Computational Approaches
to Protein Dynamics

SERIES IN COMPUTATIONAL BIOPHYSICS
Nikolay V. Dokolyan, Series Editor

Molecular Modeling at the Atomic Scale: Methods and Applications in Quantitative Biology
Edited by Ruhong Zhou

Computational Approaches to Protein Dynamics: From Quantum to Coarse-Grained Methods
Edited by Mónika Fuxreiter

Forthcoming titles

Multiscale Methods in Molecular Biophysics
Cecilia Clementi

Coarse-Grained Modeling of Biomolecules
Garegin A. Papoian

SERIES IN COMPUTATIONAL BIOPHYSICS
Nikolay V. Dokolyan, Series Editor

Computational Approaches to Protein Dynamics

From Quantum to Coarse-Grained Methods

Edited by
Mónika Fuxreiter

CRC Press
Taylor & Francis Group
Boca Raton London New York

CRC Press is an imprint of the
Taylor & Francis Group, an **informa** business

Cover Image: Structural ensemble of the cyclin-dependent kinase inhibitor p27Kip1 (shown by colored ribbons) bound to the Cdk2/cyclin complex (blue-white surface). 100 residue long p27 tail coordinates were derived from the molecular dynamics simulation of the full protein [J. Mol. Biol. 2008, vol. 376, 827-838]. Figure was prepared by the editor from the coordinates, which were kindly provided by Dr. Richard Kriwacki.

CRC Press
Taylor & Francis Group
6000 Broken Sound Parkway NW, Suite 300
Boca Raton, FL 33487-2742

First issued in paperback 2017

© 2015 by Taylor & Francis Group, LLC
CRC Press is an imprint of Taylor & Francis Group, an Informa business

No claim to original U.S. Government works

ISBN-13: 978-1-4665-6157-1 (hbk)
ISBN-13: 978-1-138-19888-3 (pbk)

Visit the Taylor & Francis Web site at
http://www.taylorandfrancis.com

and the CRC Press Web site at
http://www.crcpress.com

To my family

Contents

SECTION I Introduction

SECTION II Enzymatic Catalysis: Multiscale QM/MM Calculations

SECTION III   Protein Motions: Flexibility Analysis

SECTION IV   Approaches to Intrinsically Disordered Proteins

SECTION V   Large-Scale Dynamics

SECTION VI   Ensemble Methods

Series Preface

The 2013 Nobel Prize in Chemistry was awarded for the "development of multiscale models for complex chemical systems." This prize was particularly special to the whole computational community as it was finally recognized the role computation has played since the pioneering works of Lifson, Warshel, Levitt, Karplus, and many others.

This SERIES IN COMPUTATIONAL BIOPHYSICS has been conceived to reflect the tremendous impact of computational tools in the study and practice of biophysics and biochemistry today. The goal is to offer a suite of books that will introduce the principles and methods for computer simulation and modeling of biologically important macromolecules. The titles cover both fundamental concepts and state-of-the-art approaches, with specific examples highlighted to illustrate cutting edge methodology. The series is designed to cover modeling approaches spanning multiple scales: atoms, molecules, cells, organs, organisms, and populations.

The series publishes advanced level textbooks, laboratory manuals, and reference handbooks that meet the needs of students, researchers, and practitioners working at the interface of biophysics/biochemistry and computer science. The most important methodological aspects of molecular modeling and simulations as well as actual biological problems that have been addressed using these methods are presented throughout the series. Prominent leaders have been invited to edit each of the books, and in turn those editors select contributions from a roster of outstanding scientists.

The SERIES IN COMPUTATIONAL BIOPHYSICS would not be possible without the drive and support of the Taylor & Francis Group series manager Luna Han. All the editors, authors, and I are greatly appreciative of her support and grateful for the success of the series.

Nikolay V. Dokholyan
Series Editor
Chapel Hill, NC

Foreword

The modeling of biological systems and processes has advanced remarkably in recent years, gradually becoming a legitimate field of research. It has also started to be clear that obtaining a complete understanding of complex biomolecules cannot be fully based on experimental approaches and that the structural and kinetic findings emerging from experimental work should be augmented by computer simulations. Some of the progress in the field has been described in the scholarly articles compiled in this book. However, despite its enormous potential, the field has not yet fully developed, and some of the problems involved in choosing a proper simulation strategy are outlined below. That is, it would be comfortable to assume that modeling macromolecular function has come of age with the award of the 2013 Nobel Prize for multiscale modeling. However, in some respects, we are still at the early stages of quantifying biological function on a detailed molecular level. It is true that the tools of using force fields to describe the energetics of protein conformational changes, using molecular dynamics (MD) to obtain free energies, and using quantum mechanics/molecular mechanics (QM/MM) approaches to model chemical processes in proteins have been used and introduced successfully. However, we still do not have a broad realization of what tools should be used in different cases and how to validate the different methods. For example, some workers assume that just running nano- or even microsecond MD simulations of a given biological system can tell us how this system is working. Others may assume that enzyme catalysis is associated with dynamical effects without actually defining what such effects mean or without ever showing that the relevant rate constant can be reproduced by dynamical effects. Similarly, we see the tendency to equate targeted molecular dynamics (tMD) studies of conformational changes with consistent evaluation of the corresponding energetics and reaction path. Unfortunately, the forces applied in tMD may lead to very biased conclusions. Similarly, attempts to use QM/MM calculations can sometimes be complicated by the assumption that the use of accurate quantum methods is a license not to perform sufficient sampling for the evaluation of accurate free energies. Attempts to generate coarse-grained (CG) models, which would allow for a reduction in the time and size of problems in studies of biological functions, are encouraging but not always followed by developing models that reproduce the relevant energetics (which is primarily electrostatic).

Another problem is associated with the relatively poor understanding of what constitutes relevant validation of a model. For example, validating a model for ion transport in ion channels by studying the potential of mean force (PMF) for rotation around peptide bonds is hardly useful validation. Similarly, validating long-range electrostatic treatments by looking at energy conservation in the MD trajectory instead of the dependence of the potential on the system size is not particularly helpful. Validating QM/MM results by looking at the effect of the basis set instead of the reproduction of observables such as pK_as is also a common problem. Keeping such issues in mind is crucial, since a useful approach must actually reproduce the relevant observables rather than simply appearing formally rigorous and corresponding to a sophisticated formulation.

All of the words of caution above are mainly intended to point out that while the technological aspects are important, keeping an open eye on the key problems and on whom to represent them is crucial in the present stage of the field. Thus, the reader should look at this book as a report that reflects the current developments in the field and then make sure when picking a particular approach that this approach actually reproduces reality. Here, reality means reproducing a particular experimental observable and function rather than just a nice movie of some motions that would appear to support some hypothesis that has been put forward in an experimental paper. Of course, in the future, when the field comes to maturity, it will become clearer what method to use in different cases and different methods will provide complementary and similar results, uniting to provide powerful multiscale tools for studying biological molecules and other complex systems.

Arieh Warshel
Nobel Laureate
Member, National Academy of Sciences, HonFRSC
Distinguished Professor of Chemistry and Biochemistry
Dana and David Dornsife Chair in Chemistry
Member of the Norris Cancer Center, University of Southern California

Preface

Biology is all about adaptation to different environments. Proteins have to function under ever-changing conditions while exploiting a genetic material of limited size. As three-dimensional organization of amino acids underlies various biological activities, protein architectures must accommodate to versatile contexts to survive. Structure-function studies of proteins, especially in the last decade, showed that flexibility and structural multiplicity are essential parts of function. The discovery of proteins with a conformational ensemble as an active form underscores this point.

Dynamics can vary in a wide range, from small amplitude vibrations to large-scale domain rearrangements, which is extremely difficult and can only partly be characterized experimentally. Most of our knowledge on protein dynamics is provided by molecular simulations. Computational approaches to protein dynamics emerged in the early 1970s and by now their performance has started to reach biologically relevant timescales. The importance of these methods in biochemistry and drug discovery was recognized by a Nobel prize to Arieh Warshel, Martin Karplus, and Michael Levitt in 2013.

This book is dedicated to advanced methods that can be applied to study dynamic aspects of protein function. The book covers a wide spectrum of dynamics, from electronic structure-based to coarse-grained techniques via multiscaling at different levels. All chapters focus on a given biological problem and describe the computational approaches to tackle them. The book starts with a historical overview and description of major directions for improving biological simulations. These are detailed in the subsequent chapters to answer the following questions.

Is there a quantitative relationship between enzymatic catalysis and protein dynamics? The origin of enzymatic catalysis has long been debated. No doubt, rigid enzymes cannot do chemistry, but the exact relationships between different scales of dynamics and catalytic efficiency have yet to be established. Experimental, mostly nuclear magnetic resonance (NMR) results indicate a direct correlation between μs-ms motions and reaction rates. Chapters 2 to 4 describe how multiscale simulations can address such problems and discuss the solutions on representative systems. They also deal with methodological issues, such as how dynamics can be incorporated into quantum mechanical/

molecular mechanical (QM/MM) simulations, for example, via dynamic partitioning of the system.

Which are the functionally relevant motions of proteins? Normal mode analysis is a classical way to characterize motions of proteins. Unfortunately, the method is slow and the potential energy landscape may not justify the harmonic approximation. Chapter 5 describes possible simplifications to increase the efficiency of determining major directions of protein motions. The elastic network models have versatile applications in coarse-graining or advanced sampling, which are discussed in other chapters.

How can structural properties and partner recognition mechanisms of intrinsically disordered proteins be simulated? The discovery of proteins, which function without a well-defined three-dimensional structure, was an earthquake for classical biochemistry. Although experimental evidence demonstrates the existence of intrinsically disordered (ID) proteins *in vitro*, we still await the interpretation of their action at a molecular level. This is probably the most challenging area for biomolecular simulations (Chapters 6 to 11). Owing to the size of the available configurational space and the paucity of regular secondary structures, a number of methodological issues have to be addressed: development of force fields (Chapters 8 and 9), solvent models (Chapter 6), and sampling. Coarse-graining obviously accelerates simulations, but parametrization is far from trivial (Chapter 10). Despite these difficulties, a number of examples demonstrate the feasibility of ascribing binding and folding pathways of ID proteins to versatile partners. Characterization of ID ensembles and parametrization of respective energy-terms can also benefit from residue-based descriptors. These can be derived from thermodynamic data (Chapter 7) or directed evolution experiments (Chapter 11). The latter also illustrates evolutionary pressures on dynamic behavior of proteins.

How can we speed up molecular dynamics? A huge gap exists between the time steps of molecular dynamics simulations and the timescales of biological events. Although this can be bridged by parallelization and improvement in hardware, methodological developments are also critical. Discrete molecular dynamics (dMD) provides a general way to approach this problem. The basis of the technique as well as a wide range of applications are presented (Chapter 12).

How can conformational ensembles be described by synergistic effort of computations and experiments? Measurements of protein dynamics usually provide low-resolution information. Higher resolution results could only be obtained with the aid of computational input and/or derived databases. Chapters 13 and 14 describe experimental techniques (NMR, small-angle x-ray scattering [SAXS], hydrogen-deuterium [HD] exchange), which can deal with dynamical systems and the related structural/dynamical parameters of proteins. They discuss algorithms to incorporate computational into experimental data and vice versa, and to efficiently sample the configurational space.

These combined approaches show promise to answer questions related to very flexible systems or protein machines.

In the past decade, we realized that dynamics is essential to interpret protein action. We have to replace individual structures with ensembles and also consider multiplicity in sequence and function space. Nevertheless, there is a long way ahead for dynamics to be an integral component in structure-function relationships of proteins. Advances in computational methodology, such as those presented in this book, can largely facilitate this process and contribute to reshaping our classical views in biochemistry.

Acknowledgments

I am especially grateful for the Momentum Program of the Hungarian Academy of Sciences for providing long-term support, which allowed me to compile the book. I also thank the Department of Biochemistry and Molecular Biology of the University of Debrecen, Hungary, for their friendly, open-minded environment.

Editor

Mónika Fuxreiter received her MSc and PhD degrees from the Eötvös Loránd University of Sciences, Hungary, in 1993 and 1996, respectively. She was a postdoctoral fellow with a 2013 Nobel laureate in chemistry, Arieh Warshel, at the University of Southern California, Los Angeles, where she worked on QM/MM simulations of enzymatic catalysis. She was a senior scientist at the Institute of Enzymology, Hungarian Academy of Sciences, and is currently head of the Laboratory of Protein Dynamics at the Department of Biochemistry and Molecular Biology, University of Debrecen, Hungary. Dr. Fuxreiter has 20 years of experience in simulations of biological systems and has also contributed to the development of state-of-the-art approaches. In the last 10 years, she has been working in the field of intrinsically disordered proteins (IDPs) and has developed various models for their partner recognition mechanisms. She proposed the concept of fuzzy protein complexes, where conformational heterogeneity is functionally required for specific interactions of IDPs. She revealed unique regulatory mechanisms of fuzzy complexes and relationships to context-dependence. She is a mother of three children and a recipient of the 2014 L'Oréal–UNESCO For Women in Science award.

Contributors

Christopher M. Baker
Department of Chemistry
University of Cambridge
Cambridge, England

Pau Bernadó
Centre de Biochimie Structurale
INSERM U1054, CNRS UMR-5048
Université Montpellier I and II
Montpellier, France

Noam Bernstein
Center for Computational Materials
    Science
Naval Research Laboratory
Washington, District of Columbia

Robert B. Best
Laboratory of Chemical Physics
NIDDK, National Institutes of Health
Bethesda, Maryland

Alexandra T. P. Carvalho
Department of Cell and Molecular
    Biology
Uppsala University
Uppsala, Sweden

Wing-Yiu Choy
Department of Biochemistry
The University of Western Ontario
London, Ontario, Canada

Xiakun Chu
College of Physics
Jilin University
Changchun, Jilin, China

Elio Cino
Department of Biochemistry
The University of Western Ontario
London, Ontario, Canada

Ramon Crehuet
Institute of Advanced Chemistry of
    Catalonia, CSIC
Barcelona, Spain

Gábor Csányi
Engineering Laboratory
University of Cambridge
Cambridge, United Kingdom

Alexander Cumberworth
Centre for High-Throughput Biology
University of British Columbia
Vancouver, British Columbia, Canada

Rahul K. Das
Department of Biomedical
    Engineering
and
Center for Biological Systems
    Engineering
Washington University in St. Louis
St. Louis, Missouri

David de Sancho
Department of Chemistry
University of Cambridge
Cambridge, United Kingdom

Fernanda Duarte
Department of Cell and Molecular
    Biology
Uppsala University
Uppsala, Sweden

Agustí Emperador
Institute for Research in Biomedicine
Barcelona, Spain

Mónika Fuxreiter
MTA-DE Momentum Laboratory of
    Protein Dynamics
Department of Biochemistry and
    Molecular Biology
University of Debrecen
Debrecen, Hungary

Josep Lluis Gelpí
Department of Biochemistry and
    Molecular Biology
University of Barcelona
Institute for Research in Biomedicine
Barcelona, Spain

Jörg Gsponer
Centre for High-Throughput Biology
University of British Columbia
Vancouver, British Columbia, Canada

Florian Heinkel
Centre for High-Throughput Biology
University of British Columbia
Vancouver, British Columbia, Canada

Vincent J. Hilser
Department of Biology
and
T.C. Jenkins Department of
    Biophysics
The Johns Hopkins University
Baltimore, Maryland

Shina Caroline Lynn Kamerlin
Department of Cell and Molecular
    Biology
Uppsala University
Uppsala, Sweden

Mikko Karttunen
Department of Chemistry and
Waterloo Institute for
    Nanotechnology
University of Waterloo
Waterloo, Ontario, Canada

Jing Li
Department of Biology
and
T.C. Jenkins Department of
    Biophysics
The Johns Hopkins University
Baltimore, Maryland

Enrique Marcos
Institute of Advanced Chemistry of
    Catalonia, CSIC
Barcelona, Spain

Anuradha Mittal
Department of Biomedical
    Engineering
and
Center for Biological Systems
    Engineering
Washington University in St. Louis
St. Louis, Missouri

Letif Mones
Engineering Department
University of Cambridge
Cambridge, United Kingdom

Modesto Orozco
Department of Biochemistry and
    Molecular Biology
University of Barcelona
Institute for Research in
    Biomedicine
Barcelona, Spain

Rohit V. Pappu
Department of Biomedical
    Engineering
and
Center for Biological Systems
    Engineering
Washington University in St. Louis
St. Louis, Missouri

Tali H. Reingewertz
Institute of Human Virology
University of Maryland School of
    Medicine
Baltimore, Maryland

Melchor Sanchez-Martinez
Institute of Advanced Chemistry of
    Catalonia, CSIC
Barcelona, Spain

Yves-Henri Sanejouand
UFIP, UMR 6286 du CNRS
Faculté des Sciences et des
    Techniques
Nantes, France

Pedro Sfriso
Institute for Research in Biomedicine
Barcelona, Spain

Iván Solt
Institute of Enzymology
Hungarian Academy of Sciences
Budapest, Hungary

Eric J. Sundberg
Institute of Human Virology
Department of Medicine
Department of Microbiology and
    Immunology
University of Maryland School of
    Medicine
Baltimore, Maryland

Dmitri I. Svergun
European Molecular Biology
    Laboratory, Hamburg Outstation
Hamburg, Germany

Giancarlo Tria
European Molecular Biology
    Laboratory, Hamburg Outstation
Centre for Bioinformatics
University of Hamburg
Hamburg, Germany

Csilla Várnai
Engineering Department
University of Cambridge
Cambridge, United Kingdom

and

Warwick Systems Biology Center
University of Warwick
Warwick, United Kingdom

Konstantinos Vavitsas
Department of Cell and Molecular
    Biology
Uppsala University
Uppsala, Sweden

Andreas Vitalis
Department of Biochemistry
University of Zurich
Zurich, Switzerland

Jin Wang
Department of Chemistry, Physics &
    Applied Mathematics
State University of New York at Stony
    Brook
Stony Brook, New York

Yong Wang
State Key Laboratory of
    Electroanalytical Chemistry
Changchun Institute of Applied
    Chemistry
Chinese Academy of Sciences
Changchun, Jilin, China

Steven A. Winfield
Engineering Department
University of Cambridge
and
Cantab Capital Partners
Cambridge, United Kingdom

James O. Wrabl
Department of Biology
The Johns Hopkins University
Baltimore, Maryland

Introduction

Chapter 1

Dynamics

A Key to Protein Function

Mónika Fuxreiter

CONTENTS

INTRODUCTION

Proteins are complex entities created from 20 types of amino acids. The diverse functionality of proteins originates from the spatial organization of the protein chain. This brings residues that are far from sequence into proximity and enables the coordinated action of chemical groups. The cleavage of an amide bond, for example, requires an attacking nucleophile and a general acid to activate the nucleophile. Although the variety of three-dimensional architectures (i.e., folds) results in versatile functions, the static arrangement of amino acids cannot fulfill even the simplest biological activity. Binding of oxygen to hemoglobin would be hindered by steric conflicts unless protein sidechains change their position to create a free path toward the heme group.

An increasing body of evidence shows that proteins are not rigid. All atoms move around their equilibrium positions due to thermal energy. Protein motions could be local or collective and cover a wide range of timescales from 10^{-15} s to 1 s (Figure 1.1). Local motions allow the selection of different side chain conformers (rotamers), which are optimal for binding or chemical conversion of the substrate. Low-frequency motions (20–200 cm^{-1}) enable concerted action of domains and could be conserved across protein families. Enzymatic catalysis, for example, is coupled to distinct motions of a protein, which are also sampled in a substrate-free form.

Flexibility of proteins or domains is determined by the conformational space around their equilibrium geometries (i.e., deformability). Organisms living in different temperatures have markedly different flexibilities. Proteins or protein complexes participating in signaling might need to adopt different

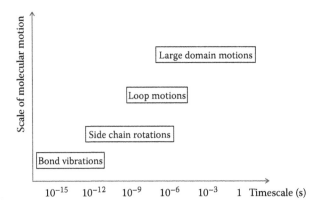

Figure 1.1 Timescales of molecular motions. Formation of individual secondary structure elements is on μs timescale, while folding of whole protein structures is on μs–ms timescale.

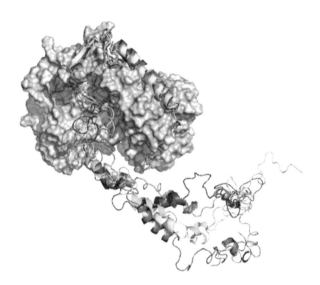

Figure 1.2 Conformational ensemble of p27^{Kip1} as bound to the Cdk-cyclin complex, as determined by NMR and SAXS. Representative coordinates are provided by Richard Kriwacki. (From Galea CA et al., *J Mol Biol* 376, 827–838, 2008.)

structures in different conditions. These proteins could be characterized as conformational ensembles rather than by variations around a given equilibrium structure (Figure 1.2). Experimental data inevitably argues that we cannot understand molecular mechanisms of proteins without characterizing the dynamics of their structure.

Solution-based spectroscopic techniques provide direct experimental information on protein flexibility and dynamics. However, even if dynamic information is available experimentally, the sampling might not be performed at functionally relevant timescales. For these reasons, computational methods are required to complement experimental data. Application of *in silico* techniques can also improve the ratio of observables and structural variables.

This chapter provides a historical overview of the computational methods to protein dynamics and a brief introduction to recent development in different areas.

BASIC APPROACHES

Energy Minimization

In order to describe a macromolecule with a physical model an energy function has to be constructed, which assigns a potential energy value of different

configurations of particles. In most cases this corresponds to different arrangements of atoms, but larger units such as whole residues could also be considered. Given the set of independent variables $(x_1, x_2, ..., x_N)$ the goal is to find
stationary points of the $V(x)$ potential energy surface (PES). In general, one is
interested in *min V(x)*, the lowest energy conformation of the PES. Localization
of the global minimum requires a nonlinear optimization [1]. To this aim, the
potential energy function $V(x)$ is expanded as Taylor series. Different optimization methods are ranked based on the highest derivatives they use. Out
of the most commonly used approaches the zeroth order is the grid search,
the first order is the steepest descent and conjugate gradient technique, and
the second order is the Newton-Raphson method [2]. These algorithms differ
in their efficiency and precision to localize the energy minimum and also in
their application range (i.e., the distance between the starting point and the
minimum).

All these optimizations are problematic above 10 or more independent variables. For macromolecules only a very small fraction of the conformational
space could be explored in this way. Hence minimization is useful to resolve
steric conflicts or to assist experimental structure determination instead of
describing the whole potential energy surface.

Vibrational Analysis

Even though energy minimization is not very helpful in revealing large conformational transitions, it can still contribute to characterization of large-scale
molecular motions. Normal mode analysis involves decomposition of the vibrational enthalpy into orthogonal vibration modes [3–5]. Atoms (units) of the
macromolecule are treated as harmonic oscillators and the potential energy is
described by a quadratic function. In this manner, the vibrational modes could
be derived from the mass-weighted Hessian matrix (v''):

$$F = M^{-1/2}v''M^{-1/2} \tag{1.1}$$

where M is a diagonal matrix of atomic masses ($3N \times 3N$). Eigenvalues and
eigenvectors of F are determined by solving the secular equations.

Low-frequency vibrational modes (below 200 cm^{-1}) could correspond to
large-scale motions and they are of potentially functional importance [6].
For example, normal mode analysis uncovered switches that induce changes
between guanosine-5′-triphosphate (GTP) and guanosine diphosphate (GDP)
bound form of Rasp21 [7]. Oncogenic mutations were also found to be more
rigid than the wild type protein hampering the structural transition between
the two functional states. Thermodynamic quantities such as entropy or heat
capacity could also be computed from vibrational analysis [8].

However, harmonic approximation of the energy surface in the vicinity of the minimum could be far from reality. Hence, quasi-harmonic approximations (QHAs) are applied in molecular dynamics (MD) simulations [9].

Molecular Dynamics

An efficient and direct method to explore configurational space of proteins is to assign velocities to each atom at a given temperature and allow them to move according to the forces derived from $V(x)$ [10,11].

$$F_i = -\frac{\partial V}{\partial r_i} \tag{1.2}$$

$$3k_B T = \sum_{i=1}^{N} m_i v_i \cdot v_i / N \tag{1.3}$$

where k_B is the Boltzmann constant, m_i and v_i are the mass and velocity of atom i, and N is the number of atoms in the system. In this classical model, atoms move deterministically according to Newton's law.

The simulation can be performed at different ensembles (canonical, isotherm-isobar, grand-canonical) depending on the problem of interest. The result of the simulation is a trajectory that stores the atomic coordinates and velocities at each time point. This could be used to compute thermodynamic quantities.

The efficiency to visit the configurational space of a protein depends on the temperature and timescale of the simulation. The latter is limited by integration errors and reversibility. The length of the simulation is quickly expanding due to the development of supercomputers and more efficient algorithms.

Free-Energy Calculations

In most biomolecular simulations, one aims to obtain quantities that could be compared to experiments. This is the work of transferring the system between two points of the configurational space e.g., between two different structural states (open and closed forms) from unbound to bound state or in between two points of the reaction pathway. Although free energy depends only on the difference between the two points/states and not on the route of how the transfer was carried out, from a computational point of view it is critical how the conversion is achieved.

In the potential of mean force (PMF) approaches the free energy is described as

$$W(\xi) = -k_B T \ln \rho(\xi) \tag{1.4}$$

where $W(\xi)$ is the free energy or work along a specific reaction coordinate and $\rho(\xi)d\xi$ is the relative probability of the system in the interval between ξ and $\xi + d\xi$. The variable ξ could be discrete or continuous. Thus free energy is evaluated in changing the variable ξ in small incremental steps. For example:

$$V_\lambda = \lambda V_1 + (1 - \lambda)V_2 \tag{1.5}$$

where V_λ is a mixed potential that is constructed by a linear combination of V_1 and V_2 and the coefficient λ is varied from 0 to 1.

Calculation of the thermodynamic quantities is limited by the sampling of the MD simulation. Biased simulations are required to visit higher energy regions of the PES, for example around a transition state of an enzymatic reaction (e.g., umbrella sampling [12]). These biases are obviously taken into account and removed when thermodynamic quantities are computed.

A simple but useful way to obtain thermodynamic quantities is to use thermodynamic cycles. The choice of a proper reference system ensures that the computed differences will represent the process of interest.

Environmental Effects

Solvation effects are critical to protein structure, dynamics, and activity. Protein regions, which are not directly involved in a chemical transformation or binding, can also be considered as solvents of the catalytic center or binding site. Computing the solvent environment, however, is challenging due to slow convergence of such interactions. Solvent configurations have to be equilibrated long before the solute is immersed and then averaged over the solute structure. Different solvent models could range from microscopic to macroscopic, including all-atom, dipolar, and implicit representation.

Most force fields are constructed from pairwise potentials, thus calculations scale with N^2, where N is the number of atoms in the system. Polarizable force fields scale with N^3. This limits the number of nonbonded atoms that can be included in evaluating the energy function. Computing solvent interactions increases the simulation time by approximately 10-fold. To reach biologically relevant timescales and achieve convergence in thermodynamic quantities, long-range forces are approximated and cutoffs are usually applied.

For explicit solvent, the two most generally used methods are the Ewald summation and spherical treatment of the environment. Infinite representation of the solvent can be achieved by applying periodic boundary conditions using sufficiently large cells so that the neighboring images do not interact. In this representation, solvent interactions could be taken into account by minimum image convention to avoid interactions over multiple cells. In the

Ewald summation method the interactions of the central box with all images are considered:

$$V = \frac{1}{2} \sum_{|n|=0}^{\infty} \sum_{i=1}^{N} \sum_{j=1}^{N} \frac{q_i q_j}{4\pi\varepsilon_0 |r_{ij} + n|} \tag{1.6}$$

where q_i and q_j are charges in different simulation boxes, r_{ij} is the distance between charges i and j, n is the multiplier of the cell length, and ε_0 is the dielectric constant. The summation is performed in reciprocal space using Fourier transformations. The Ewald summation is an efficient way to compute long-range forces [13]. The disadvantage is that the method imposes artificial constraints on charges, thus creating an external field along cellular directions.

For proteins and macromolecules of globular shape a more natural option is to apply spherical approximations [14,15]. In the explicit solvent approach, this could be primary hydration shell approximation, when only the directly coordinating water molecules are considered and are constrained to the solute during the MD simulation [16]. However, in this case determination of appropriate forces could require long preliminary tests that are not guaranteed to work under different conditions. The surface-constrained all-atom solvent (SCAAS) is an efficient option [17] when the explicit solvent is embedded into a semi-microscopic environment. In SCAAS the solvent molecules within the surface region are constrained to have polarization and radial constraint that would correspond to that in an infinite system.

Extensive efforts aim to reduce the cost of solvent treatment while using a representation that still results in reliable thermodynamic quantities. These approaches describe the solvent in a simplified manner (or lower level of theory); for example, by dipoles, such as in the protein dipole Langevin dipole (PDLD) model [18,19], or by a continuum [20]. The Poisson theory establishes a theoretical framework for computing the free energy of solvation when a set of charges are embedded in a low dielectric cavity. The Poisson equation relates the variation in the electrostatic potential ϕ within a medium of uniform dielectric constant ε to the charge density ρ:

$$\nabla^2 \phi(r) = -\frac{\rho(r)}{\varepsilon_0 \varepsilon} \tag{1.7}$$

or if the dielectric is not constant:

$$\nabla\varepsilon(r)\nabla\phi(r) = -4\pi\rho(r) \tag{1.8}$$

Obtaining the complete free energy of solvation also requires the knowledge of nonpolar terms (i.e., entropic contribution from the solvent). Solving the Poisson or Poisson–Boltzmann (PB) equation involves solving a set of non-linear differential equations [21]. Thus application of the PB theory is not very practical in MD simulations. Programs to carry out finite difference PB calculations are available and provide the solution on a grid.

Due to the complications with the PB approach, the generalized Born (GB) formalism is more common in MD simulations [22]. This is also a continuum dielectric model, where the solvent-induced reaction field is approximated by a pairwise sum of interacting charges:

$$\Delta G^{elec}_{\varepsilon_p \to \varepsilon_w} = -\frac{1}{2}\left(\frac{1}{\varepsilon_p} - \frac{1}{\varepsilon_w}\right)\sum_{i,j}\frac{q_i q_j}{\sqrt{r_{ij}^2 + \alpha_i \alpha_j \exp(-r_{ij}^2/F\alpha_i\alpha_j)}} \quad (1.9)$$

where ε_p and ε_w are the interior and exterior dielectric constants, r_{ij} is the distance between charges q_i and q_j, α_i is the GB radius, and F is an empirical factor that modulates the Born radii, which typically ranges between 2 and 10. Determination of an appropriate Born radius is the most critical point of this approximation, which could be derived from the Poisson equation or it could be parametrized using solvation free-energy values. The Born equation is used in the form of

$$\alpha_i = -\frac{1}{2}\left(\frac{1}{\varepsilon_p} - \frac{1}{\varepsilon_w}\right)\frac{1}{G^i_{pol}} \quad (1.10)$$

where G^i_{pol} is the self-energy of a unit charge in an uncharged solute when transferred between two dielectric (solvent and protein) media.

ADVANCED METHODS

Improved Sampling Techniques

The performance of computer simulations and the relevance of the results for interpretation of experimental data is influenced by the timescales, which can be achieved *in silico* and their relationship to the characteristic times of functionally important conformational changes. The first microsecond all-atom MD simulation was carried out in 1998 by the Kollman group [23], and now various trajectories of similar length are available [24–27]. However, increasing timescales might not be enough to sufficiently explore a large part of the phase space.

Three approaches to improve sampling will be discussed: (1) simulations at higher temperatures, (2) simplifying/modifying the potential energy surface, and (3) changing coordinates for optimization or reduce the degrees of freedom.

Simulations at higher temperatures. Elevating the temperature obviously increases kinetic energy and thus increases the available conformational space during a fixed amount of simulation steps. In simulated annealing, high temperatures are used to explore configurations and then gradual cooling is applied to locate the global energy minimum [28]. MD simulations run on many processors enable using different temperatures for the different runs [29]. Exchange of coordinates and temperatures is carried out between the different nodes using Metropolis criteria. This technique, called the replica exchange method (REM), became popular in the past decade [30,31]. REM is powerful for flexible proteins or peptides, but its application for large molecular systems is limited by the number of nodes (i.e., simulations run on different temperatures) that are required for efficient exchange rates. Recent developments of REM aim to reduce the number of replicas by, for example, simplified representation of the environment. Excluding solvent-solvent interactions from the exchange attempts can decrease the number of replicas by 10-fold in solute tempering (REST) [32]. Hybrid representation of the solvent by explicit and implicit models greatly increases the exchange probability [33]. Further improvements in REM could be obtained by subjecting only a small part of the system for exchange, for example in partial (RPEMDs) and local replica exchange methods (LREMDs) [34]. These two approaches have been successfully applied to large systems.

Simplifying/modifying the potential energy surface. One of the main factors that limits the sampling is the height of saddle points. Lowering the energy barriers enables exploring a larger phase space. However, the PES modifications, which need to be applied for more efficient progress of the simulation, are a priori unknown. This might require adjusting potentials in the course of the simulation (i.e., on the fly) or bias the potential so that sampling is decreased in low-energy regions. On-the-fly simulations, however, raise a series of problems for temperature control. Biasing potentials below a given threshold were demonstrated to accelerate the evolution of the MD simulation [35]. For example, accelerated molecular dynamics (AMD) represents an efficient and versatile enhanced conformational space sampling method, which requires about 1% central processing unit (CPU) time as compared to classical MD at the same level of description. In this approach a boost energy is added to modify the potential below a certain limit, which destabilizes the low-energy conformations [36]. The system moves on the modified potential at an accelerated rate (also controlled by the simulation) with a nonlinear timescale.

Selective biases—to reduce certain barriers but not others—were also attempted, but their general application is problematic due to the complexity of biomolecular systems and the corresponding configurational space. Metadynamics [37] and conformational flooding [38] sample along a selected collective coordinate.

Modifications of specific energy terms could be more effective. In a variant of REM a set of modified potentials instead of a set of temperatures is used. In Hamiltonian REM the nonbonded (e.g., Lennard–Jones) potential is altered in simulations on different nodes [39,40]. In grow-to-fit MD simulations (G2FMDs) van der Waals terms are perturbed via manipulating the radii of tightly packed side chains [41]. Another approach to modify potentials during the course of simulations is to use information from a short piece of a preceding simulation. These guiding forces are based on density of different states, for example in density-guided importance sampling (DGIS) [42]. The sampling efficiency of such approaches exceeds that of traditional Metropolis criterion.

MD simulations can also be biased toward certain directions that are derived from essential dynamics or principle component analysis. These guide the system toward the functionally interesting/important collective motions and can greatly accelerate folding simulations [43]. Elastic network methods (ENMs) are utilized to assist structure determination [44].

Reducing the degrees of freedom. Cartesian coordinates are not efficient for optimization of complex systems. Degrees of freedom could be reduced if internal coordinates such as torsion angles are used. Torsion angle molecular dynamics (TAMD) allows larger time steps without compromising the precision of the calculation as compared to conventional MD [45]. For smaller systems natural internal coordinates could be used that could be obtained from vibrational analysis of the molecule [46,47].

Various methods that explore the transition between two known states, such as umbrella sampling [12,48], nudge elastic band method [49], transition path sampling [50], or targeted molecular dynamics [51], also critically depend on the choice of coordinates used to model the conversion. For example, metadynamics was shown to be more effective in terms of localization of the transition state of a simple reaction when an energy-based rather than geometric reaction coordinate was employed [52].

The number of degrees of freedom is obviously reduced if the potential energy function is simplified. Coarse-graining is typically used for simulation of large, complex systems and is becoming more common [53–56]. Different coarse-grained (CG) models are applied depending on the level of details in the model description and also in terms of parametrization. In proteins, one residue could represent one unit/particle in the force field, but three–four atoms could also be grouped. In a CG force field for DNA, the phosphate, sugar, and nucleobase are treated separately [57–59]. Energy functions in CG techniques could be structure-based or structure-independent depending on the system of interest. Folding simulations of structured (globular proteins) are greatly accelerated by using Go potentials [60] complemented by nonnative and favorable native terms (e.g., electrostatics) [61–63]. Simulations of structural ensembles of very flexible/disordered proteins require structure-independent CG models [64].

Parametrization of CG techniques is carried out using all-atom MD results, equilibrium geometries, or thermodynamic data as a reference [65]. The obvious benefit of the CG method is enabling the use of longer time steps, which exceed at least one order of magnitude than that of classical MD. Hence CG simulations increase simulation length, which enables the study of large-scale conformational changes. The drawback is obviously losing some of the details, which could be avoided if multiscale or hybrid approaches are applied (see "Multiscale Techniques" section). The CG technique was successfully applied to study protein folding, membrane proteins [66,67], formation of large macromolecular assemblies [68,69], and complex processes [70], as well as to provide an ensemble description of very flexible or disordered systems [63,64,71,72].

Multiscale Techniques

Increasing the system size and complexity in molecular simulations is the driving force behind development of hybrid techniques. In hybrid methods different regions are described at different levels of theory.

In this way, a low-resolution description—a simplified force field or reduced degrees of freedom—could be combined with a high-resolution model; the latter would be applied only for a limited part of the system. Examples of such methods are quantum mechanics/molecular mechanics (QM/MM) [18,73–78] and all-atom/coarse-grained (AA/CG) [68,79–83].

In hybrid techniques, the division into different regions which are described by different models creates a boundary between two models that could be defined permanently (for the whole simulation) or on the fly (updated as the simulation progresses). The interaction between different models (scales) could be *embedded* coupling between the high- and low-resolution information, *iteratively* alternating/cycling between different scales [84–87], or *hierarchically* nesting the different scales (e.g., ONIOM) [88–90]. The interaction in multiscaling could be achieved in two ways: (1) by mapping, when the different scales are adjusted only once in the course of the simulation (single information transfer) [91–93] or (2) by bridging, when interaction between the different scales is dynamic (updated during the simulation) [94]. In both approaches, parametrization could use a single structure or an ensemble of structures.

For mapping CG method shape or elasticity information could be used from all-atom MD simulation. These are applied for example to viral capsids [93] or actin filaments [95]. EMNs [96,97] are useful to indicate the location of CG sites at nodes of relevant collective motions [98,99]. Bridging methods are less common, as exchanging information between regions that are described by different models could be difficult due to the boundary problem.

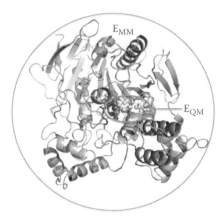

Figure 1.3   QM/MM representation of acetylcholinesterase (1ace.pdb). The QM region consists of the acetylcholine substrate (orange), Ser-200 serving as an attacking nucleophile (green), and His-440, which plays a role as a general base. Main chain dipoles of Gly-116, Gly-117, and Gly-118 at the rim of the active site provide the largest contribution to the catalytic effect. (From Fuxreiter M, Warshel A, *J Am Chem Soc* 120, 183–194, 1998.)

The hybrid QM/MM scheme was first proposed by Warshel and Levitt, where the zone of primary interest is described by quantum mechanics and the embedding environment is represented by simple empirical potentials (Figure 1.3) [18]:

$$H = H_{QM} + H_{QM/MM} + H_{MM} \qquad (1.11)$$

where H_{QM} and H_{MM} are the Hamiltonian for the quantum and molecular mechanics regions, respectively, and $H_{QM/MM}$ is the interaction between the two regions. This enables the modeling of rearrangements of chemical bonds in the center of the system, which occurs under the influence of a given, possibly very complex environment, such as an ionic solvent or enzyme.

The hybrid scheme has been successfully applied to model enzymatic reactions in good agreement with experiments [100]. A simplistic representation of the quantum region with proper representation of chemically relevant states, for example the empirical valence bond (EVB) approach, was demonstrated to be powerful in modeling and interpretation of enzymatic catalysis [101].

In the most widespread methods the coupling (i.e., interaction) between the different regions normally does not allow the interchange of particles. In a variety of processes, like membrane transport, this could present a major obstacle for modeling. Dynamic formalisms to alleviate this problem have been proposed, which are based on force-mixing rather than energy matching. The hybrid schemes raise further problems in maintaining the temperature or diffusion artifacts [102].

Representation of Solvent

Improvement of computational power allows a more accurate representation of solvent using the all-atom treatment. Simulation of large macromolecular complexes, especially to obtain a detailed description of large conformational changes and ensembles, still could only be performed under simplified or macroscopic representation of solvent. Recent methodological developments of implicit solvent models result in more accurate electrostatic solvation energies.

The second generation GB models use new sets of atomic radii that could reproduce experimental transfer energies of small molecules and free energies of charging derived from explicit solvent simulations [103,104]. A major problem with macroscopic theories is the arbitrary choice of a proper dielectric constant. The simplistic treatment with $\varepsilon = 80$ for water and $\varepsilon = 1$ or some low values (typically <10) for protein environment has to be revisited. As proteins are electrostatically inhomogenous, the dielectric constant depends on the actual location or distance from a charged group. This is partly reproduced by a distance-dependent dielectric constant [18,105], which is easy to apply, but still far from realistic representation of the electrostatic environment of proteins. Dielectric properties of the actual region or media also depend on a solvent-excluded volume and thus on the actual configuration(s) of the protein. It has been demonstrated that the dielectric constant inside the protein also varies with the actual description of the protein environment and the treatment of induced dipoles and can vary in a wide range between 1 and 15 [106]. Careful selection of ε enables the reproduction of mutation data of polar side chains [107]. Convergence of microscopic simulations could also be accelerated using ε as a scaling factor (e.g., in the semimicroscopic PDLD [PDLD/S] approach) [108,109].

Recently developed fast methods to approximate solvent-accessible surfaces [110,111] were implemented into implicit solvent models and resulted in improved accuracy:

$$G_{solv} = \sum_i \sigma_i A_i \tag{1.12}$$

where A_i is the solvent accessible surface area and σ_i is the surface tension coefficient. In addition to proteins and polar solvents, implicit solvent models were also applied to a low-dielectric medium such as membranes [112].

There are various disadvantages of implicit models. One of the major bottlenecks is still the estimation of the nonpolar solvation (entropy-related) energy term, which could play a dominant role in target selection, for example in minor groove-binding deoxyribonucleic acid (DNA) proteins.

Estimation of Free Energy

Most simulations are validated by comparing observed quantities such as binding or rate constants to the corresponding computed free energies. Free energies are state functions, which—in principle—do not depend on the pathway, just the difference between the two states. In practice, however, due to sampling problems in high-energy regions, the pathway and the sampling methods indeed matter. Free energies are crucial for drug discovery, which has an increasing demand for novel techniques for fast and efficient estimation of binding affinities [113–115]. The most frequent applications of free energy methods are docking and ligand selection.

Classical methods such as free-energy perturbation [116] or thermodynamic integration [117] offer a precise yet slow approach to obtain practically relevant quantities [118,119]. The main reason is the slow exploration of the phase space by MD calculations, especially if the solvent is explicitly taken into account. Due to growing awareness of the importance of entropic effects, conformational entropy in particular, a variety of statistical models have been developed.

Alchemical free energy methods are applicable to obtain both absolute and relative binding affinities [120]. In alchemical methods the interaction between the ligand and its surroundings is progressively switched off. Computing relative affinities could be quite straightforward if no major rotamer change takes place upon binding various ligands [121]. If experiments manage to resolve multiple binding modes, these could be taken into account as alchemical intermediates [122,123]. Obtaining absolute binding affinities on the other hand is a real challenge [113,124]. It is not only hampered by sampling, but also on the imprecision of the existing force fields. The current generation of force fields produce a moderate error range of 1–2 kcal/mol [115,125]. However, some problems are difficult to address, such as (1) describing the electrostatics of hydrophobic pockets within a protein environment [126,127] and (2) slow side chain repacking upon binding (e.g., in receptor-ligand interactions) [128].

Accelerating MD algorithms could help to overcome sampling problems, such as in NAMD [129], Gromacs [130], and Desmond. A new direction is to partition the configurational space into smaller regions along the slow degrees of freedom and these regions are sampled independently [131]. The results, which are produced by the parallel calculations are subsequently merged. If there are several slow degrees of freedom, many dimensional PMFs have to be used, which on the other hand are difficult to cope with. This problem was discussed in connection with metadynamics [37,132]. Finding slow degrees of freedom is also challenging. To this end, a Markov state model has been applied recently that was based on metastable states of the configurational space [133–135]. An alternative way is the general ensemble method, where multiple intermediate states are visited in a single alchemical simulation [136].

Solvation free energies are mostly estimated based on electrostatics, using the molecular mechanics Poisson-Boltzmann surface area (MM/PB-SA) method [137–139].

$$\Delta G_{bind} = G_{complex} - G_{free\_protein} - G_{free\_ligand} \qquad (1.13)$$

$$G = E_{gas} + G_{solvation} - TS \qquad (1.14)$$

$$G_{solvation} = G_{PB} + G_{sur} \qquad (1.15)$$

$$G_{sur} = \gamma A + b \qquad (1.16)$$

where the solvation free energy is composed of polar (G_{PB}) and nonpolar terms (G_{sur}). The polar term is obtained by the Poisson–Boltzmann equation and the nonpolar term is based on the solvent accessible surface. The MM/PB-SA method is a postprocessing technique where snapshots are collected from the MD simulation and then subsequently minimized. This method has the potential to take reorganization energy into account.

CHALLENGING BIOLOGICAL PROBLEMS FOR SIMULATIONS

Protein Folding

Understanding how protein structures are formed presents a major challenge for computer simulations. Attaining a protein structure is on the μs–ms timescale, which requires ~10^{12}–10^{15} time steps in an all-atom simulation if femtosecond time steps are used. Furthermore, multiple trajectories are needed to obtain a sufficient sampling of the configurational space. Despite the enormous computational effort, a few simulations demonstrate the realm of such an approach. Three factors are critical to achieve good agreement with the experiments: (1) force field accuracy, (2) sampling, and (3) data analysis.

The major problem with the most widely used AMBER94 [140] and CHARMM22 [141] force fields is that they are biased toward regular secondary structures α-helix and β-sheet, respectively. Although their developments were based on different approaches, they were standardized against structural data of globular proteins. Recently however, protein flexibility has gained more attention, especially with the discovery of highly dynamic or flexible proteins. For this group of proteins the native structure or conformational ensemble cannot be reproduced using classical force fields [142,143]. Even when applying enhanced sampling or high temperatures, classical force fields return to their globular forms [144]. To overcome this problem, torsion angle terms have to be modified [145–147]. In general, very small corrections could provide a good balance between helical and extended conformations. To improve sampling in nonhelical regions, AMBER ff99 torsion potentials were set to reproduce the helix-coil transition temperatures or were fitted to alanine tetrapeptide *ab initio* surfaces [147]. Surprisingly, setting torsion potentials for ϕ and ψ to zero was also quite effective to simulate flexible peptides in good agreement with nuclear magnetic resonance (NMR) data [146]. Comparison of secondary structure propensities of peptides by 12 force fields to NMR results indicated a convergence in recent force fields [148].

However, sophisticated optimization of energy functions that produces correct secondary structure propensities of peptides may not guarantee applicability to folding of real proteins, as tertiary structure interactions often overcome inherent secondary structure preferences. For GB1 and Trp cage miniprotein, however, optimization of AMBER ff03 was demonstrated to produce the correct folded structures for β-hairpin and helix-containing forms, respectively [149]. Empirical correction of torsion angles was successful to fold 12 proteins [27]. Optimization of force field parameters solely based on NMR data, however, is impossible due to the large number of variables [150]. Another perspective for folding simulations is to use polarizable force fields that require reasonable computational time. For example, atomic polarizability was shown to contribute to the energetics of helix formation [151].

A key to improve sampling of folding is to extend the timescales of the simulations. Recent developments in simulation codes in terms of parallelization and scaling (e.g., GROMACS [130] and NAMD [129]) made it possible to reach trajectories of the length of 10 μs. Combining multiple trajectories resulted in a correctly folded state for Villin HP35 [152] and fip35 WW domain [24]. Improvement in hardware exemplified by the ANTON supercomputer also largely facilitated the sampling of reversible folding events [153].

Sampling methodology has also improved especially in terms of finding/optimizing the path between two minima of the configurational space. Transition path sampling [50] for example was successfully applied to Trp cage [154]. This approach requires trajectories longer than the transition path approximately of microsecond length. Alternative transition path sampling was developed for CG folding simulations [155]. A string method was applied to optimize path between carefully chosen end structures at relatively low computational cost [156]. Other improved methods, such as replica exchange [157] or Markov state models (MSMs) [158] have been discussed before. In MSMs independent short simulations are generated and then assembled on a statistical basis. In this manner small pieces of the protein phase space are explored and then combined with proper weights, similar to solving a jigsaw puzzle. This method is easy to parallelize; independent trajectories can be produced in different machines.

A careful combination of existing individual trajectories with proper weighting could improve the reliability of the predicted native structure. To reproduce proper kinetics, however, an appropriate reaction coordinate should be found, along which there is a 50% probability for folding or unfolding events [159]. All these approaches operate with physically relevant intermediates or transition state. The advantage of such PMF-like methods is the reduction of the configurational space. MSMs on the other hand, utilize discrete structures and could be performed at various levels of resolution. This allows to discard very fast dynamics and thus to achieve long timescales. However, a considerable disagreement was found between the results of the MSMs and long trajectory-based methods, indicating that these approaches explore different folding pathways [160].

Although no general conclusions could yet be drawn, the success to fold specific proteins (~20 examples) by all-atom simulations provides a glimpse into the physical principles of how protein structures could be organized. The mechanisms and pathways are still debated, but the findings support the funnel hypothesis: the folding is dominated by native interactions. Nonnative interactions are mostly avoided along folding pathways. This imparts robustness on folding landscapes so that single mutations have a small perturbation/ impact on the funneled landscape. Capturing cooperativity along the folding pathway still presents a major bottleneck in simulations. Modeling of this phenomenon demands improvement of force fields.

Protein folding can also be studied by CG simulations, which allow investigation of longer timescales [161,162]. The benefit of using all-atom methods as compared to CG methods is the potential to study the unfolded state in detail. Representation of the highly dynamic state on the other hand could be a challenge for energy functions, as discussed above. Reliable simulation not only can deepen our understanding of how proteins reach their functional forms, but can also be useful to interpret experimental data. Low-resolution techniques (e.g., Förster resonance energy transfer [FRET]) could benefit from computational prediction of low-energy conformers.

Ensemble Description of Protein Structures

The discovery of proteins with multiple native states is a new challenge for molecular dynamics simulations. These group of proteins, termed as intrinsically disordered (ID), are devoid of a single, well-defined structure [163]. Instead, ID proteins interconvert between many conformations and exist as an ensemble of structures [164]. Protein segments with similar properties could also be functionally relevant [165]. These intrinsically disordered regions (IDRs) could impart regulatory features to proteins via embedding short interaction motifs or posttranslational modification sites [166–168]. Bioinformatics studies demonstrated the abundance of proteins with ID regions in eukaryotic organisms [169,170]. ID proteins (IDPs) are distinguished in molecular recognition [171,172] and play critical roles in signaling pathways [173].

The importance of IDPs and their involvement in various diseases [174–176] motivated experimental [177–179] and computational structural studies to identify their functional forms. Although IDPs are similar to the unfolded state of proteins, they are not unfolded [180,181]; rather, they exhibit some structural organization at the secondary or even tertiary structure level [182–185]. Similar to the unfolded state of proteins, describing these weak, variable contacts is a huge challenge for MD studies.

Owing to force field and sampling problems, ensemble modeling is applied to determine the structural ensemble of IDPs. These approaches explore the configurational space under the restraints derived from experimental parameters [186–188]. Sampling could be carried out by generating multiple MD trajectories or Monte Carlo algorithms to satisfy observed parameters. Ensemble models could be fitted to a variety of experimental data; for example various NMR parameters are sensitive to different structural and dynamic properties. Restraints based on chemical shifts are employed to determine secondary structures and characterization of the disordered state. The ensemble of tertiary structures could be determined using paramagnetic relaxation enhancement data (PREs). Residual dipolar couplings (RDCs) also contribute to the ensemble of tertiary structures. Small-angle x-ray scattering (SAXS) and hydrodynamic radius are crucial for size distribution and molecular shape [189]. In the ENSEMBLE methodology, 10–25 structures are required to fit experimental restraints [180,187]. Validations are mostly performed using 100-structure ensemble. Flexible-meccano is an efficient minimization algorithm, where different copies of the polypeptide are generated by randomly sampling of the dihedral space [190]. Orientation of each peptide plane is determined based on RDC parameters. The model is a highly simplified representation of the protein, where amino acids are represented by hard spheres and repulsive interatomic forces are applied. A statistical coil model represents the unfolded state, which is tested against experimental data (RDCs, SAXS, and chemical shifts). Owing to simplicity, huge ensembles (~100,000 conformers) could be generated even within short simulation time (~1 hour).

IDPs often adopt a well-defined structure upon binding to other proteins or DNA [191]. Mechanisms of such binding events could be described by using CG simulations on various systems [64,71,72,192]. This requires the knowledge of the native contacts, which drive the simulation via Go-potentials. Binding of ID regions are often governed by electrostatic interactions [193,194]. Hence, non-native potentials critically influence the actual pathway [63]. Parametrization of such terms, however, are far from trivial as validation against experimental binding constants requires extensive sampling efforts [195].

Most IDP simulations utilize implicit solvent. Recently, an explicit-solvent simulation has been attempted by combining explicit-solvent MD with implicit-solvent REMD data [196]. Long (>10 μs) all-atom trajectories were also generated for selected systems (e.g. α-synuclein), but these results were not compared to experimental data [197].

Simulating Enzymatic Mechanisms

Enzymes catalyze chemical reactions at $\sim 10^{10}$-fold higher rates than in solution. The origin of this catalytic power has long been investigated. Two opposing scenarios were proposed to explain rate acceleration by enzymes: (1) destabilization of the ground state of the reaction and (2) stabilization of the transition state [198]. Formally, both mechanisms contribute to reduction of the activation barrier ($\Delta g^{\ddagger}$). Various factors, such as proximity effects [199], acid-base catalysis, near attack conformation [200], strain [201], dynamics [202], and desolvation [203], contribute to lowering the activation barrier as compared to solution reactions. The individual effect of these factors is moderate and results in a rate acceleration $<10^4$ fold. The only factor with major impact on catalysis is the electrostatic preorganization [101], which can provide 10^7–10^{10}-fold rate acceleration [100].

A critical element on simulating enzymatic catalysis is the interaction between the active site and the protein environment. Computer-assisted design of enzymes, which catalyze unnatural reactions, clearly illustrates this problem [204]. Investigating a reaction mechanism in the context of a few residues may not result in optimal active site architecture for catalysis [205]. Furthermore, the reaction mechanism may not follow the pathway with lowest barrier. Owing to these problems, even if the energetics of the reaction in the theozyme (the substrate + few functional residues) is computed by high-level *ab initio* methods [206], the activity of the design is far from that of natural enzymes. In order to capture the catalytic effect, the dynamic interplay between the active site and the protein environment has to be taken into account [207]. Attempts to evaluate the catalytic effect by using gas phase models is unlikely to reproduce the correct catalytic effect since it is impossible to assess the preorganization effect without considering the protein reorganization in the simulations [208,209].

Thus it is a key point to consider the effect of protein dynamics on the energetics of enzymatic catalysis [210]. It has been argued that specific protein conformations or characteristic motions significantly contribute to catalysis [211] (e.g., in dihydrofolate reductase [212], cyclophilin A [213], purine nucleoside phosphorylase [214], and triose phosphate isomerase [215]). In the case of adenylate kinase for example, a direct link between local ps-ns motions and ms motions has been established [216]. Flexibility of a hinge region was proposed to facilitate larger-scale lid opening generating a catalytically competent state. It was hypothesized that such a connection between motions of different timescales could be an important factor in shaping protein energy landscapes [211,217]. Such correlated motions could be explored by molecular dynamics and normal mode calculations [218]. The relevance of these motions to lowering the catalytic barrier, however, has to be investigated by multiscale simulations when activation barriers in enzyme and solution are quantitatively evaluated. Such approaches demonstrated that dynamics could be of importance for substrate binding but not for the chemical step and hence it does not contribute significantly to the catalytic effect [219]. Correlated motions could increase the dynamical recrossing of the reactive trajectories (i.e., transmission coefficient) in the enzyme as compared to the uncatalyzed reaction in solution [220]. Although transition path sampling calculations indicate that specific vibrations increases the lifetime of the transition state (TS) [214], QM/MM calculations demonstrate that the transmission coefficients between the enzymatic and solution reactions do not differ significantly because correlated motions also exist in solvent [221]. In the case of dihydrofolate reductase, it was shown that it is electrostatic preorganization, and not the correlated motions, that contribute to catalysis. [222]. Specific motions might also increase the efficiency of hydrogen tunneling [223,224]. The impact of this effect on catalysis is still under debate as similar motions also exist in solution [225]. Overall, we can conclude that dynamics is required for capturing the reorganization effect, the major contribution to the reduction of the catalytic barrier [101].

Recent methodological developments enable the study of complex problems, such as chemical reactions, which are coupled to large-scale conformational changes. For example, in monomeric chorismate mutase the relationship between protein folding and catalytic efficiency could be quantified [226]. The free-energy landscape of protein configurations was determined by CG simulations and hybrid QM/MM calculations were employed to obtain the activation barriers in the respective conformations. These resulted—for the first time—in a chemical and folding free-energy landscape in the absence and presence of the substrate. It is concluded that the rate is determined by the electrostatic energy gap between the reactant and product states. This can also be explored using Langevin dynamics (LD) to estimate slow electrostatic fluctuations resulted by different protein configurations. With proper parametrization, simplified modeling by LD or CG simulations could be used to investigate the coupling between conformational changes and chemistry (e.g., in the case of ATPases) [227,228]. All computational results that agree with experimental rates demonstrate that enzymatic catalysis originates in stabilization of the transition state.

Although the toolware has been established to obtain activation free energies comparable to those that are derived from measured kinetic parameters, there are still various open issues in simulating catalysis. One of the bottle-necks is the choice of a proper reaction coordinate. Complex reaction coordinates depend on many geometrical variables of the system. Increasing the number of variables to bias the system, however, slows down the convergence and increases the computational cost [37]. Energy-based reaction coordinates, such as E_{GAP} in empirical valence bond methods can provide a solution to this problem [229,230]. Although this coordinate has been implemented within the framework of the valence bond approximation, it could also be utilized in high-level *ab initio* methods [52,231].

The other problem is to follow the dynamics of the reaction; for example, if water molecules penetrate into the active site of the enzyme in course of the reaction, they could also contribute to the reaction coordinate [232]. To take this into account, the QM region definition has to be modified during the pathway. The boundary problem makes the dynamic or on-the-fly definition of QM regions especially difficult [85,102].

PERSPECTIVES

Establishing the sequence to function relationships for proteins and nucleic acids has been a long-standing dream. The key for atomic-level interpretation of biological action is understanding the structure. A series of experimental evidence demonstrates that the structure of proteins and nucleic acids is not static and its changes are intertwined with the function of the molecule. However, characterizing the dynamics of biomolecules is still a challenge owing to force field and sampling issues. A few questions that we aim to answer are: What is the optimal structure of a biomolecule? How is the structure formed (i.e., folding pathways)? What are the characteristic changes of the structure at functionally relevant timescales? How do we interpret the experimental results, especially low-resolution data? What is the relationship between dynamics and function?

With growing interest in protein dynamics there has been a rapid development both in hardware and software (methodology) to simulate conformational rearrangements and characterize structural ensembles, which has enabled increasing timescales and molecular size for studying dynamics. For example, owing to improved sampling techniques, multiscale approaches, and parallelization, folding dynamics simulations will soon attain millisecond timescales [27]. Large molecular complexes, so-called machines could also be simulated using simplified models (e.g., ribosome) [233,234]. Experimental techniques have advanced to determine conformational ensembles of proteins lacking a well-defined three-dimensional structure [235]. However, these methods require the assistance of computational approaches to complement observed data. CG description of highly dynamic proteins enables the study of the mechanisms of partner recognition and the folding process, which is coupled to the binding. Multiscale approaches reveal mechanisms of enzymatic reactions and molecular transport [101].

What Remains to Be Solved?

We still do not have a deep insight into how molecular complexes are formed, especially if any of the partners is highly dynamic, if transient interactions contribute to binding, or if conformational heterogeneity is retained in the bound form [236]. Mechanisms enabling long-range intra- or intermolecular communications are still poorly understood. A new concept of dynamic allostery emerges, where the flexibility or dynamic properties of the interface are modulated instead of the structure of the interface (e.g., CAP [237] or p27[Kip1] [238]) [239,240]. This concept, however, still needs to be shaped to take conformational heterogeneity in bound form into account. Specificity in molecular recognition is also poorly understood at the quantitative level. Various questions of enzymatic catalysis still remain to be answered that involve the relationship between folding and catalysis, design of artificial enzymes, and the origin of promiscuous activities.

How Can We Approach Such Problems?

First, the accuracy of energy functions need to be improved. This could be achieved by incorporating missing effects such as cooperativity, polarization, and charge transfer. Although polarizable force fields are available more comparison with experimental data would be needed to improve performance. Description of the environment can critically influence the precision of the force field. Implicit solvent models are becoming more popular, but a force field that is consistent with such models is yet to be developed. Generalized born/surface area (GB/SA) models were successfully applied for studying conformational ensembles and also for *ab initio* protein folding. A mixed explicit/implicit solvent model would help to investigate atomistic events at moderate computational cost.

CG models can describe giant biomolecules and slow, large-amplitude motions with the loss of atomistic details. CG simulations are also useful for interpretation of low-resolution experimental data such as cryoelectron microscopy. Currently there is no unified CG force field for proteins, nucleic acids, or lipids. CG potentials could be improved in various ways: (1) by improving all-atom force fields, which are used for parametrization, (2) by comparison to ensemble models, which are derived from experimental data, and (3) adjusting potentials on-the-fly as the simulation proceeds. ENM-based CG potentials can characterize large-scale conformational changes and collective motions. For high-level representation of the region of interest, CG could also be combined with an all-atom description using multiscale simulations. Here the consistency between the two potentials is straightforward to achieve if the CG is derived from the AA force field. CG has a distinguished role in investigating the behavior of proteins with heterogeneous conformations in solution or upon binding to other molecules.

Multiscale simulations are becoming more widespread. QM/MM hybrid calculations are applicable to problems whereas conformational changes are coupled to chemical events such as enzymatic catalysis. Here however, additional forces can appear at the boundary resulting in undesired diffusion between the regions, which are described at different levels of theory. More complex problems, such as ion transport through a channel, would require a dynamic definition of the QM region so that it can be updated as the simulation proceeds. Consistency between the forces at the boundary are hence prerequisites for such developments.

To answer questions at biologically relevant timescales, enhanced sampling has to be performed. Parallel simulations and exchanging information between multiple trajectories significantly increased the available configurational space. For long MD simulations the relaxation time of the system has to be considered and the dependency on the initial configurations also has to be probed. For enhanced sampling techniques, the main issue is the choice of the reaction coordinate along which the system is biased. Complex systems might require too many variables to be biased. The reaction coordinate can also change during the simulation, especially if water molecules or ions also contribute to it. However, sampling techniques, which could be applied to peptides might not be transferable to proteins or larger systems. Long MD simulations could be successfully employed to simulate protein folding. Although the ultimate structures could be obtained in good agreement with the experiment, the pathways seem to depend on the sampling technology. Long MD simulations could also detect rare events or conformational states that are selected from the ensemble in signaling.

Computing free energies is probably the greatest challenge in simulations by either PMF or alchemical methods. Since most biochemical events are determined by electrostatic forces, it becomes a problem in sampling and convergence issues hinder obtaining values, which are comparable to the experimental results. Relative free-energy values (e.g., differences in binding affinities of different substrates) can be calculated at 1–2 kcal/mol precision. Evaluating free energies of binding coupled to large conformational changes however is still a challenge.

Overall, computer simulations have reached the stage when they can answer biochemical or biophysical questions at relevant timescales. There is ample room for improvement of force fields, solvent treatment, sampling, and reaction coordinate selection to increase precision and predictability of the current approaches. We must note, however, that simple models can also provide correct answers to properly asked questions. Therefore, despite of the development of computers and algorithms, the key in biological simulations is still the formulation of the problem and the proper choice of models. With all these efforts, we progress toward a holistic treatment of biomolecules, where dynamics becomes an integral component in explaining biochemical events.

ACKNOWLEDGMENT

The support of the Momentum program of the Hungarian Academy of Sciences (LP2012-41) is gratefully acknowledged.

REFERENCES

1. Levitt M, Lifson S (1969) Refinement of protein conformations using a macromolecular energy minimization procedure. *J Mol Biol* **46**, 269–279.
2. Nguyen D, Case DA (1985) On finding stationary states on large-molecule potential surfaces. *J Phys Chem* **89**, 4020–4026.
3. Noguti T, Go N (1982) Collective variable description of small amplitude conformational fluctuations in a globular protein. *Nature* **296**, 776–778.
4. Brooks B, Karplus M (1983) Harmonic dynamics of proteins: Normal modes and fluctuations in bovine pancreatic trypsin inhibitor. *Proc Natl Acad Sci U S A* **80**, 6571–6575.
5. Brooks B, Karplus M (1985) Normal modes for specific motions of macromolecules: Application to the hinge-bending mode of lysozyme. *Proc Natl Acad Sci U S A* **82**, 4995–4999.
6. Cao ZW, Chen X, Chen YZ (2003) Correlation between normal modes in the 20–200 cm-1 frequency range and localized torsio motions to certain collective motions in proteins. *J Mol Graph Model* **21**, 309–319.
7. Ma J, Karplus M (1997) Ligand-induced conformational changes in ras p21: A normal mode and energy minimization analysis. *J Mol Biol* **274**, 114–131.
8. Karplus M, Kushick JN (1981) Method for estimating configurational entropy of macromolecules. *Macromolecules* **14**, 325–332.
9. Levy RM, Srinivasan AR, Olson WK, McCammon JA (1984) Quasi-harmonic method for studying very low frequency modes in proteins. *Biopolymers* **23**, 1099–1112.
10. Rahman A (1964) Correlations in the motions of atoms in liquid argon. *Phys Rev* **136**, A405–A411.
11. Rahman A, Stillinger FH (1971) Molecular dynamics study of liquid water. *J Chem Phys* **55**, 3336–3359.
12. Torrie G, Valleau J (1977) Nonphysical sampling distributions in Monte Carlo free-energy estimation: Umbrella sampling. *J Comput Phys* **23**, 187–199.
13. Darden TA, York D (1993) Particle mesh Ewald Nlog(N) method for Ewald sums in large systems. *J Chem Phys* **103**, 8577–8593.
14. Beglov D, Roux B (1994) Finite representation of an infinite bulk system: Solvent boundary potential for computer simulations. *J Chem Phys* **100**, 9050–9063.

15. Juffer A, Berendsen HJC (1993) Dynamic surface boundary conditions: A simple model for molecular dynamics simulations. *Mol Phys* **79**, 623–644.
16. Beglov D, Roux B (1995) Dominant solvation effects from the primary shell of hydration: Approximation for molecular dynamics simulations. *Biopolymers* **35**, 171–178.
17. King G, Warshel A (1989) A surface constrained all-atom solvent model for effective simulations of polar solutions. *J Chem Phys* **91**, 3647–3661.
18. Warshel A, Levitt M (1976) Theoretical studies of enzymic reactions: Dielectric, electrostatic and steric stabilization of the carbonium ion in the reaction of lysozyme. *J Mol Biol* **103**, 227–249.
19. Lee FS, Chu ZT, Warshel A (1993) Microscopic and semimicroscopic calculations of electrostatic energies in proteins by Polaris and Enzymix programs. *J Comp Chem* **14**, 161–185.
20. Honig B, Nicholls A (1995) Classical electrostatics in biology and chemistry. *Science* **268**, 1144–1149.
21. Sharp KA, Friedman RA, Misra V, Hecht J, Honig B (1995) Salt effects on polyelectrolyte-ligand binding: Comparison of Poisson-Boltzmann, and limiting law/counterion binding models. *Biopolymers* **36**, 245–262.
22. Constanciel R, Contreras R (1984) Self-consistent field-theory of solvent effects representation by continuum models: Introduction of desolvation contribution. *Theor Chim Acta* **65**, 1–11.
23. Duan Y, Kollman PA (1998) Pathways to a protein folding intermediate observed in a 1-microsecond simulation in aqueous solution. *Science* **282**, 740–744.
24. Freddolino PL, Liu F, Gruebele M, Schulten K (2008) Ten-microsecond molecular dynamics simulation of a fast-folding WW domain. *Biophys J* **94**, L75–L77.
25. Ensign DL, Kasson PM, Pande VS (2007) Heterogeneity even at the speed limit of folding: Large-scale molecular dynamics study of a fast-folding variant of the villin headpiece. *J Mol Biol* **374**, 806–816.
26. Perez A, Luque FJ, Orozco M (2007) Dynamics of B-DNA on the microsecond timescale. *J Am Chem Soc* **129**, 14739–14745.
27. Lindorff-Larsen K, Piana S, Dror RO, Shaw DE (2011) How fast-folding proteins fold. *Science* **334**, 517–520.
28. Chou KC, Carlacci L (1991) Simulated annealing approach to the study of protein structures. *Protein Eng* **4**, 661–667.
29. Trebst S, Troyer M, Hansmann UH (2006) Optimized parallel tempering simulations of proteins. *J Chem Phys* **124**, 174903.
30. Mitsutake A, Sugita Y, Okamoto Y (2001) Generalized-ensemble algorithms for molecular simulations of biopolymers. *Biopolymers* **60**, 96–123.
31. Okamoto Y (2004) Generalized-ensemble algorithms: Enhanced sampling techniques for Monte Carlo and molecular dynamics simulations. *J Mol Graph Model* **22**, 425–439.
32. Liu P, Kim B, Friesner RA, Berne BJ (2005) Replica exchange with solute tempering: A method for sampling biological systems in explicit water. *Proc Natl Acad Sci U S A* **102**, 13749–13754.
33. Okur A, Wickstrom L, Layten M, Song K, Hornak V, Simmerling C (2006) Improved efficiency of replica exchange simulations through use of a hybrid explicit/implicit solvation. *J Chem Theory Comput* **2**, 420–433.
34. Cheng X, Cui G, Hornak V, Simmerling C (2005) Modified replica exchange simulation methods for local structure refinement. *J Phys Chem B* **109**, 8220–8230.

35. Hamelberg D, Mongan J, McCammon JA (2004) Accelerated molecular dynamics: A promising and efficient simulation method for biomolecules. *J Chem Phys* **120**, 11919–11929.
36. Markwick PR, McCammon JA (2011) Studying functional dynamics in bio-molecules using accelerated molecular dynamics. *Phys Chem Chem Phys* **13**, 20053–20065.
37. Laio A, Parrinello M (2002) Escaping free-energy minima. *Proc Natl Acad Sci U S A* **99**, 12562–12566.
38. Grubmuller H (1995) Predicting slow structural transitions in macromolecular systems: Conformational flooding. *Phys Rev E Stat Phys Plasmas Fluids Relat Interdiscip Topics* **52**, 2893–2906.
39. Affentranger R, Tavernelli I, Di Iorio E (2006) A novel Hamiltonian replica exchange MD protocol to enhance protein conformational space sampling. *J Chem Theory Comput* **2**, 217–228.
40. Kwak W, Hansmann UH (2005) Efficient sampling of protein structures by model hopping. *Phys Rev Lett* **95**, 138102.
41. Zhang W, Duan Y (2006) Grow to fit molecular dynamics (G2FMD): An ab initio method for protein side-chain assignment and refinement. *Protein Eng Des Sel* **19**, 55–65.
42. Thomas GL, Sessions RB, Parker MJ (2005) Density guided importance sampling: Application to a reduced model of protein folding. *Bioinformatics* **21**, 2839–2843.
43. Chen C, Xiao Y, Zhang L (2005) A directed essential dynamics simulation of peptide folding. *Biophys J* **88**, 3276–3285.
44. Suhre K, Sanejouand YH (2004) ElNemo: A normal mode web server for protein movement analysis and the generation of templates for molecular replacement. *Nucleic Acids Res* **32**, W610–W614.
45. Chen J, Im W, Brooks CL, III (2005) Application of torsion angle molecular dynamics for efficient sampling of protein conformations. *J Comput Chem* **26**, 1565–1578.
46. Pulay P, Fogarasi G, Pang F, Boggs J (1979) Systematic ab initio gradient calculation of molecular geometries, force constants, and dipole moment derivatives. *J Am Chem Soc* **101**, 2550–2560.
47. Koslover EF, Wales DJ (2007) Geometry optimization for peptides and proteins: Comparison of Cartesian and internal coordinates. *J Chem Phys* **127**, 234105.
48. Beutler T, van Guntersen W (1994) The computation of a potential of mean force: Choice of the biasing potential in the umbrella sampling technique. *J Chem Phys* **100**, 1492–1497.
49. Henkelman G, Jonsson H (2000) Improved tangent estimate in the nudged elastic band method for finding minimum energy paths and saddle points. *J Chem Phys* **113**, 9978–9985.
50. Bolhuis PG, Chandler D, Dellago C, Geissler PL (2002) Transition path sampling: Throwing ropes over rough mountain passes, in the dark. *Annu Rev Phys Chem* **53**, 291–318.
51. Ma J, Sigler PB, Xu Z, Karplus M (2000) A dynamic model for the allosteric mechanism of GroEL. *J Mol Biol* **302**, 303–313.
52. Mones L, Kulhanek P, Simon I, Laio A, Fuxreiter M (2009) The energy gap as a universal reaction coordinate for the simulation of chemical reactions. *J Phys Chem B* **113**, 7867–7873.

53. Takada S (2012) Coarse-grained molecular simulations of large biomolecules. *Curr Opin Struct Biol* **22**, 130–137.
54. Hinczewski M, Gebhardt JC, Rief M, Thirumalai D (2013) From mechanical folding trajectories to intrinsic energy landscapes of biopolymers. *Proc Natl Acad Sci U S A* **110**, 4500–4505.
55. Hyeon C, Thirumalai D (2011) Capturing the essence of folding and functions of biomolecules using coarse-grained models. *Nat Commun* **2**, 487.
56. Saunders MG, Voth GA (2013) Coarse-graining methods for computational biology. *Annu Rev Biophys* **42**, 73–93.
57. Knotts TA, IV, Rathore N, Schwartz DC, de Pablo JJ (2007) A coarse grain model for DNA. *J Chem Phys* **126**, 084901.
58. Voltz K, Trylska J, Tozzini V, Kurkal-Siebert V, Langowski J, Smith J (2008) Coarse-grained force field for the nucleosome from self-consistent multiscaling. *J Comput Chem* **29**, 1429–1439.
59. Savelyev A, Papoian GA (2009) Molecular renormalization group coarse-graining of polymer chains: Application to double-stranded DNA. *Biophys J* **96**, 4044–4052.
60. Go N (1983) Theoretical studies on protein folding. *Annu Rev Biophys Bioeng* **12**, 183–210.
61. Clementi C, Nymeyer H, Onuchic JN (2000) Topological and energetic factors: What determines the structural details of the transition state ensemble and "en-route" intermediates for protein folding? An investigation for small globular proteins. *J Mol Biol* **298**, 937–953.
62. Ding F, Guo W, Dokholyan NV, Shakhnovich EI, Shea JE (2005) Reconstruction of the src-SH3 protein domain transition state ensemble using multiscale molecular dynamics simulations. *J Mol Biol* **350**, 1035–1050.
63. Turjanski AG, Gutkind JS, Best RB, Hummer G (2008) Binding-induced folding of a natively unstructured transcription factor. *PLoS Comput Biol* **4**, e1000060.
64. De Sancho D, Best RB (2012) Modulation of an IDP binding mechanism and rates by helix propensity and non-native interactions: Association of HIF1alpha with CBP. *Mol Biosyst* **8**, 256–267.
65. Chu JW, Voth GA (2006) Coarse-grained modeling of the actin filament derived from atomistic-scale simulations. *Biophys J* **90**, 1572–1582.
66. Reynwar BJ, Illya G, Harmandaris VA, Muller MM, Kremer K, Deserno M (2007) Aggregation and vesiculation of membrane proteins by curvature-mediated interactions. *Nature* **447**, 461–464.
67. Yao XQ, Kenzaki H, Murakami S, Takada S (2010) Drug export and allosteric coupling in a multidrug transporter revealed by molecular simulations. *Nat Commun* **1**, 117.
68. Gorfe AA, Chang CE, Ivanov I, McCammon JA (2008) Dynamics of the acetylcholinesterase tetramer. *Biophys J* **94**, 1144–1154.
69. Trylska J, Tozzini V, McCammon JA (2005) Exploring global motions and correlations in the ribosome. *Biophys J* **89**, 1455–1463.
70. Chen J, Darst SA, Thirumalai D (2010) Promoter melting triggered by bacterial RNA polymerase occurs in three steps. *Proc Natl Acad Sci U S A* **107**, 12523–12528.
71. Terakawa T, Takada S (2011) Multiscale ensemble modeling of intrinsically disordered proteins: p53 N-terminal domain. *Biophys J* **101**, 1450–1458.
72. Chen J (2009) Intrinsically disordered p53 extreme C-terminus binds to S100B(betabeta) through "fly-casting." *J Am Chem Soc* **131**, 2088–2089.

73. Bakowies D, Thiel W (1996) Hybrid models for combined quantum mechanical and molecular mechanical approaches. *J Phys Chem* **100**, 10580–10594.
74. Field MJ, Bash PA, Karplus M (1990) A combined quantum-mechanical and molecular mechanical potential for molecular-dynamics simulations. *J Comput Chem* **11**, 700–733.
75. Gao JL, Truhlar DG (2002) Quantum mechanical methods for enzyme kinetics. *Annu Rev Phys Chem* **53**, 467–505.
76. Monard G, Merz KM (1999) Combined quantum mechanical/molecular mechanical methodologies applied to biomolecular systems. *Acc Chem Res* **32**, 904–911.
77. Mordasini TZ, Thiel W (1998) Combined quantum mechanical and molecular mechanical approaches. *Chimia* **52**, 288–291.
78. Senn HM, Thiel W (2009) QM/MM methods for biomolecular systems. *Angew Chem Int Ed Engl* **48**, 1198–1229.
79. Kamerlin SC, Vicatos S, Dryga A, Warshel A (2011) Coarse-grained (multiscale) simulations in studies of biophysical and chemical systems. *Annu Rev Phys Chem* **62**, 41–64.
80. Crowet JM, Parton DL, Hall BA, Steinhauer S, Brasseur R, Lins L, Sansom MS (2012) Multi-scale simulation of the simian immunodeficiency virus fusion peptide. *J Phys Chem B* **116**, 13713–13721.
81. Shi Q, Voth GA (2005) Multi-scale modeling of phase separation in mixed lipid bilayers. *Biophys J* **89**, 2385–2394.
82. Coluzza I, De Simone A, Fraternali F, Frenkel D (2008) Multi-scale simulations provide supporting evidence for the hypothesis of intramolecular protein translocation in GroEL/GroES complexes. *PLoS Comput Biol* **4**, e1000006.
83. Shih AY, Arkhipov A, Freddolino PL, Schulten K (2006) Coarse grained protein-lipid model with application to lipoprotein particles. *J Phys Chem B* **110**, 3674–3684.
84. Heyden A, Lin H, Truhlar DG (2007) Adaptive partitioning in combined quantum mechanical and molecular mechanical calculations of potential energy functions for multiscale simulations. *J Phys Chem B* **111**, 2231–2241.
85. Csanyi G, Albaret T, Payne MC, De Vita A (2004) "Learn on the fly": A hybrid classical and quantum-mechanical molecular dynamics simulation. *Phys Rev Lett* **93**, 175503.
86. Kerdcharoen T, Liedl KR, Rode BM (1996) A QM/MM simulation method applied to the solution of Li+ in liquid ammonia. *Chem Phys* **211**, 313–323.
87. Bulo RE, Ensing B, Sikkema J, Visscher L (2009) Toward a practical method for adaptive QM/MM simulations. *J Chem Theory Comput* **5**, 2212–2221.
88. Svensson M, Humbel S, Froese R, Matsubara T, Sieber S, Morokuma K (1996) ONIOM: A multilayered integrated MO + MM method for geometry optimizations and single point energy predictions. A test for Diels-Alder reactions and Pt(P(t-Bu)3)2 + H2 oxidative addition. *J Phys Chem* **100**, 19357–19363.
89. Kerdcharoen T, Morokuma K (2002) ONIOM-XS: An extension of the ONIOM method for molecular simulation in condensed phase. *Chem Phys Lett* **355**, 257–262.
90. Vreven T, Morokuma K, Farkas O, Schlegel HB, Frisch MJ (2003) Geometry optimization with QM/MM, ONIOM, and other combined methods. I. Microiterations and constraints. *J Comput Chem* **24**, 760–769.
91. Chen B, Tycko R (2011) Simulated self-assembly of the HIV-1 capsid: Protein shape and native contacts are sufficient for two-dimensional lattice formation. *Biophys J* **100**, 3035–3044.

92. Lu HM, Liang J (2009) Perturbation-based Markovian transmission model for probing allosteric dynamics of large macromolecular assembling: A study of GroEL-GroES. *PLoS Comput Biol* **5**, e1000526.

93. Hicks SD, Henley CL (2010) Coarse-grained protein-protein stiffnesses and dynamics from all-atom simulations. *Phys Rev E Stat Nonlin Soft Matter Phys* **81**, 030903.

94. Ayton GS, Voth GA (2010) Multiscale computer simulation of the immature HIV-1 virion. *Biophys J* **99**, 2757–2765.

95. Chu JW, Voth GA (2005) Allostery of actin filaments: Molecular dynamics simulations and coarse-grained analysis. *Proc Natl Acad Sci U S A* **102**, 13111–13116.

96. Tirion MM (1996) Large amplitude elastic motions in proteins from a single-parameter, atomic analysis. *Phys Rev Lett* **77**, 1905–1908.

97. Atilgan AR, Durell SR, Jernigan RL, Demirel MC, Keskin O, Bahar I (2001) Anisotropy of fluctuation dynamics of proteins with an elastic network model. *Biophys J* **80**, 505–515.

98. Zhang Z, Voth GA (2010) Coarse-grained representation of large biomolecular complexes from low-resolution structural data. *J Chem Theory Comput* **6**, 2990–3002.

99. Bahar I, Lezon TR, Yang LW, Eyal E (2010) Global dynamics of proteins: Bridging between structure and function. *Annu Rev Biophys* **39**, 23–42.

100. Fuxreiter M, Warshel A (1998) Origin of the catalytic power of acetylcholineesterase. Computer simulation studies. *J Am Chem Soc* **120**, 183–194.

101. Warshel A, Sharma PK, Kato M, Xiang Y, Liu H, Olsson MH (2006) Electrostatic basis for enzyme catalysis. *Chem Rev* **106**, 3210–3235.

102. Bernstein N, Varnai C, Solt I, Winfield SA, Payne MC, Simon I, Fuxreiter M, Csanyi G (2012) QM/MM simulation of liquid water with an adaptive quantum region. *Phys Chem Chem Phys* **14**, 646–656.

103. Feig M, Onufriev A, Lee MS, Im W, Case DA, Brooks CL, III (2004) Performance comparison of generalized Born and Poisson methods in the calculation of electrostatic solvation energies for protein structures. *J Comput Chem* **25**, 265–284.

104. Zhu J, Shi Y, Liu H (2002) Parametrization of a generalized Born/solvent-accessible surface area model and applications to the simulation of protein dynamics. *J Phys Chem B* **106**, 4844–4853.

105. Lazaridis T, Karplus M (1999) Effective energy function for proteins in solution. *Proteins* **35**, 133–152.

106. King G, Lee FS, Warshel A (1991) Microscopic simulations of macroscopic dielectric constants of solvated proteins. *J Chem Phys* **95**, 4366–4377.

107. Muegge I, Schweins T, Langen R, Warshel A (1996) Electrostatic control of GTP and GDP binding in the oncoprotein p21ras. *Structure* **4**, 475–489.

108. Lee FS, Warshel A (1992) A local reaction field method for fast evaluation of long-range electrostatic interactions in molecular simulations. *J Chem Phys* **97**, 3100–3107.

109. Sham YY, Chu ZT, Tao H, Warshel A (2000) Examining methods for calculations of binding free energies: LRA, LIE, PDLD-LRA, and PDLD/S-LRA calculations of ligands binding to an HIV protease. *Proteins* **39**, 393–407.

110. Ooi T, Oobatake M, Nemethy G, Scheraga HA (1987) Accessible surface areas as a measure of the thermodynamic parameters of hydration of peptides. *Proc Natl Acad Sci U S A* **84**, 3086–3090.

111. Ferrara P, Apostolakis J, Caflisch A (2002) Evaluation of a fast implicit solvent model for molecular dynamics simulations. *Proteins* **46**, 24–33.

112. Spassov VZ, Yan L, Szalma S (2002) Introducing an implicit membrane in generalized Born/solvent accessibility continuum solvent models. *J Chem Phys* **106**, 8726–8738.

113. Shirts M, Mobley D, Brown S (2010) Free Energy Calculations in structure-based drug design. In *Drug Design: Structure- and Ligand-Based Approaches.* Edited by Merz Jr KM, Ringe D, Reynolds CH. New York: Cambridge University Press.

114. Michel J, Essex JW (2010) Prediction of protein-ligand binding affinity by free energy simulations: Assumptions, pitfalls and expectations. *J Comput Aided Mol Des* **24**, 639–658.

115. Chodera JD, Mobley DL, Shirts MR, Dixon RW, Branson K, Pande VS (2011) Alchemical free energy methods for drug discovery: Progress and challenges. *Curr Opin Struct Biol* **21**, 150–160.

116. Postma JPM, Berendsen HJC (1982) Thermodynamics of cavity formation in water. A molecular-dynamics study. *Symp Chem Soc* **17**, 55–67.

117. Kirkwood JG (1935) Statistical mechanics of fluid mixtures. *J Chem Phys* **3**, 300–312.

118. Beveridge DL, DiCapua FM (1989) Free energy via molecular simulation: Applications to chemical and biomolecular systems. *Annu Rev Biophys Biophys Chem* **18**, 431–492.

119. Kollman P (1993) Free energy calculations: Applications to chemical and biochemical phenomena. *Chem Rev* **93**, 2395–2417.

120. Tembe B, McCammon J (1984) Ligand-receptor interactions. *Comput Chem* **8**, 281–283.

121. Martin YC (2009) Let's not forget tautomers. *J Comput Aided Mol Des* **23**, 693–704.

122. Boyce SE, Mobley DL, Rocklin GJ, Graves AP, Dill KA, Shoichet BK (2009) Predicting ligand binding affinity with alchemical free energy methods in a polar model binding site. *J Mol Biol* **394**, 747–763.

123. Klimovich PV, Mobley DL (2010) Predicting hydration free energies using all-atom molecular dynamics simulations and multiple starting conformations. *J Comput Aided Mol Des* **24**, 307–316.

124. Mobley DL, Dill KA (2009) Binding of small-molecule ligands to proteins: "What you see" is not always "what you get." *Structure* **17**, 489–498.

125. Wang J, Deng Y, Roux B (2006) Absolute binding free energy calculations using molecular dynamics simulations with restraining potentials. *Biophys J* **91**, 2798–2814.

126. Mehler EL, Fuxreiter M, Simon I, Garcia-Moreno EB (2002) The role of hydrophobic microenvironments in modulating pKa shifts in proteins. *Proteins* **48**, 283–292.

127. Graves AP, Shivakumar DM, Boyce SE, Jacobson MP, Case DA, Shoichet BK (2008) Rescoring docking hit lists for model cavity sites: Predictions and experimental testing. *J Mol Biol* **377**, 914–934.

128. Mobley DL, Chodera JD, Dill KA (2007) Confine-and-release method: Obtaining correct binding free energies in the presence of protein conformational change. *J Chem Theory Comput* **3**, 1231–1235.

129. Phillips JC, Braun R, Wang W, Gumbart J, Tajkhorshid E, Villa E, Chipot C, Skeel RD, Kale L, Schulten K (2005) Scalable molecular dynamics with NAMD. *J Comput Chem* **26**, 1781–1802.

130. Hess B, Kutzner C, Van der Spoel D, Lindahl E (2008) GROMACS4: Algorithms for highly efficient, load balanced and scalable molecular simulation. *J Chem Theory Comput* **4**, 435–447.

131. Jiang W, Roux B (2010) Free energy perturbation Hamiltonian replica-exchange molecular dynamics (FEP/H-REMD) for absolute ligand binding free energy calculations. *J Chem Theory Comput* **6**, 2559–2565.
132. Baftizadeh F, Biarnes X, Pietrucci F, Affinito F, Laio A (2012) Multidimensional view of amyloid fibril nucleation in atomistic detail. *J Am Chem Soc* **134**, 3886–3894.
133. Singhal N, Pande VS (2005) Error analysis and efficient sampling in Markovian state models for molecular dynamics. *J Chem Phys* **123**, 204909.
134. Jayachandran G, Shirts MR, Park S, Pande VS (2006) Parallelized-over-parts computation of absolute binding free energy with docking and molecular dynamics. *J Chem Phys* **125**, 084901.
135. Park S, Pande VS (2006) Validation of Markov state models using Shannon's entropy. *J Chem Phys* **124**, 054118.
136. Gallicchio E, Lapelosa M, Levy RM (2010) The binding energy distribution analysis method (BEDAM) for the estimation of protein-ligand binding affinities. *J Chem Theory Comput* **6**, 2961–2977.
137. Kollman PA, Massova I, Reyes C, Kuhn B, Huo S, Chong L, Lee M, Lee T, Duan Y, Wang W et al. (2000) Calculating structures and free energies of complex molecules: Combining molecular mechanics and continuum models. *Acc Chem Res* **33**, 889–897.
138. Simonson T, Archontis G, Karplus M (2002) Free energy simulations come of age: Protein-ligand recognition. *Acc Chem Res* **35**, 430–437.
139. Swanson JM, Henchman RH, McCammon JA (2004) Revisiting free energy calculations: A theoretical connection to MM/PBSA and direct calculation of the association free energy. *Biophys J* **86**, 67–74.
140. Cornell WD, Cieplak R, Bayly CL, Gould IR, Merz KM, Ferguson DM, Spellmeyer DG, Fox T, Caldwell JW, Kollman PA (1995) A second generation force field for the simulation of proteins, nucleic acids, and organic molecules. *J Am Chem Soc* **117**, 5179–5197.
141. MacKerell AD, Jr., Banavali N, Foloppe N (2000) Development and current status of the CHARMM force field for nucleic acids. *Biopolymers* **56**, 257–265.
142. Freddolino PL, Park S, Roux B, Schulten K (2009) Force field bias in protein folding simulations. *Biophys J* **96**, 3772–3780.
143. Piana S, Lindorff-Larsen K, Shaw DE (2011) How robust are protein folding simulations with respect to force field parameterization? *Biophys J* **100**, L47–L49.
144. Solt I, Magyar C, Simon I, Tompa P, Fuxreiter M (2006) Phosphorylation-induced transient intrinsic structure in the kinase-inducible domain of CREB facilitates its recognition by the KIX domain of CBP. *Proteins* **64**, 749–757.
145. Duan Y, Wu C, Chowdhury S, Lee MC, Xiong G, Zhang W, Yang R, Cieplak P, Luo R, Lee T et al. (2003) A point-charge force field for molecular mechanics simulations of proteins based on condensed-phase quantum mechanical calculations. *J Comput Chem* **24**, 1999–2012.
146. Garcia AE, Sanbonmatsu KY (2002) α-Helical stabilization by side chain shielding of backbone hydrogen bonds. *Proc Natl Acad Sci U S A* **99**, 2782–2787.
147. Simmerling C, Strockbine B, Roitberg AE (2002) All-atom structure prediction and folding simulations of a stable protein. *J Am Chem Soc* **124**, 11258–11259.
148. Best RB, Buchete NV, Hummer G (2008) Are current molecular dynamics force fields too helical? *Biophys J* **95**, L07–L09.

149. Best RB, Mittal J (2010) Balance between alpha and beta structures in ab initio protein folding. *J Phys Chem B* **114**, 8790–8798.
150. Li D-W, Brüschweiler R (2011) Iterative optimization of molecular mechanics force fields from NMR data of full-length proteins. *J Chem Theory Comput* **7**, 1220–1230.
151. Morozov AV, Tsemekhman K, Baker D (2006) Electron density redistribution accounts for half the cooperativity of alpha helix formation. *J Phys Chem B* **110**, 4503–4505.
152. Freddolino PL, Schulten K (2009) Common structural transitions in explicit-solvent simulations of villin headpiece folding. *Biophys J* **97**, 2338–2347.
153. Shaw DE, Maragakis P, Lindorff-Larsen K, Piana S, Dror RO, Eastwood MP, Bank JA, Jumper JM, Salmon JK, Shan Y et al. (2010) Atomic-level characterization of the structural dynamics of proteins. *Science* **330**, 341–346.
154. Juraszek J, Bolhuis PG (2008) Rate constant and reaction coordinate of Trp-cage folding in explicit water. *Biophys J* **95**, 4246–4257.
155. Best RB, Hummer G (2005) Reaction coordinates and rates from transition paths. *Proc Natl Acad Sci U S A* **102**, 6732–6737.
156. Weinan E, Ren W, Vanden-Eijnden E (2005) Finite temperature string method for the study of rare events. *J Phys Chem B* **109**, 6688–6693.
157. Day R, Paschek D, Garcia AE (2010) Microsecond simulations of the folding/unfolding thermodynamics of the Trp-cage miniprotein. *Proteins* **78**, 1889–1899.
158. Prinz JH, Wu H, Sarich M, Keller B, Senne M, Held M, Chodera JD, Schutte C, Noe F (2011) Markov models of molecular kinetics: Generation and validation. *J Chem Phys* **134**, 174105.
159. Hummer G (2004) From transition paths to transition states and rate coefficients. *J Chem Phys* **120**, 516–523.
160. Lane TJ, Shukla D, Beauchamp KA, Pande VS (2013) To milliseconds and beyond: Challenges in the simulation of protein folding. *Curr Opin Struct Biol* **23**, 58–65.
161. Thirumalai D, O'Brien EP, Morrison G, Hyeon C (2010) Theoretical perspectives on protein folding. *Annu Rev Biophys* **39**, 159–183.
162. Zheng W, Schafer NP, Davtyan A, Papoian GA, Wolynes PG (2012) Predictive energy landscapes for protein-protein association. *Proc Natl Acad Sci U S A* **109**, 19244–19249.
163. Wright PE, Dyson HJ (1999) Intrinsically unstructured proteins: Re-assessing the protein structure-function paradigm. *J Mol Biol* **293**, 321–331.
164. Dyson HJ, Wright PE (2005) Intrinsically unstructured proteins and their functions. *Nat Rev Mol Cell Biol* **6**, 197–208.
165. Romero P, Obradovic Z, Kissinger CR, Villafranca JE, Garner E, Guillot S, Dunker AK (1998) Thousands of proteins likely to have long disordered regions. *Pac Symp Biocomp* **3**, 437–448.
166. Iakoucheva LM, Radivojac P, Brown CJ, O'Connor TR, Sikes JG, Obradovic Z, Dunker AK (2004) The importance of intrinsic disorder for protein phosphorylation. *Nucleic Acids Res* **32**, 1037–1049.
167. Fuxreiter M, Tompa P, Simon I (2007) Local structural disorder imparts plasticity on linear motifs. *Bioinformatics* **23**, 950–956.
168. Davey NE, Van Roey K, Weatheritt RJ, Toedt G, Uyar B, Altenberg B, Budd A, Diella F, Dinkel H, Gibson TJ (2012) Attributes of short linear motifs. *Mol Biosyst* **8**, 268–281.

169. Ward JJ, Sodhi JS, McGuffin LJ, Buxton BF, Jones DT (2004) Prediction and functional analysis of native disorder in proteins from the three kingdoms of life. *J Mol Biol* **337**, 635–645.

170. Tompa P, Dosztanyi Z, Simon I (2006) Prevalent structural disorder in E. coli and S. cerevisiae proteomes. *J Proteome Res* **5**, 1996–2000.

171. Dunker AK, Garner E, Guillot S, Romero P, Albrecht K, Hart J, Obradovic Z (1998) Protein disorder and the evolution of molecular recognition: Theory, predictions and observations. *Pac Symp Biocomput* **3**, 473–484.

172. Tompa P (2005) The interplay between structure and function in intrinsically unstructured proteins. *FEBS Lett* **579**, 3346–3354.

173. Fuxreiter M, Tompa P, Simon I, Uversky VN, Hansen JC, Asturias FJ (2008) Malleable machines take shape in eukaryotic transcriptional regulation. *Nat Chem Biol* **4**, 728–737.

174. Xie H, Vucetic S, Iakoucheva LM, Oldfield CJ, Dunker AK, Obradovic Z, Uversky VN (2007) Functional anthology of intrinsic disorder. 3. Ligands, post-translational modifications, and diseases associated with intrinsically disordered proteins. *J Proteome Res* **6**, 1917–1932.

175. Mohan A, Sullivan WJ, Jr., Radivojac P, Dunker AK, Uversky VN (2008) Intrinsic disorder in pathogenic and non-pathogenic microbes: Discovering and analyzing the unfoldomes of early-branching eukaryotes. *Mol Biosyst* **4**, 328–340.

176. Uversky VN, Oldfield CJ, Dunker AK (2008) Intrinsically disordered proteins in human diseases: Introducing the D2 concept. *Annu Rev Biophys* **37**, 215–246.

177. Bussell R, Jr., Eliezer D (2001) Residual structure and dynamics in Parkinson's disease-associated mutants of alpha-synuclein. *J Biol Chem* **276**, 45996–46003.

178. Ferguson N, Becker J, Tidow H, Tremmel S, Sharpe TD, Krause G, Flinders J, Petrovich M, Berriman J, Oschkinat H et al. (2006) General structural motifs of amyloid protofilaments. *Proc Natl Acad Sci U S A* **103**, 16248–16253.

179. Wells M, Tidow H, Rutherford TJ, Markwick P, Jensen MR, Mylonas E, Svergun DI, Blackledge M, Fersht AR (2008) Structure of tumor suppressor p53 and its intrinsically disordered N-terminal transactivation domain. *Proc Natl Acad Sci U S A* **105**, 5762–5767.

180. Marsh JA, Forman-Kay JD (2009) Structure and disorder in an unfolded state under nondenaturing conditions from ensemble models consistent with a large number of experimental restraints. *J Mol Biol* **391**, 359–374.

181. Bernado P, Blanchard L, Timmins P, Marion D, Ruigrok RW, Blackledge M (2005) A structural model for unfolded proteins from residual dipolar couplings and small-angle x-ray scattering. *Proc Natl Acad Sci U S A* **102**, 17002–17007.

182. Daughdrill GW, Hanely LJ, Dahlquist FW (1998) The C-terminal half of the anti-sigma factor FlgM contains a dynamic equilibrium solution structure favoring helical conformations. *Biochemistry* **37**, 1076–1082.

183. Bienkiewicz EA, Adkins JN, Lumb KJ (2002) Functional consequences of preorganized helical structure in the intrinsically disordered cell-cycle inhibitor p27(Kip1). *Biochemistry* **41**, 752–759.

184. Fuxreiter M, Simon I, Friedrich P, Tompa P (2004) Preformed structural elements feature in partner recognition by intrinsically unstructured proteins. *J Mol Biol* **338**, 1015–1026.

185. Salmon L, Nodet G, Ozenne V, Yin G, Jensen MR, Zweckstetter M, Blackledge M (2010) NMR characterization of long-range order in intrinsically disordered proteins. *J Am Chem Soc* **132**, 8407–8418.

186. Jensen MR, Salmon L, Nodet G, Blackledge M (2010) Defining conformational ensembles of intrinsically disordered and partially folded proteins directly from chemical shifts. *J Am Chem Soc* **132**, 1270–1272.
187. Marsh JA, Forman-Kay JD (2011) Ensemble modeling of protein disordered states: Experimental restraint contributions and validation. *Proteins* **80**, 556–572.
188. Bernado P, Svergun DI (2012) Structural analysis of intrinsically disordered proteins by small-angle X-ray scattering. *Mol Biosyst* **8**, 151–167.
189. Bertini I, Giachetti A, Luchinat C, Parigi G, Petoukhov MV, Pierattelli R, Ravera E, Svergun DI (2010) Conformational space of flexible biological macromolecules from average data. *J Am Chem Soc* **132**, 13553–13558.
190. Ozenne V, Bauer F, Salmon L, Huang JR, Jensen MR, Segard S, Bernado P, Charavay C, Blackledge M (2012) Flexible-meccano: A tool for the generation of explicit ensemble descriptions of intrinsically disordered proteins and their associated experimental observables. *Bioinformatics* **28**, 1463–1470.
191. Wright PE, Dyson HJ (2009) Linking folding and binding. *Curr Opin Struct Biol* **19**, 31–38.
192. Vuzman D, Levy Y (2010) DNA search efficiency is modulated by charge composition and distribution in the intrinsically disordered tail. *Proc Natl Acad Sci U S A* **107**, 21004–21009.
193. Ganguly D, Otieno S, Waddell B, Iconaru L, Kriwacki RW, Chen J (2012) Electrostatically accelerated coupled binding and folding of intrinsically disordered proteins. *J Mol Biol* **422**, 674–684.
194. Chu X, Wang Y, Gan L, Bai Y, Han W, Wang E, Wang J (2012) Importance of electrostatic interactions in the association of intrinsically disordered histone chaperone Chz1 and histone H2A.Z-H2B. *PLoS Comput Biol* **8**, e1002608.
195. Knott M, Best RB (2012) A preformed binding interface in the unbound ensemble of an intrinsically disordered protein: Evidence from molecular simulations. *PLoS Comput Biol* **8**, e1002605.
196. Naganathan AN, Orozco M (2013) The conformational landscape of an intrinsically disordered DNA-binding domain of a transcription regulator. *J Phys Chem B* **117**, 13842–13850.
197. Sethi A, Tian J, Vu DM, Gnanakaran S (2012) Identification of minimally interacting modules in an intrinsically disordered protein. *Biophys J* **103**, 748–757.
198. Pauling L (1948) Chemical achievement and hope for the future. *Am Sci* **36**, 51–58.
199. Page MI, Jencks WP (1971) Entropic contributions to rate accelerations in enzymic and intramolecular reactions and the chelate effect. *Proc Natl Acad Sci U S A* **68**, 1678–1683.
200. Hur S, Bruice TC (2003) The near attack conformation approach to the study of the chorismate to prephenate reaction. *Proc Natl Acad Sci U S A* **100**, 12015–12020.
201. Blake CC, Johnson LN, Mair GA, North AC, Phillips DC, Sarma VR (1967) Crystallographic studies of the activity of hen egg-white lysozyme. *Proc R Soc Lond B Biol Sci* **167**, 378–388.
202. Cameron CE, Benkovic SJ (1997) Evidence for a functional role of the dynamics of glycine-121 of Escherichia coli dihydrofolate reductase obtained from kinetic analysis of a site-directed mutant. *Biochemistry* **36**, 15792–15800.
203. Lee JK, Houk KN (1997) A proficient enzyme revisited: The predicted mechanism for orotidine monophosphate decarboxylase. *Science* **276**, 942–945.

204. Baker D (2010) An exciting but challenging road ahead for computational enzyme design. *Protein Sci* **19**, 1817–1819.

205. Ruscio JZ, Kohn JE, Ball KA, Head-Gordon T (2009) The influence of protein dynamics on the success of computational enzyme design. *J Am Chem Soc* **131**, 14111–14115.

206. Tantillo DJ, Chen J, Houk KN (1998) Theozymes and compuzymes: Theoretical models for biological catalysis. *Curr Opin Chem Biol* **2**, 743–750.

207. Fuxreiter M, Mones L (2014) The role of reorganization energy in rational enzyme design. *Curr Opin Chem Biol* **21**, 34–41.

208. Frushicheva MP, Cao J, Chu ZT, Warshel A (2010) Exploring challenges in rational enzyme design by simulating the catalysis in artificial kemp eliminase. *Proc Natl Acad Sci U S A* **107**, 16869–16874.

209. Frushicheva MP, Cao J, Warshel A (2011) Challenges and advances in validating enzyme design proposals: The case of kemp eliminase catalysis. *Biochemistry* **50**, 3849–3858.

210. Kamerlin SC, Warshel A (2010) At the dawn of the 21st century: Is dynamics the missing link for understanding enzyme catalysis? *Proteins* **78**, 1339–1375.

211. Henzler-Wildman KA, Lei M, Thai V, Kerns SJ, Karplus M, Kern D (2007) A hierarchy of timescales in protein dynamics is linked to enzyme catalysis. *Nature* **450**, 913–916.

212. Agarwal PK, Billeter SR, Rajagopalan PT, Benkovic SJ, Hammes-Schiffer S (2002) Network of coupled promoting motions in enzyme catalysis. *Proc Natl Acad Sci U S A* **99**, 2794–2799.

213. Eisenmesser EZ, Millet O, Labeikovsky W, Korzhnev DM, Wolf-Watz M, Bosco DA, Skalicky JJ, Kay LE, Kern D (2005) Intrinsic dynamics of an enzyme underlies catalysis. *Nature* **438**, 117–121.

214. Saen-Oon S, Quaytman-Machleder S, Schramm VL, Schwartz SD (2008) Atomic detail of chemical transformation at the transition state of an enzymatic reaction. *Proc Natl Acad Sci U S A* **105**, 16543–16548.

215. Kempf JG, Jung JY, Ragain C, Sampson NS, Loria JP (2007) Dynamic requirements for a functional protein hinge. *J Mol Biol* **368**, 131–149.

216. Henzler-Wildman KA, Thai V, Lei M, Ott M, Wolf-Watz M, Fenn T, Pozharski E, Wilson MA, Petsko GA, Karplus M et al. (2007) Intrinsic motions along an enzymatic reaction trajectory. *Nature* **450**, 838–844.

217. Nashine VC, Hammes-Schiffer S, Benkovic SJ (2010) Coupled motions in enzyme catalysis. *Curr Opin Chem Biol* **14**, 644–651.

218. Wong KF, Selzer T, Benkovic SJ, Hammes-Schiffer S (2005) Impact of distal mutations on the network of coupled motions correlated to hydride transfer in dihydrofolate reductase. *Proc Natl Acad Sci U S A* **102**, 6807–6812.

219. Pisliakov AV, Cao J, Kamerlin SC, Warshel A (2009) Enzyme millisecond conformational dynamics do not catalyze the chemical step. *Proc Natl Acad Sci U S A* **106**, 17359–17364.

220. Schwartz SD, Schramm VL (2009) Enzymatic transition states and dynamic motion in barrier crossing. *Nat Chem Biol* **5**, 551–558.

221. Warshel A, Parson WW (2001) Dynamics of biochemical and biophysical reactions: Insight from computer simulations. *Q Rev Biophys* **34**, 563–679.

222. Adamczyk AJ, Cao J, Kamerlin SC, Warshel A (2011) Catalysis by dihydrofolate reductase and other enzymes arises from electrostatic preorganization, not conformational motions. *Proc Natl Acad Sci U S A* **108**, 14115–14120.

223. Cha Y, Murray CJ, Klinman JP (1989) Hydrogen tunneling in enzyme reactions. *Science* **243**, 1325–1330.
224. Kohen A, Cannio R, Bartolucci S, Klinman JP (1999) Enzyme dynamics and hydrogen tunnelling in a thermophilic alcohol dehydrogenase. *Nature* **399**, 496–499.
225. Cui Q, Karplus M (2002) Quantum mechanics/molecular mechanics studies of triosephosphate isomerase-catalyzed reactions: Effect of geometry and tunneling on proton-transfer rate constants. *J Am Chem Soc* **124**, 3093–3124.
226. Roca M, Messer B, Hilvert D, Warshel A (2008) On the relationship between folding and chemical landscapes in enzyme catalysis. *Proc Natl Acad Sci U S A* **105**, 13877–13882.
227. Strajbl M, Shurki A, Warshel A (2003) Converting conformational changes to electrostatic energy in molecular motors: The energetics of ATP synthase. *Proc Natl Acad Sci U S A* **100**, 14834–14839.
228. Mukherjee S, Warshel A (2011) Electrostatic origin of the mechanochemical rotary mechanism and the catalytic dwell of F1-ATPase. *Proc Natl Acad Sci U S A* **108**, 20550–20555.
229. Warshel A, Florian J (2004) The empirical valence bond (EVB) method. In *The Encyclopedia of Computational Chemistry*. Edited by Schleyer PvR, Jorgensen WL, Schaefer HF, III, Schreiner PR, Thiel W, Glen R. New York: John Wiley & Sons.
230. Warshel A, Weiss RM (1981) Empirical valence bond calculations of enzyme catalysis. *Ann N Y Acad Sci* **367**, 370–382.
231. Rosta E, Klahn M, Warshel A (2006) Towards accurate ab initio QM/MM calculations of free-energy profiles of enzymatic reactions. *J Phys Chem B* **110**, 2934–2941.
232. Solt I, Kulhanek P, Simon I, Winfield S, Payne MC, Csanyi G, Fuxreiter M (2009) Evaluating boundary dependent errors in QM/MM simulations. *J Phys Chem B* **113**, 5728–5735.
233. Ziv G, Haran G, Thirumalai D (2005) Ribosome exit tunnel can entropically stabilize alpha-helices. *Proc Natl Acad Sci U S A* **102**, 18956–18961.
234. Rychkova A, Mukherjee S, Bora RP, Warshel A (2013) Simulating the pulling of stalled elongated peptide from the ribosome by the translocon. *Proc Natl Acad Sci U S A* **110**, 10195–10200.
235. Kragelj J, Ozenne V, Blackledge M, Jensen MR (2013) Conformational propensities of intrinsically disordered proteins from NMR chemical shifts. *Chemphyschem* **14**, 3034–3045.
236. Fuxreiter M, Simon I, Bondos S (2011) Dynamic protein-DNA recognition: Beyond what can be seen. *Trends Biochem Sci* **36**, 415–423.
237. Tzeng SR, Kalodimos CG (2009) Dynamic activation of an allosteric regulatory protein. *Nature* **462**, 368–372.
238. Galea CA, Nourse A, Wang Y, Sivakolundu SG, Heller WT, Kriwacki RW (2008) Role of intrinsic flexibility in signal transduction mediated by the cell cycle regulator, p27 Kip1. *J Mol Biol* **376**, 827–838.
239. Fuxreiter M (2012) Fuzziness: Linking regulation to protein dynamics. *Mol Biosyst* **8**, 168–177.
240. Ma B, Tsai CJ, Haliloglu T, Nussinov R (2011) Dynamic allostery: Linkers are not merely flexible. *Structure* **19**, 907–917.

Enzymatic Catalysis: Multiscale QM/MM Calculations

Chapter 2

Adaptive and Accurate Force-Based QM/MM Calculations

Noam Bernstein, Iván Solt, Letif Mones, Csilla Várnai, Steven A. Winfield, and Gábor Csányi

CONTENTS

Dedicated to Martin Karplus, Michael Levitt, and Arieh Warshel, on the occasion of winning the Nobel Prize in Chemistry (2013) for laying the foundations of QM/MM.

INTRODUCTION

The QM/MM method, which couples a quantum mechanical (QM) description of bonding in a localized region to a larger molecular mechanics (MM) simulation, is widely used for simulations of proteins [1–5]. The combination allows it to benefit from the strengths of both QM and MM approaches. The QM description gives accuracy on bond-breaking energetics and structures where it is needed, but it is too computationally expensive to be used beyond a few hundred atoms at most. The MM description is much faster computationally and makes it possible to describe large-scale structure and motion, solvation, and entropic effects. In its usual formulation, the set of atoms described by QM is fixed during the simulation, but under some circumstances this requirement can be overly restrictive. One possible scenario is if protein residues move over a distance larger than the feasible QM region size or if describing the reaction requires reactants, products, or solvent molecules to be transported over large distances into or out of the QM region. Another possible scenario is if a large QM region, including one or more solvation shells around a reaction site, is needed simply for accuracy. If such a large QM region is needed, solvent molecules near its edge may diffuse away, and without adaptivity the diffusion would lead to a fragmented QM region. Instead of using adaptivity it would be possible to keep the QM molecules confined using some sort of ad hoc restraining potential, but that would require individual tuning for each system to maintain correct density in the QM region. The simulations we present here pertain to this latter situation.

To begin addressing this problem, we have developed the buffered-force QM/MM (bf-QM/MM) method based on a similar approach used for solid state systems [6,7]. The bf-QM/MM method allows for adaptivity of the set of QM atoms, letting functional groups and molecules move into and out of the QM region. The challenge in adaptive simulations is to prevent the inevitable errors at the interface between the two methods from building up over time by systematically transporting atoms into or out of the QM region. By reducing the errors in the forces on atoms near the interface, we reduce the size of artifacts in observables computed from the molecular dynamics trajectories. The resulting method produces stable, long-running simulations with adaptively changing QM regions. We note that we do not present dynamic simulation results for proteins here, only an initial evaluation of the performance and accuracy of the bf-QM/MM method when applied to small solutes in water. For proteins we only present calculations of force errors at the center of the QM region for configurations taken from an equilibrium MM trajectory as a function of QM region size, to show the effect of the QM-MM interface in this type of system and to guide the selection of appropriate QM regions in future work.

The coupling of two methods in different regions of a single atomistic simulation can be formulated in a number of ways. In one class of approaches, which we call energy mixing, one defines a total energy with contributions from the two methods, each applied to its own region, and a coupling energy. In the other class, which we call force mixing, the energy is not explicitly defined (and in fact may not be well defined at all), but instead one defines a set of atomic forces that can be used to propagate molecular dynamics and calculate free energies. While the advantages of a defined total energy for stable dynamics and calculation of free energies are obvious, there are also some disadvantages that are hard to overcome in practice. One is that the chemical potentials of various species in both methods will not in general be equal. Trying to smoothly join the raw potential energy surfaces corresponding to different partitionings between the QM and MM regions will thus result in artificial shifts in the global potential energy surface that cannot in practice be corrected for all molecular configurations at all positions in the system, leading to unphysical mass transport across the interface. Another important issue is that both methods will, in general, be affected by the presence of the unphysical interface between the two regions, and the coupling term, which must compensate for these effects, cannot do so perfectly. The presence of the interface will therefore perturb the system. Force mixing approaches take advantage of the fact that unlike the potential energy, atomic forces are inherently local properties of the atoms, and it is therefore possible to construct a global set of forces from linear combinations of forces computed with the two methods separately. As we explain in more detail in the following, the use of buffers makes it possible to calculate accurate QM and MM forces for use in force mixing.

The magnitude of the artifacts caused by the QM-MM chemical potential mismatch and perturbations caused by the presence of the QM-MM interface will depend on the individual QM and MM methods as well as the coupling scheme. Our buffered-force method was developed to be used in conjunction with electrostatic embedding where the QM atoms interact in a self-consistent way with the electrostatic potential of classical charge distributions corresponding to the MM atoms [8], and our tests use this embedding approach. It is possible that with other methods or in other chemical systems the effects of the QM-MM interface would be different, whether on the forces themselves or on the structures and free energies that are calculated from the trajectories. Our use of buffers to ensure that the dynamics follow accurate QM and MM forces is independent of the particular QM and MM models employed, and may be used to reduce, if deemed necessary, the effects of any perturbations to the dynamical forces caused by the presence of the artificial QM-MM interface.

When evaluating the errors of a QM/MM coupling scheme due to the presence of the interface, it is important to remember that results should be compared to a large QM calculation, in which artificial interfaces are either not present (by using periodic interface conditions), or are very far away from the reaction under investigation. Direct comparison with an experiment can be misleading due to unintended cancellation of error. It is possible that errors introduced by the coupling method will cancel with errors inherent to the QM model and give good agreement with experiment for the test case, but without being transferable to other systems the method would not have much predictive power. Thus changing to a different molecular system or a different QM model (perhaps even a more accurate one) can change the degree of error cancellation and lead to worse than expected results. The otherwise extensive and varied QM/MM literature does not have many direct tests against large or periodic QM calculations.

THE ADAPTIVE BUFFERED-FORCE QM/MM METHOD

For an adaptive simulation to be stable, there must not be a net driving force to transport atoms or molecules from one region to another. In energy mixing methods this goal is difficult to achieve because there is no practical way to eliminate all potential energy surface mismatch, and therefore chemical potential differences, between the QM and MM regions. By using force mixing instead, we eliminate the need to match potential energy surfaces but require that the forces governing the dynamics be accurate. Accurate forces also enable the calculation of free-energy differences using methods that require only forces and not energies [9–12].

In a conventional QM/MM calculation, even when using sophisticated electrostatic embedding, forces near the QM-MM interface are modified by the presence of the interface and can be quite different from what a full QM or full MM calculation would predict. This effect can be seen when the force on an atom in the center of the QM region is calculated as a function of QM region size. A plot of the force error (i.e., the magnitude of the difference in force components compared to a reference calculation) in a density functional theory (DFT) calculation relative to a full QM calculation for a water molecule O atom in a system consisting of liquid water as a function of QM region size is shown in Figure 2.1. Without a buffer (i.e., a single QM water molecule coupled to the MM waters by electrostatics and nonbonded interactions), the force on the

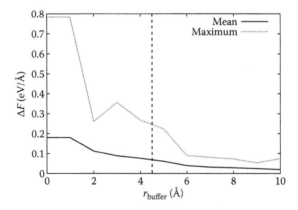

Figure 2.1   Error relative to a fully QM calculation in the force on an oxygen atom at the center of the QM region embedded in the external electric field of the MM atoms as a function of QM region size. Mean and maximum values are computed over a set of 10 atoms taken from a periodic box containing 1222 water molecules using the QM/MM implementation in CP2K [13–16]. Each QM region includes all water molecules whose O atom is within r_{buffer} of the central O atom and no covalent bonds cross the QM-MM interface. (From N. Bernstein et al., *Phys. Chem. Chem. Phys.* **14**, 646, 2012. Reproduced by permission of The Royal Society of Chemistry.)

O atom differs by up to about 1 eV/Å from the full QM calculation. Adding a buffer improves the agreement, with the mean force error going down to about 0.1 eV/Å when the buffer is about 4.5 Å. This indicates that in liquid water, atoms within 4.5 Å of the QM-MM interface of a conventional QM/MM calculation will have forces that are significantly perturbed by the interface. The simulations we present here for the purpose of benchmarking the bf-QM/MM method against pure QM calculation all use buffers. It is not clear whether other systems in general will show a strong interface effect on the forces and whether it will significantly affect the dynamics and observables derived from the trajectories, thus necessitating the use of buffers.

Note that the way we use the term "buffer" in the bf-QM/MM scheme has nothing to do with how the forces in the different regions are combined after they are calculated; it is purely a way to obtain more accurate QM and MM forces for a given geometry. In particular, our buffers are distinct from transition regions or other similar techniques that are often used to smoothly combine the various QM and MM forces (or indeed QM/MM forces corresponding to various partitionings). Other research groups using force mixing, with somewhat different formulations, systems, and force models in the two regions, have shown that at least in some cases buffers are not essential [17–19].

In the bf-QM/MM method, atoms in the QM_{dyn} region do follow accurate QM forces and atoms in the MM_{dyn} region follow accurate MM forces. By accurate we mean that these forces are close to those that would come from a full QM or MM simulation, respectively. The complete set of accurate forces needed for dynamics are generated by combining the results of two conventional QM/MM calculations, as illustrated in Figure 2.2.

Forces on atoms in QM_{dyn} are calculated in one conventional QM/MM force calculation with an enlarged QM region, consisting of QM_{dyn} surrounded by a buffer region QM_{buf}, which we call QM_{calc}^{large}. The purpose of the buffer region is to move the conventional QM/MM calculation's QM-MM interface away from QM_{dyn}, so that QM_{dyn} atoms have forces that are minimally affected by the interface. Forces on atoms in QM_{buf} are affected by their proximity to the QM-MM interface and are therefore not used. Forces on atoms in MM_{dyn} are calculated in another conventional QM/MM force calculation, with the QM region, which we denote QM_{calc}^{small}, reduced to the smallest possible size. This moves the QM-MM interface of the calculation away from any MM_{dyn} atoms, ensuring that they are subject to accurate MM forces.

The two sets of forces are combined to give a force on each atom:

$$\vec{F}_i = \begin{cases} \text{force from calculation with } QM_{calc}^{large} \text{ region treated as QM} & i \in QM_{dyn} \\ \text{force from calculation with } QM_{calc}^{small} \text{ region treated as QM} & i \in MM_{dyn} \end{cases}$$

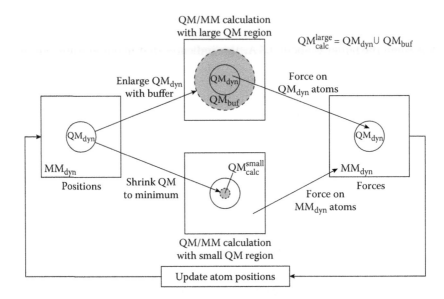

Figure 2.2   Diagram of buffered-force QM/MM method. A QM/MM calculation with large QM region $\mathrm{QM}_{\mathrm{calc}}^{\mathrm{large}}$ is created by enlarging the $\mathrm{QM}_{\mathrm{dyn}}$ region with a buffer region $\mathrm{QM}_{\mathrm{buf}}$ (top branch), and another QM/MM calculation is created with the QM region made as small as possible $\mathrm{QM}_{\mathrm{calc}}^{\mathrm{small}}$ (bottom branch). Forces for atoms in $\mathrm{QM}_{\mathrm{dyn}}$ are taken from the large QM region QM/MM calculation and forces for atoms in $\mathrm{MM}_{\mathrm{dyn}}$ are taken from the small QM region QM/MM calculation, and are used to propagate the dynamics. The regions are defined so that $\mathrm{QM}_{\mathrm{calc}}^{\mathrm{small}} \subset \mathrm{QM}_{\mathrm{dyn}} \subset \mathrm{QM}_{\mathrm{calc}}^{\mathrm{large}}$. (Reprinted with permission from C. Várnai et al. 2013, 12202. Copyright 2013, American Chemical Society.)

Because these forces are not the gradient of a single energy function, they do not obey momentum or energy conservation. We restore momentum conservation by adding a small and equal acceleration to every atom in the $\mathrm{QM}_{\mathrm{dyn}}$ and $\mathrm{QM}_{\mathrm{buf}}$ regions so that the sum of the forces over all atoms is zero. Energy conservation, on the other hand, is not a local property and cannot be restored exactly. Instead we use massive thermostats [20,21] (i.e., a separate thermostat applied to each atom) that are appropriately chosen to deal with the lack of energy conservation. While some of our simulations have used the Nosé-Hoover-Langevin thermostat [22–24], an even better choice is the adaptive Langevin thermostat of Jones and Leimkuhler [25,26]. It includes a Nosé-Hoover term that can adapt to energy conservation violation as well as a Langevin term that ensures good ergodicity. With these thermostats the bf-QM/MM method produces long-running, stable simulations at a constant temperature. Note that because there is no total energy defined, methods that require the energy (e.g., Monte Carlo sampling and some free energy calculation methods such as free energy perturbation [27] and Bennett acceptance ratio [28]) cannot be used with bf-QM/MM. However, it is still possible to carry out free-energy calculation methods that require only forces or trajectories, whether for sampling or to calculate the potential of mean force (e.g., constrained dynamics [9], meta-dynamics [10], adaptive biasing force [29], and umbrella integration [11]).

The choice of QM_{dyn} and QM_{buf} depends on the system. QM_{dyn} needs to include all the atoms that must follow the QM potential energy surface, and is analogous to the choice of QM region in a conventional QM/MM calculation. QM_{buf} should be chosen to ensure some level of accuracy for forces on QM_{dyn} atoms. Force error calculations, such as those in Figure 2.1, can be used to determine the minimum buffer size. These can be carried out on a few configurations representing important parts of the configuration space and compared to the largest possible QM region size if a full QM calculation is not feasible.

We use distance- and covalent bond connectivity based criteria to determine the precise set of atoms in the various QM regions. To make the method adaptive, we reevaluate the sets of atoms in each region throughout the simulation. The QM_{dyn} region is defined as the set of atoms within some distance of a fixed set of seed atoms where the QM description is needed. Using a hysteretic selection criterion, where the distance at which an atom becomes part of the QM_{dyn} region is smaller than the distance at which it leaves the region, helps reduce fluctuations [22]. The QM_{buf} region is defined similarly based on distance from QM_{dyn} atoms. Including and excluding entire molecules, if they are small enough for this to be efficient, eliminates errors due to the QM-MM interface cutting across covalent bonds.

STRUCTURAL PROPERTIES

As a first test for the bf-QM/MM method, we simulated the structure and dynamics of pure water and a hydrated Cl^- ion [22]. Pure water is a stringent test for an adaptive QM/MM method: the interactions between the molecules are dominated by relatively weak hydrogen bonds, and so even relatively small errors at the QM-MM interface can overwhelm the physical interactions. Since our goal is to reproduce the full QM simulation in the QM_{dyn} region and a full MM simulation in the MM_{dyn} region, we compare the bf-QM/MM results to full QM and full MM simulations.

The simulated system consisted of 84 (full QM) or 1222 (all other methods) water molecules, with DFT and the Becke-Lee-Yang-Parr (BLYP) exchange-correlation functional for the QM model [30–32] and flexible TIP3P for the MM model [33]. Molecular dynamics, with a constant temperature of 300 K applied using a massive Nosé-Hoover-Langevin thermostat, was used to propagate the system in time. Simulations were carried out at different QM_{dyn} region sizes and a buffer region size of 4 Å, and compared to full QM and full MM simulations as well as to an implementation of the hot spot method [17,34] which is adaptive but uses no buffer region. The QM_{dyn} was centered around one water molecule, and the cutoff distances were chosen to include 1–2 hydration shells.

The radial distribution function (RDF) for pure water measured for full QM, full MM, unbuffered QM/MM, and bf-QM/MM simulations at the center of the QM region are shown in Figure 2.3. The full QM and full MM models agree on the height and position of the first peak, which includes the first hydration shell, but deviate for larger distances. Without a buffer region the adaptive QM/MM simulations give poor results. For the smaller QM region (~3 Å) the first peak is pushed out to the QM-MM interface. For the larger region (~5 Å)

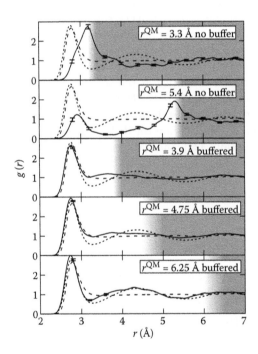

Figure 2.3   Radial distribution functions measured at the center of the QM_{dyn} region in pure water. Each panel shows reference full QM (dotted) and full MM (dashed) results as compared with a hybrid simulation with a particular choice of QM_{dyn} and QM_{buf} region sizes (solid). The top two panels show results of simulations with no buffer region, labeled by QM_{dyn} region size. The panels below show results from bf-QM/MM simulations with varying QM_{dyn} region size (the midpoint of the hysteresis range is shown). The unshaded parts of the plots represent the QM_{dyn} regions, with the gradient in shading corresponding to the hysteresis range. (From N. Bernstein et al., *Phys. Chem. Chem. Phys.* **14**, 646, 2012. Reproduced by permission of The Royal Society of Chemistry.)

a low density bubble is formed around the center of the QM region, and many of the molecules that belong in the first hydration shell are pushed out to the QM-MM interface. These artifacts are caused by the inaccurate forces at the QM-MM interface, which lead to net transport of molecules out of the QM region and into the MM region. The bf-QM/MM results, on the other hand, show very good agreement with the full QM simulations in the QM_{dyn} region and with the full MM simulations in the MM_{dyn} region. The first peak of the RDF is well reproduced even for the smaller QM_{dyn} region of about 3.9 Å, and when the cutoff distance is large enough, 6.25 Å, the second peak is reproduced as well.

The RDF centered on molecules at various distances from the QM_{dyn} center is plotted in Figure 2.4. The RDF transforms smoothly from full QM-like when it is calculated for a center molecule at the center of QM_{dyn} to full MM-like when it is calculated for a center molecule in MM_{dyn}. This smooth transition gives additional confirmation that the use of a buffer region to get accurate QM and MM forces is sufficient for obtaining a liquid water structure without artifacts at the QM-MM interface.

In addition to these results, we also examined hydrogen bond angle distribution functions and the time autocorrelation of the O–H bond length for pure

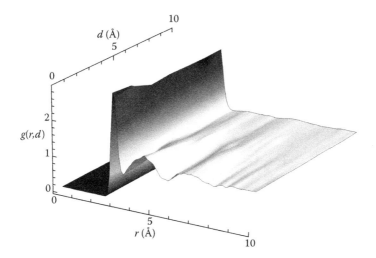

Figure 2.4 Radial distribution function in pure water for molecules at different distances from the center of the QM region d from a buffered QM/MM simulation using the largest (~6 Å) QM region. The RDF shows a smooth transition from QM-like at the center to MM-like far from the QM region, without any jumps or discontinuities through the transition. (From N. Bernstein et al., *Phys. Chem. Chem. Phys.* **14**, 646, 2012. Reproduced by permission of The Royal Society of Chemistry.)

water, as well as repeating all of the calculations for a solvated Cl⁻ ion [22]. In all cases the bf-QM/MM showed good agreement with the full QM simulation in the QM_{dyn} region.

REACTION FREE ENERGIES

The bf-QM/MM method can be used to calculate free energies using methods that require only the forces, not the energies. We have used constrained dynamics [9] and umbrella integration [11] to calculate the free-energy profiles as a function of geometric reaction coordinates for two reactions in an aqueous solvent: the nucleophilic substitution of a Cl in methyl chloride (Figure 2.5), and the deprotonation of the tyrosine side chain phenolic OH [25] (Figure 2.6).

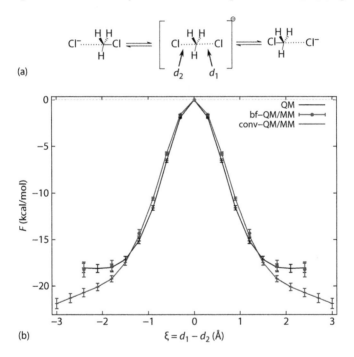

Figure 2.5 Nucleophilic substitution of a Cl in methyl chloride reaction diagram (a) and free energy as a function of the distance difference as the reaction coordinate (b). For the conventional QM/MM method, the free energy derivative does not go to zero as ξ increases, incorrectly predicting the reactant state not to be stable, while the bf-QM/MM method correctly reproduces the QM free energy and its derivative. (Reprinted with permission from C. Várnai et al. 2013, 12202. Copyright 2013, American Chemical Society.)

For the nucleophilic substitution the reaction coordinate was the distance difference between the methyl C and the bonded and attacking Cl atoms. For the tyrosine deprotonation the reaction coordinate was the distance between the phenolic O and transfering H atoms. The QM_{dyn} regions were based on a fixed set of seed atoms and a hysteretic distance criterion of about 5.5 Å for both reactions, containing about two hydration shells. All selections included only whole molecules so that no covalent bonds were cut by the QM-MM interface. The buffer regions were selected based on a hysteretic distance criterion of about 4.5 Å from the QM_{dyn} atoms. Full adaptivity was achieved by reevaluating the selection criteria every MD time step, allowing solvent molecules to diffuse into and out of the reaction region. All calculations with bf-QM/MM

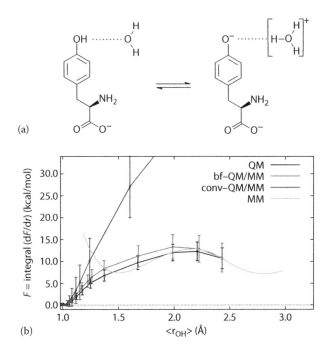

Figure 2.6 Deprotonation of the tyrosine side-chain phenolic OH reaction diagram (a) and values for the integrated free energy F for various methods as a function of mean O–H distance (b). Black symbols and dashed lines are reference full QM calculation, black symbols and solid lines are bf-QM/MM calculations, gray symbols and solid lines are conventional QM/MM calculations, and gray dashed line is nonreactive MM calculations. Error bars for the gradient indicate standard error within each sampled trajectory. Error bars for the integral indicate standard error assuming each sampling trajectory is a single independent sample. (Reprinted with permission from C. Várnai et al. 2013, 12202. Copyright 2013, American Chemical Society.)

were compared to reference full QM calculations and to conventional QM/MM calculations with only the reacting species in the QM region. Additional technical details can be found in [21].

The free energy as a function of the distance difference ξ between the C and two Cl for the methyl chloride nucleophilic substitution generated by integrating the gradient of the potential of mean force (PMF) that comes directly from the constrained dynamics is plotted in Figure 2.5. The free-energy barrier between the transition state with $\xi = 0$ and the minimum at $\xi = 2$ Å is 18.5 ± 1.5 kcal/mol, in excellent agreement with the full QM value of 18.0 ± 0.5 kcal/mol. The conventional QM/MM simulations, on the other hand, fail to reproduce the minimum, and our simulations give a lower bound on the barrier of 21 ± 0.5 kcal/mol. The actual barrier may be much higher if the conventional QM/MM results were calculated to larger values of ξ. The bf-QM/MM calculation improves on conventional QM/MM mainly by providing a better description of the free energy of the departing Cl as it separates from the methyl chloride and becomes solvated by the water. This improvement may be due to a better description of the solvated Cl, which the bf-QM/MM method surrounds by 1–2 hydration shells described by QM. It may also be due to the fact that the larger QM region, facilitated by the stable adaptivity of the bf-QM/MM method, does not become discontiguous (as the conventional QM/MM method's QM region does) as water molecules insert themselves between the methyl chloride and the departing Cl.

The free energy as a function of the tyrosine side-chain O–H distance calculated by integrating the umbrella integration PMF gradient for bf-QM/MM shown in Figure 2.6 is in good agreement with full QM. It is also in reasonable agreement with a nonreactive MM simulation at distances where the MM is expected to be applicable (i.e., after the O–H covalent bond is broken). There is a sharp covalently bonded minimum at about 1 Å and a broad maximum around 2 Å. The conventional QM/MM calculation, in contrast, overestimates the PMF gradient as soon as the covalent bond is broken, and leads to a vastly overestimated free-energy difference. Its description of the departing H, modeled as a QM proton solvated by MM water molecules, is apparently quite inaccurate. The bf-QM/MM description, which surrounds the phenolic O and departing H by 1–2 hydration shells described by the QM model, is needed to capture the energetics of deprotonation in the aqueous solvent.

FORCES AND BUFFERS IN PROTEINS

To apply the bf-QM/MM method to proteins, it is first necessary to determine how to properly select QM_{dyn} and QM_{buf} regions. Since proteins have extended covalently bonded structures, it is not practical, as it was in our previous applications to small molecules and water, to select these regions so that no covalent bonds are cut. To test the effects of the buffer on forces in the QM region, we calculated force errors for two atoms in hen-egg lysozyme (PDB code 4lzt, additional technical details can be found in [35]) shown in Figure 2.7: the OE1 atom in the active site residue Glu35, which is in a polar region, and the CG1 atom of the buried Ile55 region [35]. The force errors were computed relative to the forces for the largest QM region feasible, with a radius of about 11 Å, expanded to contain whole residues and containing 700–750 QM atoms. The forces were calculated for 150 configurations from a well-equilibrated MM simulation at 300 K.

Force and charge errors for representative atoms in the polar and nonpolar regions are plotted in Figure 2.8. Force errors are generally larger in the polar region and decay slowly and not monotonically with QM region size. For the mean force error to be less than 0.1 eV/Å, the QM region must include over 200 atoms, and for the maximum force error to be this small requires over 500 atoms or about a 9-Å selection criterion. In the nonpolar region the force errors for the calculation without a buffer are also substantial, but they go down quickly once the QM region becomes larger. Even for 50 atoms the force error mean is already less than 0.1 eV/Å.

The comparison of the two regions shows that the bf-QM/MM approach, which adds buffers to the atoms where QM forces are thought to be necessary for dynamics, is essential for atoms in proteins, especially in polar regions. The conventional QM/MM treatment of the QM-MM interface leads to inaccurate forces near the interface, and buffers can eliminate this inaccuracy. The

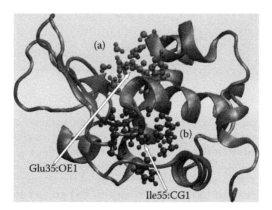

Figure 2.7   Three-dimensional structure of lysozyme (PDB code: 4lzt) and the two regions considered for QM force error calculations: (a) the polar region is centered around the OE1 atom of active site residue Glu35, and (b) the apolar region is selected around the CG1 atom of the buried Ile55. (Reprinted with permission from Solt et al. 2009, 5728. Copyright 2009, American Chemical Society.)

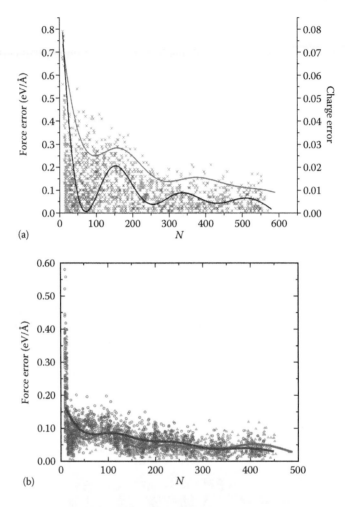

Figure 2.8 (a) Mean force errors (gray) and mean charge errors on the central OE1 of Glu35 atom of the polar QM region (black) as a function of the size of the QM region using a residue-based definition of the QM region in the polar region of the protein. The raw charge error data is shown as symbols. (b) Force errors on the CG1 atom of Ile55 in the apolar region as a function of the size of the QM region using two QM region definitions—residue-based (open triangles) and group-based (open circles). The mean errors are shown by the continuous lines; the raw errors are shown by symbols. (Reprinted with permission from Soft et al. 2009, 5728. Copyright 2009, American Chemical Society.)

importance of this in polar regions, where strong electrostatic interactions are present, is unsurprising.

SUMMARY

Adaptive QM/MM simulations are particularly challenging because the QM and MM potential energy surfaces are quite different, and allowing atoms or molecules to switch from one to the other can lead to severe artifacts. Force-mixing methods can avoid some of these artifacts because they do not need to smoothly connect the potential energy surfaces, but they require accurate QM and MM forces to be mixed. However, conventional QM/MM simulations using electrostatic embedding result in forces that are significantly affected by proximity to the (unphysical) QM-MM interface. The bf-QM/MM method uses force-mixing to ensure that every atom in the QM/MM simulation follows forces that are accurate (i.e., close to those from a full QM or full MM simulation) by using buffers to make sure that the QM-MM interface does not affect the forces used in the dynamics. These accurate forces are sufficient for achieving stable adaptive simulations with large QM regions.

We have used the bf-QM/MM method to simulate pure water and shown that the resulting structure is in excellent agreement with a full QM simulation, while an adaptive QM/MM method without buffers is unstable. We have also used the method to calculate two reaction free-energy profiles: nucleophilic substitution of Cl in methyl chloride, and the deprotonation of the tyrosine side chain phenolic OH. The bf-QM/MM method is in good agreement with a full QM simulation, while a conventional, nonadaptive QM/MM simulation with only the reactants in the QM region yields incorrect free-energy profiles. In proteins, the effects of the QM-MM interface on the forces calculated using an electrostatic embedding QM/MM simulation are significant and depend strongly on the presence of nearby polar groups or where the QM region is in relation to the surface of the protein, or both. The use of buffers to calculate accurate forces in the bf-QM/MM can reduce these effects. It remains to be seen how strong these effects remain overall for computing reaction free energies and free-energy differences involving proteins. The algorithmic tool kit and software implementations in the CP2K [13] and Amber [36] packages are now in hand to perform these studies.

ACKNOWLEDGMENTS

G.C., C.V., and L.M. acknowledge support from the EU-FP7-NMP grant 229205 ADGLASS. G.C. acknowledges support from the Office of Naval Research under award no. N000141010826. N.B. acknowledges funding for this project by the Office of Naval Research (ONR) through the Naval Research Laboratory's Basic Research Program. Some computation was carried out at the DOD MHPCC Distributed Center and ERDC and AFRL Major Shared Resource Centers.

REFERENCES

1. A. Warshel, *Biochemistry* **20**, 3167 (1981).
2. A. Warshel, P. K. Sharma, M. Kato, Y. Xiang, H. Liu, and M. H. M. Olsson, *Chem. Rev.* **106**, 3210 (2006).
3. J. Gao, S. Ma, D. T. Major, K. Nam, J. Pu, and D. G. Truhlar, *Chem. Rev.* **106**, 3188 (2006).
4. H. M. Senn, and W. Thiel, in *Atomistic Approaches in Modern Biology*, edited by M. Reiher (Springer Berlin, Heidelberg, 2007), Vol. 268 of *Topics in Current Chemistry*, pp. 173–290.
5. T. Steinbrecher and M. Elstner, in *Methods in Molecular Biology*, edited by L. Monticelli and E. Salonen, (Springer, New York, 2012) Vol. 924.
6. G. Csányi, T. Albaret, M. C. Payne, and A. De Vita, *Phys. Rev. Lett.* **93**, 175503 (2004).
7. N. Bernstein, J. R. Kermode, and G. Csányi, *Rep. Prog. Phys.* **72**, 026501 (2009).
8. A. Laio, and J. VandeVondele, *J. Chem. Phys.* **116**, 6941 (2002).
9. E. Carter, G. Ciccotti, J. T. Hynes, and R. Kapral, *Chem. Phys. Lett.* **156**, 472 (1989).
10. A. Laio, and M. Parrinello, *Proc. Natl. Acad. Sci. U.S.A.* **99**, 12562 (2002).
11. J. Kastner, and W. Thiel, *J. Chem. Phys.* **123**, 144104 (2005).
12. E. Darve, D. Rodriguez-Gomez, and A. Pohorille, *J. Chem. Phys.* **128**, 144120 (2008).
13. Available at http://cp2k.org/.
14. J. VandeVondele, M. Krack, F. Mohamed, M. Parrinello, T. Chassaing, and J. Hutter, *Comput. Phys. Commun.* **167**, 103 (2005).
15. T. Laino, F. Mohamed, A. Laio, and M. Parrinello, *J. Chem. Theory Comput.* **1**, 1176 (2005).
16. T. Laino, F. Mohamed, A. Laio, and M. Parrinello, *J. Chem. Theory Comput.* **2**, 1370 (2006).
17. T. Kerdcharoen, K. R. Liedl, and B. M. Rode, *Chem. Phys.* **211**, 313 (1996).
18. R. E. Bulo, B. Ensing, J. Sikkema, and L. Visscher, *J. Chem. Theory Comput.* **5**, 2212 (2009).
19. K. Park, A. W. Goetz, R. C. Walker, and F. Paesani, *J. Chem. Theory Comput.* **8**, 2868 (2012).
20. D. J. Tobias, G. J. Martyna, and M. L. Klein, *J. Phys. Chem.* **97**, 12959 (1993).
21. J. Schmidt, J. VandeVondele, I.-F. W. Kuo, D. Sebastiani, J. I. Siepmann, J. Hutter, and C. J. Mundy, *J. Phys. Chem. B* **113**, 11959 (2009).
22. N. Bernstein, C. Varnai, I. Solt, S. A. Winfield, M. C. Payne, I. Simon, M. Fuxreiter, and G. Csanyi, *Phys. Chem. Chem. Phys.* **14**, 646 (2012).
23. S. A. Winfield, *Hybrid Multiscale Simulation of Liquid Water*, Ph.D. thesis (Pembroke College, University of Cambridge, Cambridge, 2009).
24. B. Leimkuhler, E. Noorizadeh, and F. Theil, *J. Stat. Phys.* **135**, 261 (2009).
25. C. Várnai, N. Bernstein, L. Mones, and G. Csányi, *J. Phys. Chem. B* **117**, 12202 (2013).

26. A. Jones, and B. Leimkuhler, *J. Chem. Phys.* **135**, 084125 (2011).
27. R. W. Zwanzig, *J. Chem. Phys.* **22**, 1420 (1954).
28. C. H. Bennett, *J. Comput. Phys.* **22**, 245 (1976).
29. E. Darve, and A. Pohorille, *J. Chem. Phys.* **115**, 9169 (2001).
30. A. D. Becke, *Phys. Rev. A* **38**, 3098 (1988).
31. C. Lee, W. Yang, and R. G. Parr, *Phys. Rev. B* **37**, 785 (1988).
32. B. Miehlich, A. Savin, H. Stoll, and H. Preuss, *Chem. Phys. Lett.* **157**, 200 (1989).
33. A. D. MacKerell, D. Bashford, M. Bellott, R. L. Dunbrack, J. D. Evanseck, M. J. Field, S. Fischer, J. Gao, H. Guo, S. Ha et al., *J. Phys. Chem. B* **102**, 3586 (1998).
34. G. W. Marini, K. R. Liedl, and B. M. Rode, *J. Phys. Chem. A* **103**, 11387 (1999).
35. I. Solt, P. Kulhánek, I. Simon, S. Winfield, M. C. Payne, G. Csányi, and M. Fuxreiter, *J. Phys. Chem. B* **113**, 5728 (2009).
36. D. A. Case, V. Babin, J. T. Berryman, R. M. Betz, Q. Cai, D. S. Cerutti, T. E. Cheatham, III, T. A. Darden, R. E. Duke, H. Gohlke et al., *AMBER* **14**, University of California, San Francisco (2014).

Chapter 3

Conformational and Chemical Landscapes of Enzyme Catalysis

Alexandra T. P. Carvalho, Fernanda Duarte,
Konstantinos Vavitsas, and Shina Caroline Lynn Kamerlin

CONTENTS

INTRODUCTION AND OVERVIEW

Enzymes are nature's catalysts, reducing the timescales of the chemical reactions that make biology possible from millions of years to fractions of seconds (Warshel et al. 2006b; Wolfenden and Snider 2001). On the one hand, they are "merely" large molecules: polymers of amino acids folded into a particular three-dimensional (3-D) structure in order to solve specific chemical problems. However, despite their deceptive simplicity and over a century's worth of extensive research effort, the molecular basis for their tremendous catalytic proficiencies remains highly controversial and, to some extent, poorly understood. Tying in with this, a large number of hypotheses have been put forward over the years in order to explain how enzymes really work (e.g., Benkovic et al. 2008; Hammes-Schiffer 2013; Kamerlin and Warshel 2010a; Kamerlin, Mavri et al. 2010; Nagel and Klinman 2009; Villali and Kern 2010; Warshel et al. 2006b; Zalatan and Herschlag 2009).

In this chapter, we will focus on one proposal that has gained significant popularity over the past three decades (particularly in the last few years), namely the idea that dynamical contributions play a significant role in enzyme catalysis (Bhabha et al. 2011; Cannon et al. 1996; Careri et al. 1979; Daniel et al. 2003; Fan et al. 2013; Gavish and Werber 1979; Henzler-Wildman and Kern 2007; Henzler-Wildman, Lei et al. 2007; Henzler-Wildman, Thai et al. 2007; Kale et al. 2008; Karplus and McCammon 1983; Klinman 2013; Kohen et al. 1999; Kurkcouglu et al. 2012; McCammon et al. 1979; McGowan and Hamelberg 2013; Neria and Karplus 1997; Radkiewicz and Brooks 2000; Rajagopalan and Benkovic 2002; Saen-Oon et al. 2008; Vendruscolo and Dobson 2006; Wolf-Watz et al. 2004; Zavodszky et al. 1998). In its simplest form, this proposal is qualitatively quite attractive (enzymes are dynamic entities, and this dynamic motion can be apparently linked to enzyme function; e.g., Bhabha et al. 2011; Eisenmesser et al. 2002; Henzler-Wildman, Lei et al. 2007; Henzler-Wildman, Thai et al. 2007). However, it suffers from the fact that there is no clear definition of what precisely is *meant* by a dynamical effect, which makes this proposal very hard to formulate in a clear and precise way (Hammes-Schiffer and Benkovic 2006; Kamerlin and Warshel 2010a; Warshel et al. 2006a) and results in multiple different usages of the same term in the literature. In addition to this, as we will discuss in the following, even what is actually meant when using the term "enzyme catalysis" is different from a biological and a chemical perspective, creating substantial semantic confusion in the field. Finally, the most challenging aspect of this issue is the fact that, at present, there is no experiment that can clearly and unambiguously link enzyme dynamics as being causative for the chemical step of catalysis. Therefore, all dynamical proposals have been based on *indirect* connections such as similar timescales for conformational changes and the chemical step (Henzler-Wildman, Lei et al. 2007) or mutations that impair catalysis also appearing to freeze out local conformational flexibility (Bhabha et al. 2011). Finally, the microsecond-to-second timescales (Frey and Hegeman 2006) of such conformational changes make addressing this problem intractable using standard computational approaches.

We will begin this chapter by providing the reader with a clear definition of what is meant by the term dynamical effect based on the different incarnations it currently takes in the literature. We will also outline the distinction between the biological and chemical usage of the expression catalysis in the context of enzymes. Following from this, we will discuss recent computational advances that allow us to actually explore the energy landscape for enzyme catalysis as a function of both conformational changes and the chemical step. We will also illustrate the applications of these approaches to studying more complex biological problems, such as deoxyribonucleic acid (DNA) translocation and the molecular machinery of ATPases. Finally, we will complete this chapter by touching on alternative computational approaches to probe the role of dynamics in enzyme catalysis that do not involve examining the full energy landscape, and present our perspective for future directions in this field.

DEFINING DYNAMICAL EFFECTS IN THE CONTEXT OF ENZYME CATALYSIS

In simplest terms, a dynamical effect on catalysis can be defined as a situation in which an enzyme has evolved to optimize a specific vibrational mode, to either move the system to the transition state (TS), or to convert a system that is already at the TS to the product state (Olsson et al. 2006). One of the biggest challenges in any discussion of a potential role for dynamical effects in driving enzyme catalysis arises from the fact that the concept of enzyme dynamics is used to mean many different things by scientists with different research backgrounds, ranging from local fluctuations of amino acid side chains (Bhabha et al. 2011) through to large-scale conformational changes (Grant et al. 2010). In addition to this, the term catalysis is used interchangeably to refer to both the chemical step of the catalytic cycle (Warshel et al. 2006a), as well as the full cycle itself from start to finish (Nagel and Klinman 2009; Villali and Kern 2010). That is, the background (uncatalyzed) reaction in aqueous solution can be as simple as a single-step process going directly from reactants to products. However, for the catalyzed counterpart, the reacting species also have to associate with and dissociate from the catalyst, the overall rates of which may be limited by both local and more global conformational changes. If one wants to consider this in terms of the kinetics, this can (in its simplest form) be described as:

$$E+S \underset{k_{-1}}{\overset{k_1}{\rightleftharpoons}} ES \xrightarrow{k_{cat}} ES^* \rightarrow EP \rightarrow E+P \tag{3.1}$$

Here, E, S, and P denote free enzyme, substrate, and product, respectively. ES denotes the Michaelis complex in the ground state (after substrate binding), ES^* denotes the enzymatic transition state, and EP denotes the bound enzyme-product complex. Finally, k_1, k_{-1}, and k_{cat} denote the corresponding rate constants for (reversible) substrate binding, and theoretically the chemical step, although in principle chemistry need not be rate-limiting, and k_{cat} could, in fact, be reflecting a different nonchemical step of the catalytic cycle. Note that k_{cat} (s^{-1}) is also sometimes referred to as a turnover number (Michaelis et al. 2011; Nelson et al. 2008), as it measures the number of substrate molecules each enzyme produces per second. Therefore, the inverse of k_{cat} (s) provides the time required by one enzyme to turn one molecule of substrate into one molecule of product. To an approximation (Nelson et al. 2008), the rate of the formation of product, υ, can be given by the Michaelis-Menten equation (Michaelis et al. 2011):

$$\upsilon = \frac{d[P]}{dt} = V_{max} \frac{[S]}{K_m + [S]} \tag{3.2}$$

Here, V_{max} is the maximum rate of the reaction under saturating substrate conditions and the Michaelis constant, K_m, is the substrate concentration at which the reaction rate is half of V_{max} (defined by $K_m = \frac{k_{-1} + k_{cat}}{k_1}$ (for a full derivation, see Nelson et al. 2008). Now as $V_{max} = k_{cat}[E]_0$ (where $[E]_0$ is enzyme concentration), therefore, Equation 3.2 can be rewritten as:

$$\upsilon = \frac{k_{cat}[E]_0[S]}{K_m + [S]} \tag{3.3}$$

where, under ideal (saturating substrate) conditions, k_{cat} describes catalytic turnover rate of the enzyme and the ratio k_{cat}/K_m describes how efficiently the enzyme converts a substrate with a given starting concentration to product.

While it can superficially appear to merely be a philosophical issue, the problem, which lies to some extent at the heart of the discussion surrounding the role of dynamical effects in enzyme catalysis, is whether one considers enzyme catalysis as referring to just the chemical step ($ES \rightarrow EP$) or the full catalytic cycle of the enzyme ($E + S \rightarrow E + P$). That is, as illustrated in Figure 3.1, even Equation 3.1 provides a hugely simplified view of enzyme catalysis. The 2Å crystal structure of hen egg-white lysozyme in 1965 (Blake et al. 1965) marked a turning point in enzymology: this was the first enzyme to have its structure solved, and having access to a physical structure from which to draw functional conclusions jump-started the field of mechanistic enzymology (Johnson 1998). However, the static nature of x-ray structures led, in turn, to a static view of enzymes during catalysis, and it was not until a decade later, with the advent of computer simulations of biological systems (Levitt and Warshel 1975; McCammon et al. 1977) as well as increasingly advanced nuclear magnetic resonance (NMR) spectroscopy studies of biological systems (Fiaux et al. 2002; Henzler-Wildman and Kern 2007) or other spectroscopic methods, such as neutron spin echo spectroscopy (Bu et al. 2005; Farago et al. 2010), electron

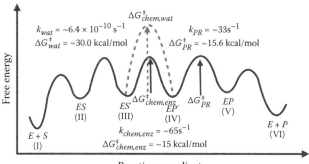

Figure 3.1   Free-energy landscape for a hypothetical enzyme catalyzed reaction. $E + S$, ES, ES', EP', EP, and $E + P$ denote (I) the free enzyme and free substrate, (II) the enzyme-substrate complex, (III) the Michaelis complex after potential conformational changes, (IV) the product complex, (V) the product complex after potential conformational changes, and (VI) free enzyme and free product, respectively. $\Delta G^{\ddagger}_{chem,wat}$ and $\Delta G^{\ddagger}_{chem,enz}$ denote the activation barriers for the reference reaction and the chemical step of the enzyme's catalytic cycle respectively. Similarly, k_{wat} and k_{enz} denote the rates of the chemical step for the reference and the enzyme catalyzed reaction, respectively. Finally, k_{PR} and $\Delta G^{\ddagger}_{PR}$ denote the rate and free energy of product release, respectively. Shown here are typical rates for such processes, and the main purpose of this image is to illustrate that while the chemical step is not necessarily the rate-limiting step in the catalytic cycle, it *is* where the main rate acceleration occurs. Once the enzyme has evolved to accelerate the reaction by (sometimes quite extreme) orders of magnitude (Wolfenden and Snider 2001), it is then trivial to fine-tune the overall rate by making a different step such as a conformational change rate-limiting. (Reprinted with permission from Adamczyk, A. J., J. Cao, S. C. L. Kamerlin, and A. Warshel. 2011. *Proc. Natl. Acad. Sci. U. S. A.* 108:14115–14120. Copyright 2011, National Academy of Sciences, U.S.A.)

paramagnetic resonance (EPR) (Kline et al. 2013), circular dichroism (CD) (Karabencheva-Christova et al. 2013), Föster resonance energy transfer (FRET) (Truong and Ikura 2001), and dual polarization interferometry (DPI) (Coan et al. 2012) that it became increasingly clear that enzymes are, in fact, fluctuating dynamical entities.

Tying in with this came the first arguments that such dynamics plays a role in enzyme catalysis, dating back to the late 1970s and early 1980s (Careri et al. 1979; Gavish and Werber 1979; Karplus and McCammon 1983; McCammon et al. 1979). However, in addition to local conformational dynamics, a wealth of kinetic, biochemical, and structural data has demonstrated that enzymes can undergo (sometimes substantial) conformational changes during their catalytic cycles (Hammes 2002), and these clearly play an important role in enzyme function and regulation and add further complexity to the catalytic landscape (Figure 3.1). It is these conformational changes in systems such as adenylate kinase (Arora and Brooks 2007; Hanson et al. 2007; Lu and Wang 2008; Maragakis and Karplus 2005; Muller et al. 1996; Pisliakov et al. 2009; Vonrhein et al. 1995; Whitford et al. 2007; Wolf-Watz et al. 2004), dihydrofolate reductase (Adamczyk et al. 2011; Arora and Brooks 2013; Benkovic and Hammes-Schiffer 2003; Bhabha et al. 2011; Boehr et al. 2006; Fan et al. 2013; Hammes 2002; Liu and Warshel 2007b; Loveridge et al. 2011; McElheny et al. 2005; Radkiewicz and Brooks 2000; Schnell et al. 2004), and cyclophillin A (Bosco and Kern 2004; Bosco et al. 2002, 2010; Eisenmesser et al. 2002, 2005; Fraser et al. 2009; Homme et al. 2005; Kern et al. 1993, 1994, 1997, 2005; Kern, Kern, Scherer et al. 1995; Kern, Kern, Schmid et al. 1995; McGowan and Hamelberg 2013; Thai et al. 2008) that have been central to the more recent debate about the role of dynamics in enzyme catalysis (Benkovic and Hammes-Schiffer 2003; Glowacki et al. 2012b; Hammes-Schiffer 2013; Henzler-Wildman and Kern 2007; Kamerlin and Warshel 2010a; Villali and Kern 2010).

This is why it is crucial to start any discussion of enzyme catalysis with a clear definition of what one actually *means* by the term catalysis in this context. For example, on the one hand and as is often done (Eisenmesser et al. 2002; Henzler-Wildman and Kern 2007; Nagel and Klinman 2009; Villali and Kern 2010), one could take the term catalysis to refer to the entire cycle shown in Figure 3.1. However, this would, in our opinion, be conflating the overall function of the enzyme with the actual catalytic step. That is, it has been clearly and unambiguously demonstrated that such conformational changes are an important part of the catalytic cycle and are critical to enzyme function, for example by bringing the enzyme to a catalytically active state that has been preorganized for optimal chemistry (McGeagh et al. 2011; Prasad and Warshel 2011; van der Kamp et al. 2013) as well as regulating the rates of the overall catalytic cycle (Stadtman 2006). Therefore, anything that impairs such functionally important conformational changes will also impair the function of the catalyst. However, while any functional conformational changes are an important prerequisite for efficient catalysis, this does not mean that they necessarily play a role in driving the catalysis itself. That is, by definition, chemical catalysis describes the acceleration of the rate of a chemical reaction compared to the uncatalyzed counterpart, and therefore, in the strictest sense, the process of actual catalysis refers only to the acceleration of the chemical step, not the full cycle (Warshel et al. 2006a). This can be described as the ratio of k_{cat}/k_{uncat}, where k_{uncat} is the rate of the corresponding uncatalyzed reaction. Therefore, the issue becomes not whether such conformational fluctuations occur and are important, but whether they directly affect the enzymatic rate constants k_{cat} and k_{cat}/K_M and therefore by extension the chemical step.

There have been many recent works that have explored the potential role of dynamics in enzyme catalysis (e.g., see the discussion in Benkovic and Hammes-Schiffer 2003; Glowacki et al. 2012b; Hammes-Schiffer 2013; Henzler-Wildman and Kern 2007; Kamerlin and Warshel 2010a; Villali and Kern 2010). However, as it is not possible to prove a direct causal link between the observed dynamics and the catalytic efficiency of the enzyme, most experimental conclusions have instead been drawn on indirect inferences such as similar time-scales for the conformational and chemical steps (Henzler-Wildman, Lei et al. 2007; Henzler-Wildman, Thai et al. 2007). The closest we have been able to come to direct experimental evidence for a dynamical contribution to enzyme catalysis is a recent elegant study of dihydrofolate reductase (DHFR), which showed that mutations that freeze out the millisecond thermal fluctuations of the active site residues also impair the catalytic activity of the enzyme (Bhabha et al. 2011). As we will discuss in the section titled "Thermophilic/Mesophilic DHFR," we have demonstrated (Adamczyk and Warshel 2011) through detailed computational work that the origin of detrimental effect of these mutations on the catalytic activity is purely electrostatic in origin. More important, we have also demonstrated that the reduced flexibility of the enzyme is *due to* and not the *cause of* the changing catalytic landscape for this enzyme upon mutation. This therefore leads to the next semantic issue, which is to define what is actually *meant* by a dynamical effect in the context of enzyme catalysis. As we have elaborated on elsewhere (Kamerlin and Warshel 2010a), it is not sufficient to say that dynamical effects are linked to enzyme motion, as atoms move in *any* chemical process occurring at above 0 K. Therefore, any meaningful discussion of dynamical effects on catalysis has to examine the actual rate of the chemical process (defined through k_{cat} and k_{cat}/K_M) and then verify that such effects indeed impact the chemical step. As a starting point, one can take into account the relationship:

$$k = \kappa k_{TST} \tag{3.4}$$

Here k_{TST} is the rate constant from transition state theory (TST), and κ denotes a transmission factor (or coefficient) introduced to account for the fact that not every vibrational mode responsible for converting reactant to product is necessarily productive. This factor also allows for the quantum mechanical phenomenon of tunneling (i.e., molecules that have insufficient energy to pass over the TS may tunnel through the barrier) (Jensen 2007). Normally, κ is taken to be roughly equivalent to 1 (Jensen 2007). Additionally, after the introduction of a number of approximations and simplifications (e.g., see Glowacki et al. 2012b; Kamerlin and Warshel 2010a), k_{TST} can be related to the activation barrier for the process, $\Delta g^{\ddagger}$, using the Eyring–Polanyi equation (Evans and Polanyi 1935):

$$k_{TST} \approx \frac{k_B T}{h} \exp\left(-\frac{\Delta g^{\ddagger}}{RT}\right) \qquad (3.5)$$

Therefore, clearly, for any dynamical contribution to have a measurable effect on either the reaction rate or the corresponding activation barrier, it is necessary that they cause a deviation from transition state theory. Following from this, all dynamical contributions can essentially be grouped into the transmission factor (Bennett 1977; Grimmelmann et al. 1981; Keck 1966), and for a dynamical contribution to affect the reaction rate, this would require a transmission factor of substantially smaller than 1. Another valid definition of a dynamical effect would be the existence of coherent, non-Boltzmann motions that affect the probability of reaching a transition state. Additionally, coupled motions could theoretically indicate a dynamical contribution, *provided* that they are substantially different from the corresponding motions in the uncatalyzed reaction in aqueous solution and that the coupled motions in the enzyme do not follow the Boltzmann distribution. However, if one wants to directly address the main challenge of whether enzyme conformational flexibility (or full conformational changes) directly affect the chemical step of catalysis, this is most easily achieved using the representation shown in Figure 3.2. This figure defines the catalytic landscape as a function of two distinct reaction coordinates, one of which is a conformational (or flexibility) coordinate and the other of which corresponds to the chemical step. This figure illustrates two limiting cases: a diffusive model in which the chemical step has no memory of the preceding conformational change, and an inertial model in which there is a true dynamical effect, and memory of the conformational transition helps drive the chemical step. This is shown for the example situations in which the barrier to the conformational change $\left(\Delta g^{\ddagger}_{conf}\right)$ is much smaller than that of the chemical step ($\Delta g^{\ddagger}_{chem}$, Figure 3.2a and c), and one in which the two barriers are roughly equal (Figure 3.2b and d). From this figure, it can be seen that in the

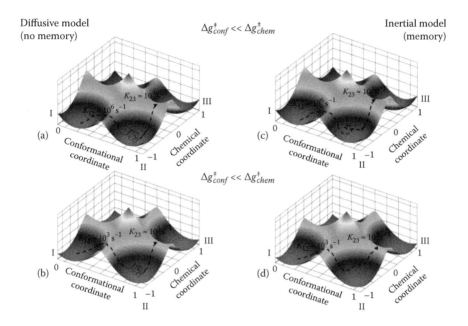

Figure 3.2   Energy landscapes for enzyme catalysis as a function of both conformational changes and chemistry. This figure describes four potential scenarios, in two of which the barrier to the conformational change, $\Delta g^{\ddagger}_{conf}$, is much smaller than the chemical step, $\Delta g^{\ddagger}_{chem}$ (i.e., the chemistry is rate-limiting, a and c), and a second two in which the two barriers are roughly equal (b and d). In the first extreme (a and b), we illustrate a diffusive model in which the chemical step has no memory of the preceding conformational change and the energy of the conformational change is completely dissipated in the Michaelis complex. In the second extreme (c and d) we present an intertial model, in which memory of the conformational transition is carried across and helps drive the chemical step. Such a situation would constitute a true dynamical effect, where conformational flexibility is helping drive the rate acceleration. (Reprinted with permission from Pisliakov, A. V., J. Cao, S. C. L. Kamerlin, and A. Warshel. 2009. *Proc. Natl. Acad. Sci. U. S. A.* 106:17359–17364. Copyright 2009, National Academy of Sciences, U.S.A.)

diffusive model the system moves randomly in the reactant state and eventually acquires enough thermal energy to reach the TS. In contrast, in the inertial model the conformational transition helps to drive the chemical step. If this inertial model were to exist in a real system, it would describe a true dynamical contribution to catalysis, and addressing this issue remains a substantial challenge for biophysics.

The purpose of this chapter is *not* to convince the reader of whether dynamical effects play a role in enzyme catalysis, for which we refer the reader to other sources (e.g., Benkovic and Hammes-Schiffer 2003; Glowacki et al. 2012b; Hammes-Schiffer 2013; Henzler-Wildman and Kern 2007; Kamerlin and Warshel 2010a; Villali and Kern 2010). Rather, we believe that the *only* way to address this problem is with a clear definition such as that shown in Figure 3.2, and by exploring the actual chemical landscape. Having defined both what we mean when referring to enzyme catalysis and also what can be constituted to be true dynamical effects, we will therefore proceed to illustrate advances in computational approaches that allow us to actually explore the multidimensional landscapes of enzyme catalysis by using an appropriate reaction coordinate and computational models with varying degrees of simplification.

MODELING FREE-ENERGY LANDSCAPES OF ENZYME ACTIVITY

General Considerations

An idea that has gained increasing popularity in recent years is that of energy landscapes for enzyme folding and function (e.g., Benkovic et al. 2008; Boehr et al. 2006; Cheung et al. 2004; Frauenfelder et al. 1991; Hammes-Schiffer 2013; Hammes et al. 2011; Heath et al. 2007; Henzler-Wildman, Lei et al. 2007; Kuhlman and Baker 2004; Kumar et al. 2000; Lu and Wang 2008; Messer et al. 2010; Min et al. 2008; Okazaki et al. 2006; Onuchic et al. 1997). This has been based on extensive experimental and theoretical work (Balabin et al. 2009; Hatzakis et al. 2012; Hubbard et al. 2013; Liu et al. 2012; Mauldin and Sauer 2013; Seco et al. 2012; Sulkowska et al. 2012; Williams 2010; Yu et al. 2012), which has shown that enzymes are dynamic entities that can exist in multiple conformations, through which they fluctuate on various timescales that can range from femtoseconds to seconds (Benkovic and Hammes-Schiffer 2003; Henzler-Wildman and Kern 2007; Henzler-Wildman, Lei et al. 2007). The existence of such conformations has led to the energy landscape hypothesis of enzyme catalysis (Benkovic et al. 2008; Hammes et al. 2011; Hammes-Schiffer 2013). That is, the catalytic landscape for an enzymatic reaction can, in principle, be reduced to two dimensions (Benkovic et al. 2008; Xiang et al. 2008), as illustrated in Figure 3.2. Here, one dimension is a conformational coordinate, which refers to any sort of environmental reorganization (local fluctuations or large-scale conformational changes) relevant to the chemical step. The second dimension is a chemical coordinate, which responds to the chemical step of enzyme catalysis. It has been argued (Benkovic et al. 2008) that this energy landscape is rugged, with multiple minima and transition states, and where the multiple conformations that can occur during the catalytic mechanism form catalytic networks.

Such a hypothesis is particularly attractive in light of work that has shown that the activities of individual enzyme molecules can vary drastically in a bulk sample (Engelkamp et al. 2006; Tan and Yeung 1997) (a condition referred to as static disorder). For example, as highlighted in Engelkamp et al. (2006), studies have shown that the activities of individual lactate dehydrogenase molecules can vary by a factor of four (Xue and Yeung 1995), and in the case of alkaline phosphatase, activities can vary by a whole order of magnitude (Craig et al. 1996). Extending from this, single-molecule studies such as English et al. (2006), Lu et al. (1998, 1999), Schenter et al. (1999), and Xie (2002) have pointed to the concept of dynamic disorder. In such works, time tracers have been used in order to measure enzymatic turnovers in real time, and it has been demonstrated that not only do enzymatic rates fluctuate between individual enzyme molecules, but they also fluctuate over time. This has in turn been argued to be a widespread property of enzyme catalysis (Engelkamp et al. 2006). Such fluctuations in activities would be perfectly consistent with a rugged landscape, with local activation barriers separating the different minima, and with multiple trajectories across the surface, giving rise to different kinetics depending on the starting point. Additionally, and as we clearly demonstrated in a recent study of DHFR (Adamczyk et al. 2011), external factors such as mutations can and will change the topology of this landscape.

Now, while a two-dimensional (2-D) representation of the catalytic landscape such as that shown in Figure 3.2 may seem oversimplified, it is nevertheless very effective, and such 2-D representations have routinely been used, for example, in folding studies (Frauenfelder et al. 1991; Zhuravlev et al. 2009). Recent experimental studies have also highlighted the importance of catalytic landscapes for understanding enzymatic function and catalysis. For instance, Hilvert and coworkers (Pervushin et al. 2007; Vamvaca et al. 2004) have been able to generate an engineered monomeric form of chorismate mutase that appears to be a molten globule, but at the same time is still as effective a catalyst as the native enzyme. Similarly, to take this argument even further, recent NMR studies on T4 lysozyme (Pervushin et al. 2007) have shown that targeted mutations can change the conformational landscape of the enzyme sufficiently to cause it to adopt different functions in different conformations.

While such experimental studies provide valuable insight, if one wants to directly probe the relationship between such landscapes and the observed enzymatic activities, it is clearly important to be able to move beyond qualitative inference to clear physical descriptions. As we will showcase in the section titled "Showcase Systems," computer power has reached a stage where it is in fact possible to generate both folding and conformational landscapes using reduced models with ease and both we and our coworkers have, in fact, done so for a number of systems over the past few years (e.g., Adamczyk and Warshel 2011; Adamczyk et al. 2011; Braun-Sand et al. 2008; Pisliakov et al. 2009; Roca et al. 2008; Xiang et al. 2008, among others). However, before proceeding to discuss these works, we would like to take some time to discuss the challenges involved in addressing this problem computationally. Essentially, the problem can be reduced into two parts: reliably modeling the chemical step of catalysis, and reliably modeling the corresponding conformational flexibility. One could address these problems individually, generating (as much as possible of) the full free-energy landscape through "simply" examining the chemical step at different protein conformations, taking slices through the landscape (Braun-Sand et al. 2008; Roca et al. 2008). Alternately, one could use reduced representations of the system in order to generate the full landscape directly as a function of the two relevant coordinates (Adamczyk and Warshel 2011; Adamczyk et al. 2011; Pisliakov et al. 2009). The challenge in both approaches of course is to determine what the best way to define the relevant coordinates actually *is*, and here, we will discuss these two issues separately for simplicity.

Modeling the Chemical Dimension of Enzyme Catalysis

In order to break down the problem, we will start by discussing the chemical coordinate corresponding to the chemical step of enzyme catalysis. That is, in recent years, there has been significant progress in the use of *ab initio* methods (Cremer 2011; de Jong et al. 2010; Del Ben et al. 2012) and approaches based on density functional theory (DFT) (Ehrlich et al. 2013; Sena et al. 2011) to study chemical reactivity in the gas and condensed phases. One advantage when dealing with such smaller systems is that, given the advances in the scaling of DFT approaches (Siegbahn and Himo 2009; te Velde et al. 2001; Liao and Thiel 2012), one can take the full system into account. Additionally, in such cases, as the chemical step is usually well defined, it is possible to use a simple geometric reaction coordinate to describe reaction progress. The main challenge, however, is moving from this to the complex enzyme energy landscape characterized by several minima. It is becoming increasingly popular to use cluster calculations on model systems to describe enzyme reactivity (Siegbahn and Himo 2009; Sousa et al. 2012). However, for a proper description of long- range electrostatics and the system's dynamical behavior it is necessary to take into account the full system. Now, in principle, one could instead perform hybrid quantum mechanical/molecular mechanical (QM/MM) calculations (Gao and Truhlar 2002; van der Kamp et al. 2013; Warshel and Levitt 1976) in which the part of the system one is interested in is treated at a higher level of theory (*ab initio* or DFT) with the remainder of the system being treated using MM. However, such approaches are still very computationally expensive, making it challenging to perform the extensive conformational sampling required to obtain meaningful convergent free energies (Klähn et al. 2005), and therefore one can easily end up in a situation where different starting conformations give completely different results due to the complexity of the landscape (Kamerlin et al. 2009; Klähn et al. 2005).

One way to sidestep this problem when studying chemical reactivity is simply by use of a valence-bond-based description of the system that reflects bond properties rather than a molecular orbital (atomic) description of the system. This in turn allows for the use of the energy gap reaction coordinate, which we will expand on later in this section. This can be achieved using approaches such as the empirical valence bond (EVB) approach (Kamerlin and Warshel 2010b, 2011a; Warshel 1991), which is a semiempirical QM/MM approach that is on the one hand fast enough to allow for the extensive conformational sampling required to obtain convergent free energies, while at the same time carrying sufficient chemical information to describe bond cleavage and formation processes in a chemically meaningful way. A schematic representation of the EVB approach for the simple case of a two-state S_N2 reaction is shown in Figure 3.3. That is, in the EVB approach, the individual species involved in the reaction (in this case just reactants and products, although an unlimited number of intermediate states can be introduced) are described as being at the center of parabolic surfaces corresponding to different zero-order diabatic states. These diabatic states correspond to the classical valence-bond structures that would describe each of these species, and the actual ground state adiabatic energy surface can be obtained by mixing the effective off-diagonal terms of an $N \times N$ (in this case 2×2) EVB Hamiltonian. The ground state energy of this system, E_g, corresponds to the lowest eigenvalue of this Hamiltonian. The potential energies of each of these

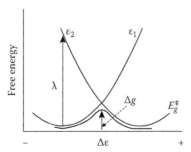

Figure 3.3 Simplified description of the relationship between the diabatic parabola and the adiabatic ground state free energy of a hypothetical two-state reaction using an EVB description. Also shown here is the reorganization energy, λ, and the energy gap reaction coordinate ($\Delta\varepsilon$). For further discussion, see main text.

diabatic states is represented by classical MM force fields, which take the form:

$$\varepsilon_i = \alpha_{gas}^i + U_{intra}^i(\mathbf{R},\mathbf{Q}) + U_{Ss}^i(\mathbf{R},\mathbf{Q},\mathbf{r},\mathbf{q}) + U_{ss}(\mathbf{r},\mathbf{q}) \qquad (3.6)$$

Here, $\mathbf{R}$ and $\mathbf{Q}$ denote the coordinates and charges of the diabatic states, respectively, and $\mathbf{r}$ and $\mathbf{q}$ denote the corresponding values for the surrounding protein and/or solvent. $U_{intra}^i(\mathbf{R},\mathbf{Q})$ and $U_{Ss}^i(\mathbf{R},\mathbf{Q},\mathbf{r},\mathbf{q})$ denote the intramolecular potential of the solute system (relative to its minimum) and the solute (S) solvent (s) interactions, respectively. $U_{ss}(\mathbf{r},\mathbf{q})$ denotes the surroundings-surrounding (ss) interactions between the protein atoms and surrounding solvent, and finally, α_{gas}^i designates the gas-phase energy of the ith diabatic state, when all other fragments are taken to be at infinite separation. The first term on the left-hand side of Equation 3.6, ε_i, forms the diagonal elements of the EVB Hamiltonian (H_{ii}) for the relevant state:

$$\hat{H}^{EVB} = \begin{pmatrix} H_{ii} & H_{ij} \\ H_{ij} & H_{jj} \end{pmatrix} = \begin{pmatrix} \varepsilon_i & H_{ij} \\ H_{ij} & \varepsilon_j \end{pmatrix} \qquad (3.7)$$

(here the indices i and j denote the reactant and product states, respectively, for the example illustrated in Figure 3.3). The corresponding off-diagonal terms, H_{ij}, can be obtained using the rigorous definition below:

$$H_{ij} = \sqrt{(\varepsilon_i - E_g)(\varepsilon_j - E_g)} \qquad (3.8)$$

Note as an aside that one of the underlying assumptions of the EVB approach is that the H_{ij} term is same in the gas phase, solution, and proteins, and a detailed study has established the validity of this assumption (Hong et al. 2006). In any case, once the EVB diagonal and off-diagonal elements have been defined, the relevant activation energies ($\Delta g^{\ddagger}$) can be obtained by mapping from one diabatic state to another. In its simplest form involving two diabatic states, this can be represented as a linear combination of the potentials of the two diabatic states; that is

$$\varepsilon_m = (1 - \lambda_m)\,\varepsilon_i + \lambda_m \varepsilon_j \qquad (3.9)$$

where ε_m denotes the corresponding mapping potential. Here, ε_i and ε_j denote the potentials of the two diabatic states i and j, and λ_m is a weighting factor that is changed from 0 to 1 in $n + 1$ fixed increments (where $\lambda_m = 0/n$, $1/n$, $2/n$, ..., m/n) and forces the system to fluctuate in a specific place along the reaction coordinate depending on the weighting of the potentials. The corresponding free energy associated with changing λ_m from 0 to m/n (ΔG_m) is evaluated using standard free-energy perturbation (FEP) (Warshel 1991). The subsequent free-energy functional that provides the free energy corresponding to the adiabatic ground state surface along the reaction coordinate, x, can be obtained by FEP umbrella sampling (US) (Torrie and Valleau 1977; Warshel 1991), which can also be used to obtain the free-energy functional of the individual diabatic states by (Warshel 1991):

$$\Delta g_i(x') = \Delta G_m - \beta^{-1} \ln\langle \delta(x-x')\exp[-\beta(\varepsilon_i(x)-\varepsilon_m(x))]\rangle_{\varepsilon_m} \qquad (3.10)$$

Here, $\Delta g_i(x')$ corresponds to the free-energy functional of the diabatic state i at position x' along the reaction coordinate x. The potentials ε_i and ε_m have already been defined above, and here ε_m is simply used as a potential to keep x in the region of x'. Additionally, β corresponds to $1/k_B T$, where k_B is Boltzmann's constant and T is the temperature. Provided that the incremental changes in ε_m are sufficiently small, the resulting $\Delta g(x')$ obtained with several values of m overlap over a range of different x' values, and patching together all $\Delta g(x')$ obtained across the reaction coordinate (moving from $\lambda_m = 0$ to m/n) should give the full adiabatic free-energy curve for the reaction.

The reason we have presented such a lengthy introduction to the EVB approach is to be able to introduce two key concepts that are central to our discussion of modeling dynamical effects in enzyme catalysis; namely, the energy gap reaction coordinate, and more critically, the concept of reorganization energy. As it is a simpler concept, we will start by discussing the energy gap reaction coordinate. That is, a reaction coordinate is simply a qualitative descriptor used to describe the process of a chemical event from one state to another (although not strictly semantically correct, we will use the concept "reaction coordinate" in this chapter to also denote the conformational dimension). Clearly, the correct choice of reaction coordinate is *crucial* to any study of complex chemical or biochemical processes, as it introduces bias into the system and will therefore to some extent determine the outcome (Gao et al. 2006; Rohrdanz et al. 2013; Warshel 1982). Now for the simple case of a small model reaction occurring either in the gas phase or modeled using an implicit solvent model, the two reacting states (ground and product states) can be clearly defined and reaction progress can normally be described in terms of changes in 2–3 degrees of freedom (c.f. More O'Ferrall Jencks plots [Jencks 1972; O'Ferrall 1970] or by using reaction cubes [Grunwald 1985; Guthrie 1990; Scudder 1990; Trushkov et al. 1990]). However, the situation becomes much more complicated when modeling the system using either explicit solvation or (even more challengingly) in a complex biological system such as an enzyme, as one has many more degrees of freedom and a simple geometrical coordinate is no longer an adequate descriptor of the system. This problem can be resolved in a valence bond framework, as one can use the energy gap between the two diabatic states as the reaction coordinate (which is defined as $x = \varepsilon_i - \varepsilon_j$) (Warshel 1991, 1982). This is a particularly powerful choice of reaction coordinate when dealing with multidimensional reaction coordinates, as it allows for the projection of the full multidimensional space onto a single reaction coordinate (Hwang et al. 1988) without assuming a predefined reaction path, while at the same time accounting for the full environmental reorganization along the reaction coordinate. Clearly, this is especially relevant in the case of examining conformational space (see the section titled "Modeling the Conformational Dimension of Enzyme Catalysis"), where it is otherwise much more difficult (if not impossible) to define a clear and unambiguous reaction trajectory in one dimension. In addition to fully accounting for the system reorganization, the energy gap reaction coordinate also takes into account the solvent response to the solute polarization. Finally, it should be noted that it was independently demonstrated that an additional advantage of the energy gap reaction coordinate is that it greatly accelerates the convergence of free-energy calculations (Mones et al. 2009) and the power of the energy gap reaction coordinate is now becoming increasingly appreciated by some of the key workers in the field (Kamerlin and Warshel 2010b, 2011a; Mones et al. 2009; Watney et al. 2003), thus greatly increasing its applications.

Having defined the reaction coordinate not in terms of geometries but rather in terms of the energy gap between the two diabatic states, this then allows us to address the concepts of the reorganization energy and electrostatic preorganization. That is, as illustrated in Figure 3.4, in the case of the uncatalyzed reaction in aqueous solution, as the reaction progresses from ground state to transition state the solute charges also change and the solvent becomes polarized by the field of the solute (reorienting its dipoles to optimally stabilize the charged solute). There is a free energy cost to this reorientation, which is referred to as the reorganization energy (Warshel 1978). This reorganization energy, λ_r, can be obtained by rigorously evaluating the Marcus parabola (corresponding to the different diabatic states) (Warshel 1991), as outlined in Figure 3.3. Mathematically, λ_r can be described using the relationship:

$$\lambda_r = 0.5 \left(\langle \Delta\varepsilon \rangle_j - \langle \Delta\varepsilon \rangle_i \right) \tag{3.11}$$

Here, ε_i and ε_j are the two diabatic states originally introduced in Equations 3.6 and 3.9. $\Delta\varepsilon$ denotes the energy difference between these two diabatic states, and $\langle \Delta\varepsilon \rangle_i$ and $\langle \Delta\varepsilon \rangle_j$ denote averages obtained over trajectories run using the potentials of states i and j, respectively. Note the subscript r, which we use to distinguish between the reorganization energy (λ_r) and the frames of the mapping potential in Equation 3.6 (λ_m). Additionally, we would like to point out again that in order to be physically meaningful, the reorganization energy should *not* be evaluated using, for example, simply the energy along the least

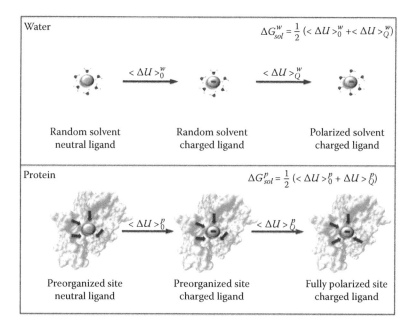

Water

$$\Delta G_{sol}^{w} = \frac{1}{2} (< \Delta U >_0^{w} + < \Delta U >_Q^{w})$$

$< \Delta U >_0^{w}$

$< \Delta U >_Q^{w}$

Random solvent
neutral ligand

Random solvent
charged ligand

Polarized solvent
charged ligand

Protein

$$\Delta G_{sol}^{p} = \frac{1}{2} (< \Delta U >_0^{p} + \Delta U >_Q^{p})$$

$< \Delta U >_0^{p}$

$< \Delta U >_Q^{p}$

Preorganized site
neutral ligand

Preorganized site
charged ligand

Fully polarized site
charged ligand

Figure 3.4   Illustrating the electrostatic preorganization concept. This figure highlights the environmental response to the change in solute charges upon moving from the ground to the transition state for a hypothetical reaction in aqueous solution and in an enzyme active site. It can be seen, in aqueous solution, that the reaction involves substantial reorganization of the water molecules upon moving to the transition state, which corresponds to a large reorganization energy. In the enzyme active site, however, the active site dipoles are already preorganized to optimally stabilize the transition state, and therefore there is much less change upon moving to transition state and a correspondingly smaller reorganization energy. (Reprinted with permission from Kamerlin, S. C. L., S. K. Pankaz, Z. T. Chu, and A. Warshel. 2010. *Proc. Natl. Acad. Sci. U. S. A.* 107:4075. Copyright 2010, National Academy of Sciences, U.S.A.)

energy path, but rather has to be evaluated using the pure, unmixed diabatic states (Kamerlin and Warshel 2010b). Figure 3.5 shows a schematic illustration of the relationship between the two diabatic states, ε_i and ε_j, the activation energy $\Delta g^{\ddagger}$, the ground state energy, E_g, and the reorganization energy, λ_r, for a hypothetical reaction occurring in aqueous solution as well as its counterpart in an enzyme active site. From this figure, it can be seen that in crude terms the reorganization energy provides a quantitative measure of the extent to which a system changes upon moving from state i to state j. In addition to this, it can also be seen that λ_r is substantially smaller in the enzyme than in solution. This is due to the electrostatic preorganization of the enzyme's active site. That is, while there is a large free-energy penalty associated with reorienting

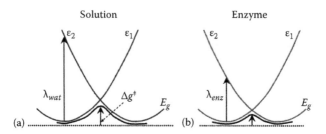

Figure 3.5   Schematic illustration of the relationship between the relative positions of the two diabatic states, ε_i and ε_j, and the activation energy, $\Delta g^{\ddagger}$, for a hypothetical two-state reaction occurring in (a) aqueous solution and (b) in an enzyme active site. It can be seen that as the reaction in the enzyme has a much smaller reorganization penalty, this alters the position of the two parabolas, thus reducing the activation energy. (Adapted from Roca, M. et al., *Biochemistry* 48:3046, 2009.)

the randomly oriented solvent dipoles in aqueous solution, while not perfectly aligned, the protein active site already has partially oriented dipoles. Therefore, in the case of the protein, the system has to pay a much lower reorganization penalty upon moving to the transition state (note that while related to the concept of Marcus' reorganization energy [Marcus and Sutin 1985], the key difference is that in the present case, the protein has to have fixed dipoles to be able to reach both small reorganization and a stable transition state). In any case, this phenomenon, which has been numerous simulation (Adamczyk et al. 2011; Prasad and Warshel 2011; Warshel et al. 2001, 2006a), is arguably the key determining factor in enzyme catalysis (as well as rationalizing the effect of mutations [Adamczyk et al. 2011; Liu and Warshel 2007b], even driving both natural and artificial protein evolution [Labas et al. 2013; Liu and Warshel 2007b; Luo et al. 2012]). As was seen in the discussion above, rigorously calculating the shift in the Marcus parabola within a valence bond description allows for a quantitative determination of this effect and actually probing the precise determinants of chemical catalysis.

Modeling the Conformational Dimension of Enzyme Catalysis

The previous section, "Modeling the Chemical Dimension of Enzyme Catalysis," introduced the concepts of the energy gap reaction coordinate and the reorganization energy in the context of modeling the *chemical* step of enzyme catalysis. Here we will start by illustrating that these concepts are equally (if not even more) powerful when exploring enzyme conformational changes. That is, there is extensive debate in the literature about the choice of appropriate reaction coordinate to describe complex, multidimensional processes (e.g., see

Nymeyer et al. 2000; Rohrdanz et al. 2011, 2013, among others). Also as shown in "Modeling the Chemical Dimension of Enzyme Catalysis," the energy gap has the advantage of being able to project this entire multidimensional space onto a single, nongeometric reaction coordinate while still taking into account the full system reorganization during the course of the process of interest. Now if one's main interest is in only examining the isolated conformational/folding coordinate, this becomes less of an issue—as of 2013, computational power reached a stage where microsecond (Mayor et al. 2003; Schlick et al. 2011) (and sometimes even millisecond [Dror et al. 2010; Shaw 2013]) simulations of protein conformational dynamics are now quickly becoming routine. There exists a plethora of approaches in order to perform biased and guided simulations of protein conformational dynamics (Rohrdanz et al. 2013; Schlick 2009), such as targeted molecular dynamics (Isralewitz et al. 2001; Schlitter et al. 1993), string methods (Weinan and Vanden-Eijnden 2010), and Markov state models (Pande et al. 2010). Things become complicated if one wants to not only examine the conformational dimension but also examine its coupling to the corresponding chemical coordinate (i.e., examining part or all of the full catalytic landscape for the system).

In cases where the system proceeds through clearly defined conformational states (for which structures exist), exploring the coupling to the chemical steps becomes an easier problem. That is, one can in principle simply run EVB (or similar) trajectories at different points in the conformational landscape and stack the resulting slices through the surface to obtain a qualitative approximation of the topology of the surface, as was done for example in Figure 3.6 in the case of chorismate mutase (Roca et al. 2008). If one wants, this can also be taken one step further by using the linear response approximation (LRA) in order to evaluate the energetics of moving between these conformational states, providing a more (semi-)quantitative description of the catalytic landscape. Such an approach has been effectively applied to a number of biological systems, including bacteriorhodopsin (Braun-Sand et al. 2008), F_1 ATPase (Mukherjee and Warshel 2012) and cytochrome c oxidase (Muegge et al. 1997). This can be achieved using a variant of the standard LRA treatment (Braun-Sand et al. 2008):

$$\Delta\varepsilon\left[i(r_i) \rightarrow j(r_j)\right] \cong \frac{1}{2}\left[\left\langle \Delta\varepsilon_{i \rightarrow j}\right\rangle_{i,r_i} + \left\langle \Delta\varepsilon_{i \rightarrow j}\right\rangle_{j,r_j}\right] \qquad (3.12)$$

Here, $\Delta\varepsilon_{i \rightarrow j} = \varepsilon_j - \varepsilon_i$, and $\left\langle \Delta\varepsilon_{i \rightarrow j}\right\rangle_{i,r_i}$ and $\left\langle \Delta\varepsilon_{i \rightarrow j}\right\rangle_{j,r_j}$ designate trajectories run on the relevant state (i or j) with a constraint that keeps the system near the relevant protein conformation (r_j or r_i) (see also discussion in Braun-Sand et al. 2008; Strajbl et al. 2003). This approach is described schematically in Figure 3.7 and is discussed in greater length in the section titled "Bacteriorhodopsin."

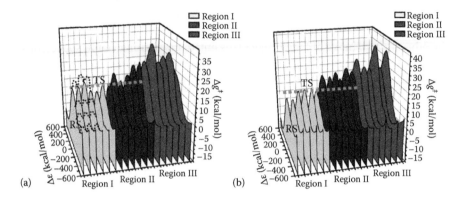

Figure 3.6 Energy landscapes for (a) monomeric (engineered) and (b) dimeric (native) forms of chorismate mutase. As outlined in the main text, the definition of different regions is based on the root mean square deviation (RMSD) from the native structure. RS and TS denote the reactant and product state, respectively, $\Delta\varepsilon$ denotes the energy gap reaction coordinate, and $\Delta g^{\ddagger}$ denotes the activation barrier for the process. The dashed line is drawn at 16 kcal/mol, and it can be seen that the monomeric form (a) has several catalytic configurations with energies below this cutoff, whereas the dimeric form (b) has none. Therefore, there are more catalytically active conformations accessible to the monomer than to the dimer. (Reprinted with permission from Roca, M., B. Messer, D. Hilvert, and A. Warshel. 2008. *Proc. Natl. Acad. Sci. U. S. A.* 105:13877–13882. Copyright 2008, National Academy of Sciences, U.S.A.)

Finally (and more recently), we have also been using a renormalization approach to explore the coupling between enzyme conformational dynamics and the chemical step of enzyme catalysis (Adamczyk et al. 2011; Pisliakov et al. 2009). The purpose of this approach, which is illustrated in Figure 3.8, is to allow one to move between simplified/reduced models of complex systems and full explicit models of the same system in a physically meaningful way, thus letting us model energetics and dynamics long timescale processes in complex biological systems. In this hierarchical approach, we start with an explicit all-atom model that is subjected to molecular dynamics (MD) simulations on which a series of restraints are imposed to guide the system along the reaction coordinate on different timescales (Figure 3.8a). The aim here is to use the smallest possible restraint that allows the relevant transition to occur on a computationally tractable timescale. We then move to increasingly simplified representations of the full system (Figure 3.8b and c) and run Langevin dynamics (LD) rather than MD, adjusting the friction in the LD treatment until the time dependence of the process of interest becomes identical in the explicit and simplified models for a given set of constraints. This process has been discussed in detail elsewhere (e.g., in Kamerlin et al. 2011; Kamerlin and Warshel

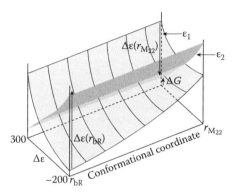

Figure 3.7   A schematic overview of the LRA procedure used in (Braun-Sand et al. 2008) in order to generate a combined surface for the proton transfer in bacteriorhodopsin, which also takes into account the protein conformational coordinate. The conformational coordinate is described by the transition between different known conformations of bacteriorhodopsin, and is taken as the energy gap between the two conformations ($\Delta\varepsilon$) where the proton is still on the Schiff base when $\Delta\varepsilon \approx -200$ kcal/mol. (Reprinted from Braun-Sand, S., P. K. Sharma, Z. T. Chu, A. V. Pisliakov, and A. Warshel. 2008. The energetics of the primary proton transfer in bacteriorhodopsin revisited: It is a sequential light induced charge separation after all. *Biochim. Biophys. Acta* 1777:441–452, Copyright 2008, with permission from Elsevier.)

2011b; Pisliakov et al. 2009). Once the system has been fully validated in order to obtain the optimal friction, it is then possible to obtain atomic detail of the process of interest using restrained all-atom molecular dynamics simulations and then run long timescale unbiased simulations of the system using the simplified model. In this way, one retains the physics of the system while also obtaining insight into the effect of removing the constraints in contrast to performing only biased simulations (see, e.g., Dryga and Warshel 2010 for further discussion). Such an approach has been successfully validated for the test cases of the gramicidin A ion channel (Dryga and Warshel 2010) and lid closure in adenylate kinase (Pisliakov et al. 2009) and then extended to a range of problems from modeling protein conformational changes (Adamczyk et al. 2011; Pisliakov et al. 2009; Prasad and Warshel 2011) to the activation of voltage gated ion channels (Dryga et al. 2012) and proton translocation across membranes (Rychkova and Warshel 2013; Rychkova et al. 2010). Finally, what is particularly relevant to the present work is the fact that once the system has been fully validated, it is possible to simplify the problem as far as representing the model using two effective coordinates corresponding to the conformational and chemical steps, respectively. From this, it is possible to obtain a 2-D free-energy landscape such as that shown in Figure 3.8c, from the ground state

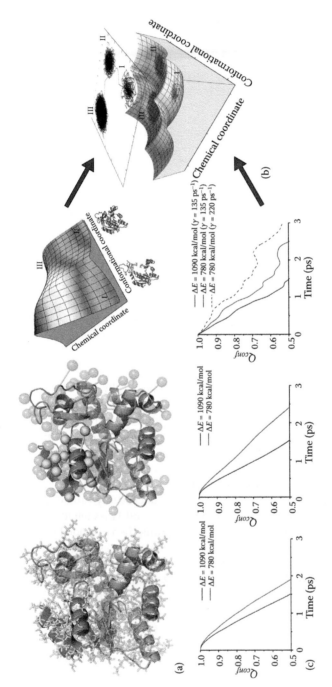

Figure 3.8 Illustration of different possible representations used in the renormalization approach, including (a) and all atom explicit model, (b) a coarse-grained model in which the side chains are described as sphere, and (c) a simplified 2-D model in which an energy landscape is generated by describing the system using two dimensionless effective coordinates, Q_1 and Q_2. (Reprinted with permission from Pisliakov, A. V., J. Cao, S. C. L. Kamerlin, and A. Warshel. 2009. *Proc. Natl. Acad. Sci. U. S. A.* 106:17359–17364. Copyright 2009, National Academy of Sciences, U.S.A.; From Kamerlin, S. C. L. et al., *Annu. Rev. Phys. Chem.* 62:41–64. Copyright 2011.)

energy (E_g) of a four-state EVB-type Hamiltonian (Pisliakov et al. 2009), the diagonal elements of which are given by

$$H_{lm,lm} = \varepsilon_{lm} = \frac{\hbar\omega_1}{2}\left(Q_1 - F_{coupl}\delta_1^{lm}\right)^2 + \frac{\hbar\omega_2}{2}\left(Q_2 - \delta_2^{lm}\right)^2 + \alpha_{lm} \tag{3.13}$$

In this equation, the two effective coordinates corresponding to the chemical and conformational steps are described by Q_1 and Q_2, respectively. Each of these coordinates can exist in one of two states (reactant/product [R/P] vs. open/closed [O/C]) or some intermediary thereof, and these states are described by the indices l and m. These indices have values of 1 and 2, depending on whether one is in the starting (open/reactant) or final (closed/product) state along the relevant reaction coordinate. α_{lm} denotes the free energy in each of the four OR, CR, CP, and OP states (compared to the gas-phase shift of Equation 3.6). The two coordinates, Q_1 and Q_2 are dimensionless effective coordinates, which are defined as $Q_1 = -(\varepsilon_{21} - \varepsilon_{11})\hbar\omega_1\delta_1$ and $Q_2 = -(\varepsilon_{12} - \varepsilon_{22})\hbar\omega_2\delta_2$, respectively. The shifts of the different states, δ_i, are related to each other by the relationships $\delta_1^{21} = \delta_1^{22} = -\delta_1^{11} = \delta_1$, $\delta_2^{12} = \delta_2^{22} = -\delta_2^{11} = \delta_2$, $\delta_1^{12} = \delta_1^{11}$, and $\delta_2^{21} = \delta_2^{11}$. The function F_{coupl} describes the change in the chemical reorganization energy upon changing the donor acceptor distance provided that this is coupled to the conformational coordinate (for a detailed definition see [Pisliakov et al. 2009]), and the effective frequency, ω_{eff}, is evaluated by

$$\omega_{eff} = \int \omega P(\omega) d\omega \tag{3.14}$$

where $P(\omega)$ is the normalized power spectrum of the corresponding contribution to the term $(\varepsilon_{2m} - \varepsilon_{1m})$. Additionally, the different δ terms are related to the reorganization energy using the relationship:

$$\lambda_1 = \left(\frac{\hbar}{2}\right)\omega_1\delta_1^2 = \left(\frac{\hbar}{2}\right)\omega_2\delta_2^2 \tag{3.15}$$

Finally, as in the simpler 2-D case described in "Modeling the Chemical Dimension of Enzyme Catalysis," the off-diagonal terms can be modulated in order to force the barriers along each coordinate to reproduce known experimental values and once again the potential surface is obtained by diagonalizing the system Hamiltonian. This gives rise to a smooth surface such as that shown in Figure 3.8c. However, in the event that one wants a more realistic representation of the actual landscape, as we demonstrated in Pisliakov et al. (2009), one can also modulate the surface in the Q_2 direction by introducing an extra contribution to ε_{lm}. This is a combination of several periodic functions that have adjustable amplitudes and frequencies, which can be described using:

$$F_{\text{fast}} = a_1 \cos(b_1 Q_2) + a_2 \cos(b_2 Q_2) + a_3 \cos(b_3 Q_2) \tag{3.16}$$

The introduction of these functions makes the landscape along the conformational coordinate more corrugated, which provides a better description of the protein surface, which can be rugged and have multiple minima. Although simplified, this renormalization approach provides a powerful projection of the actual landscape onto two coordinates and has successfully been used to describe a wide range of biological problems (Liu et al. 2009; Mukherjee and Warshel 2012; Pisliakov et al. 2009; Prasad and Warshel 2011), several of which will be discussed in "Showcase Systems."

SHOWCASE SYSTEMS

As can be seen from the section above, "Modeling Free-Energy Landscapes of Enzyme Activity," obtaining multidimensional free-energy landscapes for enzyme-catalyzed reactions as a function of both conformational flexibility and the chemical steps is far from trivial even with current computational power. As a result, in the first attempts to achieve this (Florián et al. 2005; Xiang et al. 2008), only a few key structures were calculated along the conformational landscape, and a preliminary "idea" of the resulting surface was extrapolated from this. This then evolved to the use of coarse-grained (CG) models in order to obtain better sampling of the conformational coordinate (Roca et al. 2007, 2008), since full models are still too computationally expensive to allow for sufficiently extensive sampling on the relevant (micro-, milli-, and even second timescales [Frey and Hegeman 2006]). These simulations then provide starting points in order to take a range of structures from the conformational landscape in order to calculate the chemical step, either by full evaluation of the relevant free-energy profiles using approaches such as the EVB (described in the section titled "Modeling the Chemical Dimension of Enzyme Catalysis") or by calculating the energies of key reaction structures. Once one is able to calculate the energetics of the chemical step at different conformations of the enzyme, one can obtain a more complete idea of the surface by stacking the obtained slices of information to obtain an approximation of the catalytic landscape (Roca et al. 2008). Finally, methods such as the renormalization approach allow one to obtain full surfaces for a range of processes, including enzyme catalysis (Adamczyk et al. 2011; Pisliakov et al. 2009), ion transport through ion channels (Dryga et al. 2012) and protein insertion into membranes (Rychkova and Warshel 2013), to name a few examples.

In this section, we will highlight a number of showcase systems to illustrate the evolution of the field and how it has been possible to move from highly simplified representations of the catalytic landscapes to representations providing ever more detailed information. This transition is best illustrated by DNA polymerases. The energy surface for this system, as a function of conformational/chemical coordinates, was initially studied by simply using stacking slices and more recently through the use of the renormalization approach, where detailed insights into the fidelity of DNA replication were obtained (Prasad and Warshel 2011). We will then present some test cases in order to give the reader a more detailed idea of the evolution of different approaches to explore catalytic landscapes as well as the methods employed.

DNA Polymerases and Replication Fidelity

We will start this section by discussing computational studies of the control of DNA replication fidelity, as this system provides many good examples of works that have elegantly explored the energy landscapes for enzyme catalysis as a function of both chemical and conformational changes. DNA polymerases are attractive systems to study because in all of them (as well as in the similar ribonucleic acid [RNA] polymerases), substrate binding induces a large conformational change in the enzyme leading to a structural transition from an open to a closed state (Doublie et al. 1999; Patel et al. 1991; Tsai et al. 1999). This observation has led to the proposal of an induced fit mechanism, in which the incorporation of a correct deoxyribonucleotide triphosphate (dNTP) substrate (R) would trigger the conformational transition, while the binding of a misaligned substrate (W) will hamper this closure and consequently play a major role in establishing the fidelity (Beard and Wilson 2003; Johnson 1993; Joyce and Benkovic 2004). More recently, this induced fit concept has been extended to argue in favor of the importance of prechemical conformational changes (Radhakrishnan and Schlick 2004, 2005, 2006; Radhakrishnan et al. 2006). This prechemistry proposal is, essentially, a more focused variant of the induced fit proposal, and argues that subtle side chain rearrangements can play an important role in bringing the system to a catalytically active state, thus guiding the subsequent replication fidelity (Radhakrishnan and Schlick 2004; Radhakrishnan et al. 2006). A counterargument against this proposal has been that while such conformational changes do exist, they do not play a role in determining the catalytic efficiency or fidelity *as long as the corresponding free-energy barriers are not rate-limiting* (Prasad and Warshel 2011) (due to the fact that the enzyme is not carrying any memory of the previous steps; see Figure 3.2).

There has been substantial discussion surrounding these positions; for instance in Mulholland et al. (2012), Prasad et al. (2012), and Wang and Schlick (2008). Therefore, we will not go into great detail and repeat these arguments here. However, we would like to point out that, critically, the role of the conformational change in the catalytic process or its relationship to DNA replication fidelity can only be assessed by a proper exploration of the relevant free-energy landscape. The first attempt to use an energy landscape to probe the molecular origin for DNA replication fidelity comes from a study of T7 DNA polymerase (Florián et al. 2005). In this work, the free-energy pathways for the incorporation of R and W nucleotides along the chemical coordinate in the closed protein conformation were plotted along with the approximate free-energy pathways for the R and W open conformations (Figure 3.9). A combination of the LRA and EVB approaches were then used in order to tease out the individual contribution of the binding and chemical steps to the overall fidelity. It is also worth noting that the free-energy pathways for the chemical step in the fully open conformation of the polymerase were approximated rather than calculated and were assumed to be similar to the corresponding energetics in aqueous solution (Florián et al. 2005). This work reasonably reproduced experimental trends in the fidelity and demonstrated that while the G:dATP mispair is discriminated against in both the substrate binding and chemical steps, the T:dGTP mispair was only discriminated against in the binding step. One explanation for this was that the incorporation of the T:dGTP mispair could occur from a partially open conformation, which would be energetically unfavorable. However, it was also pointed out that in this particular case the calculations appear to overestimate the binding step and underestimate the chemical step, and the reproduction of the experimental trend is due to the additive effect of the two errors canceling each other out.

In order to improve the description of the conformational coordinate, in a subsequent study on DNA polymerase β (Pol β), the open to closed conformational change was partially sampled using targeted molecular dynamics (TMD) simulations (Schlitter et al. 1993). Four conformations were obtained (open, closed, and two intermediate structures), and free-energy pathways for the chemical coordinate were again calculated for each conformation for both the R and W substrates (Xiang et al. 2008). This work allowed for a direct analysis of the proposal that induced fit plays an important role in fidelity, and interestingly, demonstrated that similar to the proposal put forward in an earlier study of T7 DNA polymerase (Florián et al. 2005), the W nucleotide is incorporated through a transition state in one (or more) partially open conformations of the polymerase, whereas the R nucleotide is incorporated in a closed conformation. This work could be taken to a much more detailed level with the recent elucidation of crystallographic structures that showed different binding sites for the R (Batra et al. 2006) and W (Batra et al. 2008) nucleotides in the closed conformation of Pol β. Additionally, in a recent study Prasad and Warshel (2011), examined in detail the W to R movement between binding sites

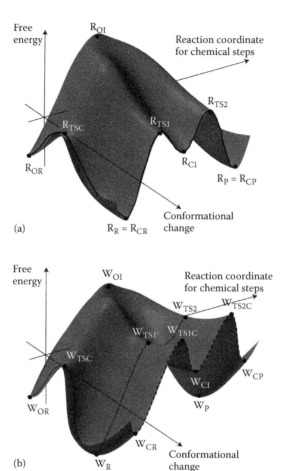

Figure 3.9 Illustration of different possible scenarios for the insertion of both correct (a) and mispairing (b) deoxynucleotides by a DNA polymerase as a function of chemical and conformational coordinates. (Reprinted with permission from Florián, J., M. F. Goodman, and A. Warshel. 2005. *Proc. Natl. Acad. Sci. U. S. A.* 102:6819–6824. Copyright 2005, National Academy of Sciences, U.S.A.)

in the closed conformation of Pol β was simulated using potential of mean force (PMF) calculations, which moved the template O3′ to the catalytic Mg^{2+}. The sampling for the closed to open conformational transition was also improved using TMD with five structures (open, closed, and three intermediates) and then subsequently refined using the renormalization approach to give the catalytic landscapes for R and W nucleotide incorporation.

Overall, these studies have culminated in an improved energy landscape, which allows us to obtain a more complete generalizable picture for the control of DNA replication fidelity by DNA polymerases. These studies have demonstrated that, because the chemical step is rate-limiting, the conformational changes do not effect flexibility, nor do dynamical effects have any influence on catalysis. Rather, the fidelity reflects an allosteric competition between the binding of the incoming ribonucleotide triphosphate (NTP) at the base binding site and the binding of the TS at the catalytic site (Prasad and Warshel 2011). In the case of the correct R substrate, the enzyme provides optimal sites for both base pairing and for the chemical step. However, in the case of the W substrate, poor preorganization in the base-binding site as well as poor binding of the base lead to structural rearrangements that are propagated to the catalytic site. These destroy the active site preorganization for the chemical step, thus yielding less TS stabilization and ultimately controlling the fidelity.

Thermophilic/Mesophilic DHFR

DHFR is a small, ~20–25 kDa enzyme (depending on variant), responsible for the reduction of 7,8-dihydrofolate (DHF) to 5,6,7,8-tetrahydrofolate (THR) through the oxidation of the cofactor coenzyme nicotinamide adenine dinucleotide phosphate (NADPH) (Fierke et al. 1987); see Figure 3.10. DHFR possesses three

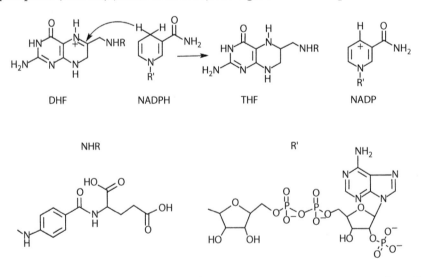

Figure 3.10 Overview of the reaction catalyzed by dihydrofolate reductase (DHFR), which is, specifically, the reduction of 7,8-dihydrofolate (DHF) to 5,6,7,8-tetrahydrofolate (THR) through the oxidation of nicotinamide adenine dinucleotide phosphate (NADPH).

interesting loops, designated the Met-20, βF-βG, and βG-βH loops (Hammes-Schiffer and Benkovic 2006; McElheny et al. 2005; Venkitakrishnan et al. 2004). In the *Escherichia coli* (*E. coli*) variant, these loops can undergo conformational transitions between open, closed, and occluded states, as shown in Figure 3.11 (Hammes-Schiffer and Benkovic 2006; McElheny et al. 2005; Venkitakrishnan et al. 2004). However, no corresponding conformational transitions have been observed in the human enzyme (Davies et al. 1990). As a result, DHFR has been used as an extensive benchmark for the study of dynamical contributions to catalysis (Adamczyk et al. 2011; Arora and Brooks 2013; Benkovic and Hammes-Schiffer 2003; Bhabha et al. 2011; Boehr et al. 2006; Fan et al. 2013; Hammes 2002; Liu and Warshel 2007b; Loveridge et al. 2011; McElheny et al. 2005; Radhakrishnan and Schlick 2005; Schnell et al. 2004), and particularly as a centerpiece for the debate surrounding whether slow motions play a catalytic role (Adamczyk et al. 2011; Arora and Brooks 2013; Benkovic and Hammes-Schiffer 2003; Bhabha et al. 2011; Boehr et al. 2006; Fan et al. 2013; Hammes 2002; Liu and Warshel 2007b; Loveridge et al. 2011; McElheny et al. 2005; Radkiewicz and Brooks 2000; Schnell et al. 2004). Arguments in favor of dynamical contributions to catalysis have been based on a variety of approaches including mutagenesis studies (Miller and Benkovic 1998a, 1998b; Miller et al. 2001), NMR measurements of the loop motions (Miller and Benkovic 1998c), classical MD simulations (Rod et al. 2003), and QM/MM calculations (Wong et al. 2005). However, as outlined in detail in the section titled "Modeling Free-Energy Landscapes of Enzyme Activity," the challenge here is that all of these studies are based on *indirect* observations because it is not possible to directly explore the dynamical proposal using experimental or standard computational techniques. Additionally,

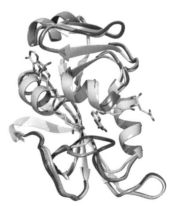

Figure 3.11 Overlay of the open (blue, from PDB ID: 1RA1), occluded (gray, from PDB ID: 1RX2), and closed (red, from PDB ID: 1RX4) conformations of the Met20 loop (residues 9–24) of dihydrofolate reductase.

there have been a number of recent computational studies that have explored both the conformational/folding and chemical coordinates and provided compelling arguments against dynamical contributions to catalysis.

Of particular relevance to the topic of this chapter is the work of Roca et al., which directly compared the thermophilic (Tm) DHFR from *Thermotoga maritima* with the mesophilic (Ms) counterpart from *E. coli* (Roca et al. 2007). The purpose of this study was to explore the argument that Tm enzymes have slow reaction rates compared to Ms enzymes due to restricted dynamical motions (Maglia et al. 2003; Wrba et al. 1990; Zavodszky et al. 1998). Specifically, this argument is based on the assumption that Tm enzymes are more stable than Ms enzymes because they have evolved to function at higher temperatures. In other words, Tm organisms live at higher temperature habitats and hence their proteins need to remain folded at these high temperatures, so the proteins must be more stable (Razvi and Scholtz 2006). In this study (Roca et al. 2007), the Tm and Ms variants of DHFR were modeled using a CG model in which the side chains of each amino acid were represented by an effective unified atom and a dummy atom. The conformational landscape was then evaluated in terms of the radius of gyration and the root-mean square deviation (RMSD) from the folded structure to the unfolded one, using FEP/US simulations (Åqvist and Warshel 1993; Hwang and Warshel 1987; Kato and Warshel 2005; Warshel 1982). However, the regions near the folded state were also explored by running 4-ns MD simulations with no restraints.

A comparison of the resulting folding free energy surfaces for the Tm and Ms enzymes (Figure 3.12) showed that the motion on the Tm surface is confined to a smaller space than that for the corresponding Ms enzyme. Furthermore, the obtained folding free-energy values are in good agreement with the experimental values. Specifically, the calculated folding energy for the Tm enzyme was −34.6 kcal/mol and the value for the Ms enzyme was −8.1 kcal/mol compared to the experimental values of −34 and −6 kcal/mol, respectively (Dams and Jaenicke 1999; Ionescu et al. 2000). The chemical coordinate was evaluated by use of the EVB approach, averaging over six trajectories, and once again, the calculated barriers were in good agreement with experiment (Maglia et al. 2003; Miller and Benkovic 1998b,c), with the Tm barrier being higher than the Ms barrier by 8.7 kcal/mol.

In order to get more insight into the origin of the change in the activation barriers, the diabatic free-energy functionals were decomposed into solvent and solute contributions and the corresponding total reorganization energies. Here, the reorganization energy represents the reduction in energy of the protein (or solvent) when the system with the equilibrium coordinate of the reactant state is placed on the potential surface of the product state and allowed to relax to the product equilibrium coordinate. It was found that the reorganization energy for the Tm enzyme was significantly larger than that for the

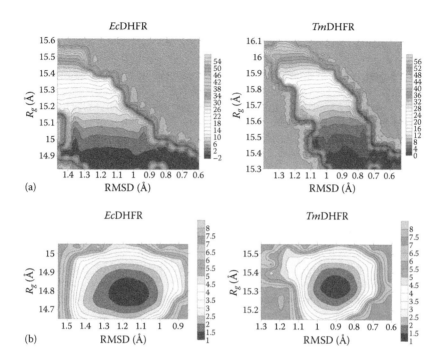

Figure 3.12 Free-energy landscapes for *Ec*DHFR (mesophile) and *Tm*DHFR (thermophile) represented as a function of the radius of gyration of the system (R_g) and the RMSD of the system from the most stable structure. All energies are expressed in kcal/mol, and all distances are expressed in Å. Shown here are surfaces obtained using either (a) free-energy perturbation umbrella sampling or (b) long molecular dynamics simulations without the imposition of any external constraints. (Reprinted with permission from Roca, M., H. Liu, B. Messer, and A. Warshel. 2007. *Biochemistry* 46:15076–15088. Copyright 2007, National Academy of Sciences, U.S.A.)

Ms enzyme. This energy was then used to construct 2-D free-energy surfaces for the chemical step (as a function of the solute, solvent coordinates, and as a function of the donor-acceptor and donor hydrogen distances). These surfaces showed that the reaction pathway is longer in the Tm; however, the curvature at the Rs is similar for both enzymes. Dispersed polaron calculations (Warshel et al. 1989) have further proven that the motions having projections onto the reaction coordinate have similar frequencies and that what changes is the *amplitude* of the motion, which is larger for the Tm than for the Ms enzyme. This means that in the Tm case, the system has to wait longer for the rare fluctuations that take the system to the transition state. This was confirmed by calculating the productive downhill trajectories from the TS, since the time reversal of these trajectories provides the reactive trajectory that moves the system from the RS to the TS (according to Muegge et al. 1997; Olsson et al. 2006).

Once again, this shows that in the Tm ground state, the system oscillates along a larger range of donor-acceptor and acceptor-hydrogen distances than in the Ms. The reaction coordinate motions are, however, not related to the folding coordinate motions. This was verified by showing that (1) the vectors between the PS and RS are nearly perpendicular to the folding vectors, (2) the average projection of the reactive trajectory onto the chemical reaction coordinate is much larger than the projection onto the folding coordinate for the Ms enzyme, and finally, (3) the normalized autocorrelation function of these projections decays in a very short timescale, indicating the unlikelihood that the fluctuation along the folding coordinate would transfer energy to the reaction coordinate. Consequently, the Tm enzymes seem to have slower rates than the Ms enzymes at "normal" temperatures because the Ms enzymes have evolved to reduce the activation barrier (improving catalysis), whereas the Tm enzyme evolved to stay folded at high temperatures, making it unable to preorganize itself to optimally provide low activation barriers. These findings corroborate the idea that proteins have to invest some folding energy in order to obtain low reorganization energy and a preorganized active site (Warshel 1978). For comparison, see also a related work that demonstrated that the effect of mutations distant from the active site of DHFR is due to changes in reorganization energies (Liu and Warshel 2007b).

A more recent elegant experimental study (Bhabha et al. 2011), which combined kinetic measurements, x-ray structure determination, and NMR, demonstrated that the N23PP, S148A, and N23PP/S148A mutations of DHFR not only led to a modest decrease in the rate constant of the hydride transfer step, but also prevent the closed to occluded conformational change of the Met20 loop (Figure 3.11). Additionally, the NMR experiments showed that in the catalytically impaired mutants, the millisecond conformational flexibility of the active site residues appears to have been frozen out. It was therefore argued that this work provides direct evidence of a dynamical contribution to catalysis through dynamical knockout mutations.

Recently, we analyzed these experimental findings in a quantitative way (Adamczyk et al. 2011). The chemical step for both WT and mutant enzymes was evaluated using the EVB approach, and both absolute activation barriers as well as changes upon mutation were reproduced with good accuracy. In order to uncover the origin of the free-energy change, we evaluated the relevant free-energy functionals (Warshel 1991), which provide the microscopic equivalent of the Marcus parabola. Here, we demonstrated that the effect of mutations appears primarily to change the solvation of the product charges (i.e., changing ΔG_0), which reflects both a change in the work function (which is associated with bringing the donor-acceptor closer together) and the reorganization energy along the solvent coordinate. Is it worth noting that based on the crystal structure, one is dealing with a not insubstantial change in donor-acceptor distance from 3.3 to 2.9Å upon mutation (Bhabha et al. 2011), which one would expect to lead to quite different active site preorganization. This was verified by examining and comparing the electrostatic contributions from each protein residue in both WT and mutant enzymes evaluated using the LRA approach. Finally, the effect of flexibility and dynamics was explored by evaluating the free-energy landscapes for the WT and mutant enzymes along the chemical and conformational coordinates (Figure 3.13). It was demonstrated that the

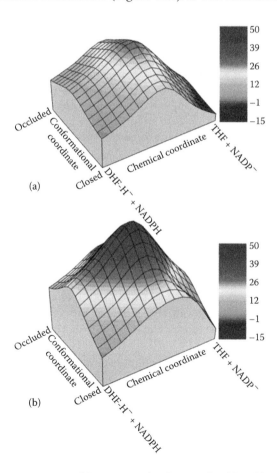

Figure 3.13 A comparison of free energy landscapes for (a) wild type DHFR, and (b) the N23PP/S148A mutant. All energies are presented in kcal/mol. (Reprinted with permission from Adamczyk, A. J., J. Cao, S. C. L. Kamerlin, and A. Warshel. 2011. *Proc. Natl. Acad. Sci. U. S. A.* 108:14115–14120. Copyright 2011, National Academy of Sciences, U.S.A.)

obtained surface for the mutant was qualitatively very similar to the wild type, but with a larger barrier and a higher free-energy difference between the occluded and closed states. Therefore, it would appear that motions in the direction of the occluded state do not help catalysis either by dynamical coupling or by an increase in flexibility, but rather, what changes the observed catalytic effect between the wild-type and mutant enzymes is again changes in reorganization free energy. This finding was corroborated by a recent experimental study (Loveridge et al. 2012) that explored the temperature dependence of the kinetic isotope effects in both WT DHFR and in the N23PP/S148A mutant. The resulting kinetic parameters again strongly suggested that the rate constant of the reaction is determined by the conformational state of the enzyme immediately before the hydride transfer occurs, which dictates the electrostatic preorganization of the active site. Tying in with this, it is worth reemphasizing, as stated in Adamczyk et al. (2011), that clearly, changes in the potential energy surface, such as that shown in Figure 3.13 will modify the reorganization free energy, and as a result, such changes will also alter the corresponding motions. However, these altered motions are simply a reflection of the topology of the energy landscape rather than driving forces for the catalytic step.

Lid Closure in Adenylate Kinase

Following from our discussion on DHFR, another system where a lot of debate has surrounded the role of a lid closure in catalysis adenylate kinase (ADK) (Arora and Brooks 2007; Hanson et al. 2007; Henzler-Wildman, Thai et al. 2007; Lu and Wang 2008; Maragakis and Karplus 2005; Muller et al. 1996; Pisliakov et al. 2009; Vonrhein et al. 1995; Whitford et al. 2007; Wolf-Watz et al. 2004). This enzyme catalyzes the reversible interconversion of adenosine triphosphate (ATP) and adenosine monophosphate (AMP) into two adenosine diphosphate (ADP) molecules (Wieland et al. 1984). It also has two highly flexible segments, designated the NMP domain and the LID domain, the latter of which closes upon substrate binding (Wolf-Watz et al. 2004). Several crystallographic structures exist that document these conformational transitions (Abele and Schulz 1995; Berry et al. 1994; Diederichs and Schulz 1991; Dreusicke et al. 1988; Muller-Dieckmann and Schulz 1994, 1995; Muller and Schulz 1992, 1993; Reuner et al. 1988; Stehle and Schulz 1992; Vonrhein et al. 1995), and there have been multiple elegant recent NMR studies that have suggested that the LID domain motion (its opening after the catalytic step) is the rate-limiting step (Wolf-Watz et al. 2004). Also, the reduced catalytic activity of the Tm enzyme at ambient temperatures was attributed to a slower LID opening conformational change (Wolf-Watz et al. 2004).

Subsequent mutational studies and NMR experiments have focused on the inverse conformational change (i.e., the open to closed transition). This study showed that the mutants had more disordered LID domains, higher ATP binding affinities, and the equilibrium between the open/closed LID conformations was more displaced toward the closed conformation. This led to the proposal that in the WT enzyme, segments within the LID unfold and fold during the open to closed conformational change—that is, at the TS for the conformational change, significant strain energy is released by the LID segments unfolding in accordance with a cracking model (Olsson and Wolf-Watz 2010).

Considering the wealth of structural and experimental data available on this system, ADK provides a perfect starting point for theoretical investigations of the role of dynamics in enzyme catalysis. Therefore, we used ADK as a model system in order to explore whether there exist coupling between the lid closure in this enzyme and the chemical step (Pisliakov et al. 2009). The chemical step was initially examined using both quantum chemical calculations in aqueous solution as well as EVB calculations in both solution and in enzyme, with ADK in its open, closed, and intermediate conformations. Having characterized the chemical step, we then used our renormalization approach to study this enzyme, simplifying the system to first a CG model (with the side chains represented only by spheres), as well as a 2-D simplified model described in the section titled "Modeling the Conformational Dimension of Enzyme Catalysis." This resulted in the catalytic landscape shown in Figure 3.8a (used as an illustration of the renormalization approach). Once this landscape was properly renormalized, it was then possible to run long timescale trajectories (with no constraints). This work (Pisliakov et al. 2009) was the first study to explore both the chemical and conformational steps on the relevant millisecond timescale using unbiased (albeit simplified) simulations.

In order to explore whether the chemical step has any memory of the preceding conformational change, we started by examining the time dependence of the 2-D simplified model, demonstrating that we can reproduce the observed experimental rates of 6500 s^{-1} (k_{open}) (Torres and Levitus 2007) and 260 s^{-1} (k_{chem}) (Wolf-Watz et al. 2004) for the conformational change and chemical steps with good accuracy. Having established the validity of our model, we then ran unconstrained trajectories using the simplified model, starting from the open conformation of ADK, and measured the average first passage time (τ_{fp}) required to reach the chemical TS and product states for the first time. We began by running the trajectories along the conformational and chemical coordinates, fitting the barriers to the experimental values for ADK (14.0 and 14.3 kcal/mol for the conformational and chemical steps). Here, it appeared that the system could reach the closed, product state in ~0.5 ms, although it spent substantial time in the reactant region before crossing to products. We then moved on to a more generalized model by fitting the barrier along the chemical coordinate to a low value of 5 kcal/mol (to accelerate sampling), changing the barrier along the conformational coordinate from 0 to 9 kcal/mol, and once again measuring τ_{fp} (see Figure 3.14). The entire procedure was repeated using

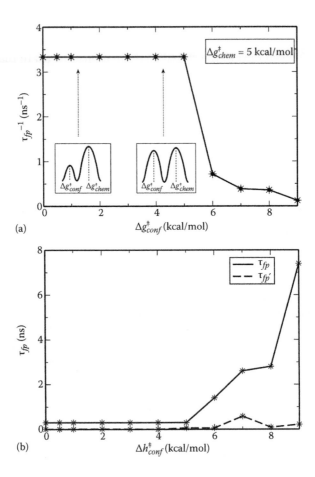

Figure 3.14 An illustration of the relationship between the height of the conformational barrier $\left(\Delta g^{\ddagger}_{conf}\right)$, and the first passage time (τ_{fp}) along the chemical barrier $\left(\Delta g^{\ddagger}_{chem}\right)$, when counting either from (a) the open ground state or (b) the Michaelis complex. This figure shows the effect on τ_{fp} of changing the conformational barrier while keeping the chemical barrier fixed at a given value (5 kcal/mol in the present case). Here it can be seen that as long as the conformational barrier is smaller than the chemical barrier there is no effect on τ_{fp} from changing this barrier, and even when $g^{\ddagger}_{conf} > g^{\ddagger}_{chem}$ the effect of changing $g^{\ddagger}_{conf}$ is negligible. (Reprinted with permission from Pisliakov, A. V., J. Cao, S. C. L. Kamerlin, and A. Warshel. 2009. *Proc. Natl. Acad. Sci. U. S. A.* 106:17359–17364. Copyright 2009, National Academy of Sciences, U.S.A.)

a slightly higher barrier of 7 kcal/mol along the chemical coordinate to ensure that this does not change the overall conclusions. Interestingly, as can be seen from Figure 3.14, it appears that as long as the barrier for the chemical step is higher than the barrier for the conformational step, changing the barrier along the conformational coordinate has no effect on τ_{fp}. This changes slightly once the conformational step is slower than the chemical step; however, if one decouples the time for the overall process (Figure 3.14a) from the time just to cross the chemical step (i.e., measuring τ_{fp} starting from the closed reactant state (Figure 3.14b), one sees that the effect of increasing the conformational barrier is negligible and might just reflect sampling effects. This effectively shows that the chemical process has no significant memory of the dynamics of the conformational motion as long as the chemical barrier is higher than a few kcal/mol (which is the case in enzymatic reactions).

Afterward, in order to further examine the proposal that the chemical reaction is partially driven by the kinetic energy of the substrate-promoted conformational change, two different tests were made: (1) the surface in the simplified 2-D model was explored by artificially increasing the energy of the open state, thus simulating a very exothermic binding process, and (2) the surface was also explored in the CG model (as before with a small chemical barrier), with either no restraints, a 0.5 kcal/(mol · Å²) restraints, or a 5 kcal/(mol · Å²) restraint to guide the conformational transition (here the strength of the restraint represents excess binding energy). Once again, both tests showed that excess kinetic energy does not help to accelerate the chemical step. Finally, we also explored the effect of introducing additional fast relaxation components to the conformational dynamics (see the section titled "Modeling the Conformational Dimension of Enzyme Catalysis") to create a more rugged landscape and explore the idea that the existence of hierarchies of timescales is relevant to enzyme catalysis (Henzler-Wildman, Lei et al. 2007). Here, we repeated the conceptual experiment of Figure 3.14 and demonstrated that even with a more rugged landscape with multiple timescales along the conformational coordinate, this does not change the overall conclusions of Figure 3.14 (see Figure 6 of Pisliakov et al. 2009). Therefore, this work provides strong theoretical evidence against dynamical coupling between the conformational and chemical trajectories for ADK and in fact suggested that the energy of the conformational change is completely dissipated in the closed Michaelis complex and not remembered during the chemical step.

Folding and Catalysis in Engineered Chorismate Mutase

While the main focus of this chapter has so far been on studying the link between smaller-scale conformational changes (such as side chain fluctuations or loop movements), clearly, the ultimate challenge still lies in trying to rationalize the relationship between the full folding landscape and an enzyme's catalytic power. That is, while there have been several studies exploring the topology of various proteins' folding landscapes, a comparatively limited number of studies have directly linked this to chemical catalysis (Adamczyk et al. 2011; Braun-Sand et al. 2008; Pisliakov et al. 2009; Prasad and Warshel 2011; Roca et al. 2007, 2008; Xiang et al. 2008). Here, chorismate mutase (CM) provides a particularly interesting case study due to the experimental finding that an intrinsically disordered engineered protein is capable of achieving the same catalytic activity as the corresponding native enzyme (Pervushin et al. 2007; Vamvaca et al. 2004). Specifically, Hilvert and coworkers (Pervushin et al. 2007; Vamvaca et al. 2004) have taken this dimeric enzyme, which is responsible for the conversion of chorismate to prephenate in the biosynthesis of tyrosine and phenylalanine, and engineered it into a monomer that basically behaves as a molten globule in the free state. Despite this, the engineered CM was demonstrated to have essentially the same catalytic activity as the native monomer. This finding clearly creates a major question mark about the nature of the complex relationship between protein folding and catalytic landscapes, and extending from this, the even more complex relationship between structural and functional evolution (Soskine and Tawfik 2010).

In order to investigate these surprising findings, Warshel and coworkers (Roca et al. 2008) used a simplified CG model in order to obtain the free-energy conformational landscape for both the native wild-type dimer and the engineered monomeric forms of CM as a function of the protein radius of gyration and the contact order. Then, the activation barriers for the chemical step were fully evaluated using structures that were directly derived from the NMR and x-ray structures (40 starting conformations, region I) as well as explicit structures taken from the CG model folding landscape, both near energy minima (region II) and far from the minima (region III); see Figure 3.6. The reasons for deriving structures from the x-ray and NMR and not only using those from the simplified model landscape was the fact that this model was not able to produce the best catalytic configurations, because it was insufficiently refined in terms of protein-substrate interactions. The final surface (shown in Figure 3.6) was generated by stacking the different reaction profiles that were obtained for the chemical step from different starting structures and spacing them equally according to their RMSD from the native structure for the three regions defined above. While this is still an incomplete approximated treatment of the conformational coordinate, nevertheless, it provides the important information that, while the disordered monomer still has several catalytic configurations in region II (low-energy structures from the CG folding landscape), the dimer has none.

In order to fully explore the nature of the catalytic effect the authors also evaluated the reorganization energy, λ, in different regions of the energy landscape. The reorganization energy was defined in detail in the context of the chemical step in the section titled "Thermophilic/Mesophilic DHFR." In the current work (Roca et al. 2008), the authors obtained the reorganization energy, λ, as a function of the free energy released if one starts at the product state (using the reactant structure), and then let the system relax to the product structure. As was discussed in the section titled "Modeling Free-Energy Landscapes of Enzyme Activity," the changes in λ should in turn manifest themselves as changes in the activation energy for the process. Therefore, the authors examined the correlation between the calculated activation barriers and λ (Figure 3.15) for both the monomeric and dimeric forms of CM and found that the regions of the landscape with low catalytic efficiencies also have large reorganization energies (as has been discussed in detail elsewhere [Warshel 1978, 1991], one of the main sources of the catalytic effect of enzymes appears

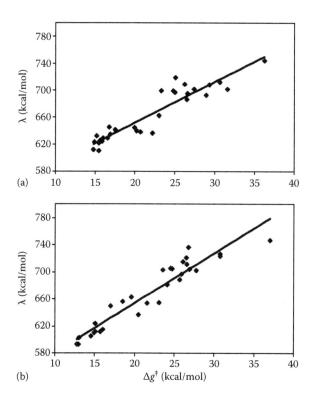

Figure 3.15 Correlation between the reorganization energy, λ, and the calculated activation barrier, $\Delta g^{\ddagger}$, for (a) monomeric and (b) dimeric chorismate mutase. (Reprinted with permission from Roca, M., B. Messer, D. Hilvert, and A. Warshel. 2008. *Proc. Natl. Acad. Sci. U. S. A.* 105:13877–13882. Copyright 2008, National Academy of Sciences, U.S.A.)

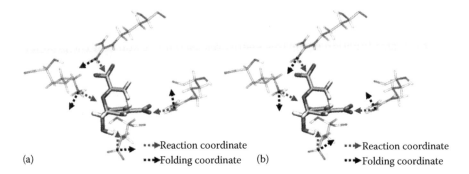

Figure 3.16 Illustration of the relationship between the reaction and folding coordinates in the reactant state of chorismate mutase, at (a) a partially unfolded and (b) an almost fully folded conformation of the enzyme. The black arrows represent the direction of the folding coordinate and the gray arrows represent the vector of motions along the reaction coordinate for residues that interact with the substrate. (Reprinted with permission from Roca, M., B. Messer, D. Hilvert, and A. Warshel. 2008. *Proc. Natl. Acad. Sci. U. S. A.* 105:13877–13882. Copyright 2008, National Academy of Sciences, U.S.A.)

to be to reduce the reorganization energy for the reaction). The key finding of this work was that although the disordered monomer appears to become more confined than the dimer upon ligand binding (Roca et al. 2008) and thus loses some of its molten globule character (Pervushin et al. 2007), it also has a wider catalytic landscape than the dimer because it has a wider region with low reorganization energy. This reduction of the reorganization energy still determines the overall rate acceleration; however, the calculations suggest that the chemical process is more complex than a single path. Finally, the authors (Roca et al. 2008) also calculated the coordinate vectors from the native to a partially unfolded structure as well as from the reactant to product EVB states, and these were shown to be nearly perpendicular (Figure 3.16). This further argues against the idea that coupled motions contribute to catalysis and strengthens the idea that preorganized active sites have evolved to minimize motions along the reaction coordinate.

Bacteriorhodopsin

The light-induced reactions of both rhodopsin (Rh) and bacteriorhodopsin (bR) are among the fastest and most efficient processes in biology (Lanyi 2004; Schoenlein et al. 1991). In both cases, the process starts with the absorption of light by the protonated Schiff base of retinal (PSBR), which then triggers a photocycle comprising an initial isomerization of the retinal chromophore, followed by a sequence of proton transfer steps that eventually lead to visual excitation in Rh and proton pumping in bR.

In addition to the general biological importance of this system, the 1976 study of the primary photochemical event in rhodopsin (Warshel 1976) is the first ever MD simulation to be performed to study the dynamics of a biological process (and its general conclusions have been corroborated by more detailed simulation studies over three decades later [Altoe et al. 2010]). Due to the lack of a crystal structure (the first structure of Rh was reported only in 2000 [Palczewski et al. 2000]), the study was performed using a model composed of PSBR and a steric constraint introduced to represent the restraining effect of the enzyme's binding pocket (Warshel 1976). In this study, the surface crossing in the photoisomerization process was simulated using a semiclassical approach that provided detailed information about the isomerization process as well as its time dependence. Impressively, despite the lack of structural information, this prediction was later proven to be in good agreement with experimental values (Mathies et al. 1988).

Subsequently, construction of a tentative 3-D model for bR (Henderson et al. 1990) as well as high-resolution crystal structures of this system (Kimura et al. 1997; Luecke et al. 1999a,b) have paved the way for many more theoretical studies of the system, which have yielded major insight into the process (Bondar et al. 2004; Braun-Sand et al. 2008; Rousseau et al. 2004; Warshel et al. 2001; Zhuravlev et al. 2009). Particular attention has been paid to the first proton transfer step from the Schiff base to Asp85. This transfer occurs between the L and M states of the bR photocycle. The L intermediate forms through the decay of the previous, photoisomerized K state (Subramaniam et al. 1999). The mechanism by which the first proton transfer occurs is extremely challenging to understand at atomic detail due to the difficulties in experimentally characterizing the associated transition state(s) required to define the process (Bondar et al. 2004).

In order to obtain more insight into the free-energy landscape for the primary proton transfer (PT) process in bR, Warshel and coworkers (Braun-Sand et al. 2008) studied the energetics of the process in terms of protein conformational changes, for which extensive structural information was available. The primary PT process was studied along the chemical and conformational coordinates using the specialized version of the LRA approach, outlined in the section titled "Modeling the Conformational Dimension of Enzyme Catalysis," to evaluate the energetics along the protein conformational coordinate (Equation 3.10). This was achieved by taking different protein intermediates (coming from different protein data bank [PDB] structures), relaxing each model, and then performing EVB free-energy mapping calculations, from which the activation free energies for the PT process were evaluated. Additional trajectories, starting from different initial configurations generated by perturbing the given PDB structures,

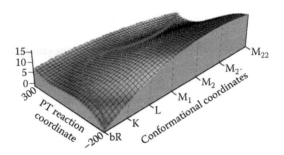

Figure 3.17 A semiquantitative description of the free-energy landscape for the primary PT in bacteriorhodopsin, defined as a function of the protein conformational change (y-axis) and the PT process (x-axis). Here, changes along the x-axis are quantitative, whereas changes along the y-axes are more qualitative. Illustrated here is the coupling between the protein structural changes and the PT, as well as the path for the overall PT process. (Reprinted from Braun-Sand, S., P. K. Sharma, Z. T. Chu, A. V. Pisliakov, and A. Warshel. 2008. The energetics of the primary proton transfer in bacteriorhodopsin revisited: It is a sequential light induced charge separation after all. *Biochim. Biophys. Acta* 1777:441–452. Copyright 2008, with permission from Elsevier.)

were also considered in order to validate the results. Taking the free-energy profiles generated from the different protein intermediates as a reasonable estimate of the conformational dependence of the PT process allowed the authors to generate a semiquantitative estimate of the catalytic landscape (Figure 3.17). This landscape describes the coupling between the protein structural changes (qualitatively described) and the primary PT pathway (estimated quantitatively, as indicated before). This study reproduced the three primary PTs and indicated that this transfer occurs at the L → M step without the need for any intervening water molecules, as suggested before (Lanyi and Schobert 2003; Schulten and Tavan 1978). The study also verified the tentative analysis proposed in earlier studies by the same authors (Warshel and Barboy 1982) and determined that rather than being driven by the intrinsic effect of the chromophore twist, the overall PT process is driven by a light-induced charge separation process between the Schiff base and its counterion Asp85. This in turn drives the subsequent structural reorganization of the protein and the proton pumping.

Helicase and F_0 ATPase

Having described a number of representative systems in great detail, we would like to complete this section by including two elegant studies on larger biological systems, both of which are responsible for controlling ATP hydrolysis. The first of these aims to understand the molecular origin of unidirectional translocation of the Ltag helicase hexamer on single-stranded DNA (Liu et al. 2009), where the DNA was simplified to ionized phosphate groups. In this work, the chemical coordinate corresponded to ATP hydrolysis, and contained 30 structures, which included the ATP bound configuration (T), the ADP bound configuration (D), and the empty configuration (E), as well as nine intermediate conformations between each of these. The conformational coordinate is a 2-D rotational translocation, and was too computationally costly to evaluate using full model, and so only the electrostatic free energy was evaluated using the PDLD/S-LRA approach (Lee et al. 1993; Warshel et al. 2006a) (note that subsequent works have performed detailed studies of protein translocation by the translocon [Rychkova and Warshel 2013; Rychkova et al. 2010]). The system was then studied using a renormalization approach in which a simplified free-energy surface was fit to the explicit surface and was subjected to LD simulations. The obtained free-energy surfaces were consistent with unidirectional translocation. Also, significantly, changing the physically based friction constant in the simplified model by two orders of magnitude (as is the case here) did not change the translocation time, suggesting that the translocation process may be driven by stochastic random motions, which are dictated by the shape of the free-energy landscape (i.e., a diffusive model such as that shown in Figure 3.2 [Liu et al. 2009]).

The second system of interest is F_0F_1-ATPase, which is a well-known ubiquitous protein that generates ATP using the pH gradient across the inner mitochondrial membrane developed by the electron transport chain (Nelson et al. 2008). This protein is made of two rotary motors (Abrahams et al. 1994), and the transmembrane one (F_0-ATPase) has been studied using a similar approach as for the systems described above (Mukherjee and Warshel 2012). Specifically, the F_0 motor consists of the subunit c-ring (the rotor part), along with the stator subunit-a and dimer subunits-b (Abrahams et al. 1994) (Figure 3.18a). In all crystal structures, the c-ring contains centrally located aspartate residues that can be protonated and that interact with an arginine of the subunit-a upon rotation (Valiyaveetil and Fillingame 1997).

In this case (Mukherjee and Warshel 2012), both the protein and the membrane were modeled using CG models; however, the protein backbone atoms were treated explicitly and the side chains are represented using a simplified united atom model. The membrane was represented through a grid of effective atoms with a width of 30Å. The c-ring central Asp was placed in the middle of the grid. The presence of a proton channel was modeled first by placing an 8Å cutoff around the Arg, and then a smaller 6Å cutoff, to test the influence of the cutoff. Both cutoffs gave similar results. The conformational coordinate, which is described as the rotation of the c-ring, was first modeled by generating several intermediate frames by rotating the c-ring gradually in the synthesis direction, while subunit-a was kept fixed. The chemical coordinate is the proton transport (PTR) from the periplasmic side (P) to the cytoplasmic side (N)

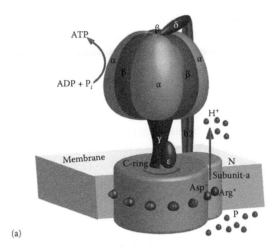

(a)

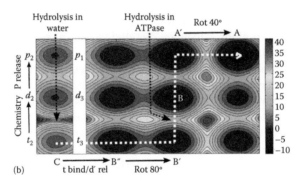

(b)

Figure 3.18   (a) A schematic overview of the structure of F_1-ATPase showing the differ-ent subunits, and (b) the conformational/chemical landscape for ATP hydrolysis in this system, with the corresponding reaction in water shown for comparison, highlighting the catalytic dwell in the enzyme. (Part a is reprinted with permission from Mukherjee, S., and A. Warshel. 2012. *Proc. Natl. Acad. Sci. U. S. A.* 109:14876–14881. Copyright 2012, National Academy of Sciences, U.S.A., and part b is reprinted with permission from Mukherjee, S., and A. Warshel. 2011. *Proc. Natl. Acad. Sci. U. S. A.* 108:20550–20555. Copyright 2011. National Academy of Sciences, U.S.A.)

and was modeled by evaluating the energetics of inserting a hydronium ion (H_3O^+) at each feasible region for all the frames of the conformational coor-dinate. Later, the free-energy landscape was explored by again resorting to the renormalization approach. The previously determined free energies of key states along the rotational coordinate and the PTR barriers were used to construct a simplified surface that was simulated with Langevin dynamics.

The obtained landscapes (Figure 3.18b) clearly show a vectorial path, where PTR from P to N drives the rotation the rotor. The detailed picture of this movement is the sequential deprotonation of the Asp residue closer to the Arg through to the N channel and reprotonation through the P channel as the c-ring rotates to the right, as rotating to the left is shown to be higher in energy (Mukherjee and Warshel 2012). This provides a clear rationale for the unidirectionality of the rotational motion during ATP synthesis. A subsequent study of F_0-ATPase has taken this one step further, using a similar approach to demonstrate the electrostatic basis for the proton gradient-driven rotary motion. Together, these works provide an excellent starting point for further studies on proton-driven mechanochemical systems.

ALTERNATIVE COMPUTATIONAL APPROACHES

Having presented a range of approaches to examine free-energy landscapes of enzyme catalysis, we would like to also briefly discuss some alternative computational approaches that have provided insightful information without needing to examine the full landscape. Note that due to the breadth of this vibrant and active field, this section is not intended to provide an exhaustive overview of the field, but rather to provide discussion of a few significant and representative works, with appropriate references as suggestions for further reading. The methods discussed in this section will include QM/MM approaches (Lodola et al. 2010), rare events MD simulations (Boekelheide et al. 2011; Hamelberg et al. 2004; McGowan and Hamelberg 2013), statistical analysis (Lodola et al. 2010; Major and Weitman 2012; McGowan and Hamelberg 2013), and kinetic approaches (Glowacki et al. 2012b; Kurkcouglu et al. 2012).

Statistical approaches such as principal component analysis (PCA) (Wold et al. 1987), allow one to identify the more relevant vibrational motions of the system, both in terms of their coupling with the reaction coordinate as well as the vibrational motions of the environment that are active during the process of interest (Castillo et al. 2008; McGowan and Hamelberg 2013). On the other hand, different types of MD simulations allow one to analyze both the ensemble of different conformations available to the system as well as the potential effect of dynamical (nonequilibrium) effects on enzyme catalysis (Boekelheide et al. 2011; Glowacki et al. 2012b; Grote and Hynes 1980; Major and Weitman 2012; Roca et al. 2006). We will summarize a few of these studies in the following section.

Studying Conformational Effects on Catalysis Using Multivariate Statistical Techniques

Multivariate statistical techniques have proven to be powerful tools in the analysis of molecular dynamics simulations. They allow the management of large and complex data sets, thus simplifying the identification of relevant information. When combined with molecular dynamics simulations, such approaches provide a powerful tool to probe dynamical effects in enzyme catalysis.

Lodola and coworkers (Lodola et al. 2010), for example, have used statistical analysis techniques to explore the effect of protein structural fluctuations on the first step of the acylation reaction between fatty acid amide hydrolase (FAAH), which is an important enzyme in the central nervous system, and oleamide. Their work illustrates the importance of conformational sampling of configurations favorable to the chemical reaction and highlights the power of statistical approaches in identifying structural features that could aid in bringing an enzyme to its catalytically active preorganized state as well as helping the subsequent reaction. That is, the use of PCA in this work allowed the authors to identify catalytically relevant enzyme-substrate complexes, and from this, to obtain an approximate estimate of the energy required to preorganize the active site for optimal catalysis.

Similarly, Hamelberg and coworkers (Doshi et al. 2012; McGowan and Hamelberg 2013) have used both normal and accelerated atomistic molecular dynamics simulations in combination with PCA analysis to study the role of enzyme conformational dynamics in catalytic function. In their accelerated MD simulations, adding a biasing potential to the true potential alters the potential energy surface such that the escape rates from the potential wells are enhanced. This accelerates the sampling and therefore extends the simulation timescale (Hamelberg et al. 2004). As a model system, the authors chose cyclophilin A (CypA), an extensively studied peptidyl-prolyl *cis-trans* isomerase that catalyzes the isomerization of the peptide bond preceding proline residues in proteins (Bosco et al. 2002; Davis et al. 2010; Eisenmesser et al. 2005; Fraser et al. 2009). Using the substrate–free enzyme as well as the enzyme–substrate complexes at three different stages along the catalytic pathway (*trans*, transition-state, and *cis* configurations), the authors carried out normal MD simulations, analyzing the ensemble of enzyme–substrate conformations along the reaction. From the statistical analysis, it was concluded that the dynamics of the active site residues of the substrate-bound CypA complexes are inherent to the substrate–free enzyme, which suggested that the enzyme conformations needed for optimal catalytic activity already exist in the ensemble of the substrate–free enzyme. Based on this, it was argued that substrate binding occurs *via* conformational sampling, as was previously suggested by NMR studies (Fraser et al. 2009; Kern, Kern, Scherer et al. 1995). Additionally, kinetic analysis, using one-dimensional Kramer's theory in the high friction regime (Hanggi et al. 1990; Kramers 1940), demonstrated that the enzymatic motions do *not* modify the free-energy landscape. Rather, they reduce the effective diffusion coefficient (as compared to the reference reaction in solution). It was therefore concluded that variations in the electrostatic and hydrophobic contacts that take place as the configuration of the substrate changes in the active site become stronger when the substrate moves from the reactant (or product) state to the transition state, where an intricate reorganized network of hydrogen bonds become much stronger than in the reactant states, thus stabilizing the transition state. This work therefore suggests that transition state stabilization is the main effect in enhancing the chemical rate (McGowan and Hamelberg 2013).

Finally, Major and Weitman (2012) studied the multiple enzymatic pathways involved in the biosynthesis of monoterpene bornyl diphosphate (BPP) from geranyl diphosphate by BPP synthase (BPPS). One of the main features of this system is the fact that it can lead to a diversity of different products from a unique starting compound, including, in addition to the main product, BPP, α-pinene, camphene, and limonene (Wise et al. 1998).

Previous computational studies in the gas-phase (Hong and Tantillo 2010; Weitman and Major 2010) have suggested the existence of numerous possible side reactions for BPPS in order to explain the product diversity, which include migrations, hydride transfers, proton transfers, or deprotonations. In these studies, it has also been suggested that the bornyl cation involved in the main pathway is not a stable intermediate, but rather a short-lived species constituting a bifurcation point on the gas-phase potential energy surface (PES) (Weitman and Major 2010). Bifurcations arise when there are sequential transition states with no intervening energy minima. In such cases, it has been suggested that the shape of the potential energy surface and dynamical effects control selectivity rather than the energetics of the transition state (Ess et al. 2008).

In order to probe the question of the dynamics of branching of the bornyl carbocation to BPP and camphene, the authors performed activated molecular dynamics simulations (Northrup et al. 1982) in combination with PCA analysis to study the complex energy landscape involved in this process. In activated molecular dynamics simulations, an ensemble of trajectories is initiated from the kinetic bottleneck and propagated downhill from there. From the analysis of these trajectories, it was suggested that the bifurcating nature of the process (which is already observed in the gas phase [Weitman and Major 2010]) is retained in the enzyme; however, involving different species. It was concluded that in order to yield selectivity, BPPS employs electrostatics as a guiding force *en route* to the final product. Indeed, the relative stability of the bornyl and camphyl cations is significantly perturbed in the enzyme compared to the gas phase in favor of the bornyl cation. Nevertheless, in addition to electrostatic interactions, some dynamical effects are also apparently required to drive the selectivity. That is, the product distribution in BPPS is not predictable on the basis of thermodynamics and kinetics alone, and in this case, specific rearrangements in the energy surface appear to guide the reaction toward product formation. These active modes include coupled motions that combine the rearrangement from the pinyl to the bornyl cation, suggesting that dynamical effects may play *some* role in defining terpene cyclase selectivity.

Dynamics and Dissipation in Enzyme Catalysis

Miller and coworkers (Boekelheide et al. 2011) have quantitatively studied the effect of enzyme vibrations in the hydride transfer catalyzed by DHFR (introduced in the section titled "Thermophilic/Mesophilic DHFR") using a combined approach based on ring polymer molecular dynamics (RPMD) (Craig and Manolopoulos 2004, 2005), a relatively recent path-integral method that enables the inclusion of nuclear quantization effects in conjunction with the EVB approach. One important goal of this work was to quantitatively analyze the effect of correlations between static (in thermal equilibrium with the environment) and dynamical (not in equilibrium) motions in enzyme catalysis. While static correlations between the active site and distal protein residues were found to be highly correlated at all reactant, transition (dividing surface), and product regions, dynamical correlations were found to be extremely local in nature and primarily confined to the enzyme, substrate, and cofactor regions, with only weak signals in the protein residues surrounding the active site. A combined measure of the dynamical correlation between a given atom and the intrinsic reaction event was obtained from the nonequilibrium ensemble average of velocities in the reactive trajectories. Donor-acceptor distances as well as atom pairs on the cofactor and on the substrate were found to be both statistically and dynamically correlated with the intrinsic reaction; however, these time-averaged quantities were limited by disparities in the relative timescales. That is, the reactive trajectories cross the transition state on a timescale that is too fast to be dynamically coupled to the protein coordinate (Boekelheide et al. 2011). These findings agree with and complement previous works on the same system (Agarwal et al. 2002; Castillo et al. 1999; Garcia-Viloca et al. 2003; Liu and Warshel 2007a) that reproduced experimental results without the need to employ nonstatistical dynamics, and found no significant coupling between the chemistry and protein dynamics.

Approaches Based on TST

As discussed previously, coupling between the reaction coordinate and the remaining coordinates may be responsible for the existence of recrossings and therefore affect the transmission coefficient. Different methods have been used to calculate the transmission coefficient of enzymatic reactions including the reactive flux trajectory method (Neria and Karplus 1997), variational transition state theory with multidimensional tunneling corrections (Alhambra et al. 2001), the centroid path integral approach (Hwang and Warshel 1996), and Grote-Hynes (GH) theory (Castillo et al. 2008; Grote and Hynes 1980; Roca et al. 2006). In the reactive flux approach, for example, one carries out unrestrained molecular dynamics simulations starting from the transition state, following the reaction path both backward and forward in time. The transmission coefficient can then be estimated from the fraction of trajectories that cross the transition state in the product direction but recross the dividing surface to return to the reactant region (Ruiz-Pernia et al. 2008). GH theory (Castillo et al. 2008; Grote and Hynes 1980; Roca et al. 2006), on the other hand, requires the evaluation of the friction kernel in the transition state. This friction kernel is obtained from the averaged forces exerted on a distinguished reaction coordinate by the remaining degrees of freedom of the system and thus gives a quantitative estimation of the coupling between the reaction coordinate and the environment (Castillo et al. 2008).

Following from this, Tuñón and coworkers (Castillo et al. 2008; Roca et al. 2006, 2010; Ruiz-Pernia et al. 2008) have shown that reaction flux theory and GH theory can be successfully applied to the analysis of enzymatic reactions combined with either rare-event reaction trajectories or QM/MM simulations. In particular, these authors applied GH theory (Nam et al. 2004; Ruiz-Pernia et al. 2008) to study the reactions catalyzed by O-methyltransferase (COMT) and chalcone isomerase (CHI) (Ruiz-Pernia et al. 2008), as well as the hydride transfer reaction catalyzed by formate dehydrogenase (FDH) (Roca et al. 2010). In these cases, GH theory provides transmission coefficients that were larger in the enzyme than in the reference reaction in solution. Using the following expression:

$$\Delta G_{dyn} = -RT \ln \frac{k_{enz}}{k_{aq}} \qquad (3.17)$$

it was found that protein dynamics' contribution to catalysis is small, representing around 5% of the total catalytic effect of Chalcone isomerase (Ruiz-Pernia et al. 2008). Additionally, the analysis of the reactive trajectories also reveals how the relative movements of some residues precede and promote the chemical reaction.

Finally, recently, Mulholland and coworkers (Glowacki et al. 2012a) combined molecular dynamics simulations and standard TST to provide a simple two-state model of enzyme kinetics for understanding tunneling and KIEs in a number of enzymes. A central feature of the model is that it assumes that multiple conformations of the enzyme-substrate complex may be involved in the reaction. In principle, one could assume a large number of conformers, each with different reactivity behavior. However, a minimal model considering just two conformations was sufficient to account for the experimentally observed kinetic behavior. This model described the enzyme-substrate reactant complex as a mixture of two enzyme-substrate conformers R1 and R2 with different reactivity and tunneling behavior, both of which may lead to a product complex, P, with rate coefficients k_1 and k_2, respectively. R1 and R2 are assumed to interconvert faster than they react, with rate coefficients k_f and k_r, and an overall equilibrium constant, $K_{eq} = k_f/k_r$. Using TST, the authors were able to provide a mathematical expression for the rate coefficients, KIEs, and their temperature dependence, for a range of different enzymes. Specifically, the model was capable of reproducing experimentally observed kinetics in enzymes such as soybean lipoxygenase (SLO-1) (Knapp et al. 2002), aromatic amine dehydrogenase (AADH) (Masgrau et al. 2006), methylamine dehydrogenase (MADH) (Basran et al. 1999), and dihydrofolate reductase (DHFR) (Loveridge et al. 2010). The model, in addition to supporting TST as an appropriate basis for understanding enzymatic catalysis, also demonstrates that there is no direct role for protein dynamics in driving enzyme reactions. Instead what should be taking into account when studying this issue is the consideration of multiple enzyme conformations.

OVERVIEW AND CONCLUSIONS

Despite their deceptive superficial simplicity, enzymes are tremendously complicated catalysts, providing catalytic amplifications that have never been observed in *any* man-made catalyst in aqueous solvent (although it has been possible to obtain enzyme-like accelerations by changing reaction conditions [Brown 2011]). Part of the complexity comes from their sheer size and their conformational flexibility, allowing them to accommodate substrates of different shapes and sizes while preorganizing their active sites for optimal chemistry (Warshel 1978). Perhaps unsurprisingly, with the development of increasingly sophisticated experimental techniques to study enzyme conformational dynamics, debate has arisen as to whether such dynamics play a role in driving the chemical step of the catalytic cycle (Benkovic and Hammes-Schiffer 2003; Glowacki et al. 2012b; Hammes-Schiffer 2013; Henzler-Wildman and Kern 2007; Kamerlin and Warshel 2010a; Villali and Kern 2010). Computational tools can play an important role in addressing this question; however, the high computational cost associated with studying conformational transitions, that can occur on timescales of as high as several seconds, still remains a bottleneck for simultaneously exploring the coupling of such dynamics to the chemical step.

In this chapter, we have presented different approaches to model energy landscapes for enzyme catalysis as a function of both conformational changes and chemistry. We have shown the evolution of such approaches, as applied to a number of different systems, starting from simple qualitative estimates of the transition between different conformations to more quantitative assessment using, for example, the renormalization approach or special implementations of the LRA approach. We have also presented some examples of alternative strategies to address this question that do not invoke examination of the full catalytic landscape. We emphasize that due to the sheer breadth of the field, this is merely a subset of the available literature and we have chosen only to highlight the most relevant key works and have guided the interested reader to more detailed reviews.

Clearly, despite constant increases in computational power, we are still at a stage in the field where we need to make substantial simplifications in order to be able to explore the full landscape, such as representing the conformational change and chemistry as a function of effective dimensionless coordinates as we have done in the case of adenylate kinase (Pisliakov et al. 2009). However, a combination of specialized infrastructure and software for massively parallelized computing (Lane et al. 2013; Pronk et al. 2011; Shaw 2013) as well as increasingly sophisticated coarse-graining approaches (Kamerlin et al. 2011) will continue to push forward the boundaries of the problems that can be addressed. In the long term, while there exist many alternative approaches to examine enzyme dynamics (some of which were presented in the section titled "Alternative Computational Approaches"), we believe that the most *direct* test of the nature of the coupling between conformational flexibility and chemistry will remain exploration of the full catalytic landscapes, ideally, and ultimately, at full atomic resolution.

ACKNOWLEDGMENTS

The authors would like to thank the Swedish Royal Academy of Sciences and the Sven and Ebba-Christina Hagberg Foundation for a generous stipend that has funded this work. The authors would also like to thank Jan Florian and Sonja Braun Sand for generously providing us with the original high-resolution versions of Figures 3.9, 3.7, and 3.17, respectively.

REFERENCES

Abele, U., and G. E. Schulz. 1995. High-resolution structures of adenylate kinase from yeast ligated with inhibitor Ap(5)a, showing the pathway of phosphoryl transfer. *Protein Sci.* 4:1262–1271.

Abrahams, J. P., A. G. W. Leslie, R. Lutter, and J. E. Walker. 1994. Structure at 2.8 Å resolution of F_1-ATPase from bovine heart mitochondria. *Nature* 370:621–628.

Adamczyk, A. J., J. Cao, S. C. L. Kamerlin, and A. Warshel. 2011. The catalytic power of dihydrofolate reductase and other enzymes arises from electrostatic preorganization, not conformational motions. *Proc. Natl. Acad. Sci. U. S. A.* 108:14115–14120.

Adamczyk, A. J., and A. Warshel. 2011. Converting structural information into an allosteric-energy-based picture for elongation factor Tu activation by the ribosome. *Proc. Natl. Acad. Sci. U. S. A.* 108:9827–9832.

Agarwal, P. K., S. R. Billeter, and S. Hammes-Schiffer. 2002. Nuclear quantum effects and enzyme dynamics in dihydrofolate reductase catalysis. *J. Phys. Chem. B* 106:3283–3293.

Alhambra, C., M. L. Sanchez, J. Corchado, J. L. Gao, and D. G. Truhlar. 2001. Quantum mechanical tunneling in methylamine dehydrogenase. *Chem. Phys. Lett.* 347: 512–518.

Altoe, P., A. Cembran, M. Olivucci, and M. Garavelli. 2010. Aborted double bicycle-pedal isomerization with hydrogen bond breaking is the primary event of bacterio-rhodopsin proton pumping. *Proc. Natl. Acad. Sci. U. S. A.* 107:20172–20177.

Åqvist, J., and A. Warshel. 1993. Simulation of enzyme-reactions using valence-bond force-fields and other hybrid quantum-classical approaches. *Chem. Rev.* 93:2523–2544.

Arora, K., and C. L. Brooks. 2007. Large-scale allosteric conformational transitions of adenylate kinase appear to involve a population-shift mechanism. *Proc. Natl. Acad. Sci. U. S. A.* 104:18496–18501.

Arora, K., and C. L. Brooks, III. 2013. Multiple intermediates, diverse conformations, and cooperative conformational changes underlie the catalytic hydride transfer reaction of dihydrofolate reductase. *Top. Curr. Chem.* 337:165–187.

Balabin, I. A., W. T. Yang, and D. N. Beratan. 2009. Coarse-grained modeling of allosteric regulation in protein receptors. *Proc. Natl. Acad. Sci. U. S. A.* 106:14253–14258.

Basran, J., M. J. Sutcliffe, and N. S. Scrutton. 1999. Enzymatic H-transfer requires vibration-driven extreme tunneling. *Biochemistry* 38:3218–3222.

Batra, V. K., W. A. Beard, D. D. Shock, J. M. Krahn, L. C. Pedersen, and S. H. Wilson. 2006. Magnesium-induced assembly of a complete DNA polymerase catalytic complex. *Structure* 14:757–766.

Batra, V. K., W. A. Beard, D. D. Shock, L. C. Pedersen, and S. H. Wilson. 2008. Structures of DNA polymerase β with active-site mismatches suggest a transient abasic site intermediate during misincorporation. *Mol. Cell.* 30:315–324.

Beard, W. A., and S. H. Wilson. 2003. Structural insights into the origins of DNA polymerase fidelity. *Structure* 11:489–496.

Benkovic, S. J., G. G. Hammes, and S. Hammes-Schiffer. 2008. Free-energy landscape of enzyme catalysis. *Biochemistry* 47:3317–3321.

Benkovic, S. J., and S. Hammes-Schiffer. 2003. A perspective on enzyme catalysis. *Science* 301:1196–1202.

Bennett, C. H. 1977. Molecular dynamics and transition state theory: The simulation of infrequent events. In *Algorithms for Chemical Computations*, R. E. Christofferson, ed., 63–97. Washington, DC: American Chemical Society.

Berry, M. B., B. Meador, T. Bilderback, P. Liang, M. Glaser, and G. N. Phillips. 1994. The closed conformation of a highly flexible protein: The structure of *Escherichia coli* adenylate kinase with bound AMP and AMPPNP. *Proteins-Struct. Func. Genet.* 19:183–198.

Bhabha, G., J. Y. Lee, D. Ekiert, J. Gam, I. A. Wilson, H. J. Dyson, S. J. Benkovic, and P. E. Wright. 2011. A dynamic knockout reveals that conformational fluctuations influence the chemical step of enzyme catalysis. *Science* 332:234–238.

Blake, C. C., D. F. Koenig, G. A. Mair, A. C. North, D. C. Phillips, and V. R. Sarma. 1965. Structure of hen egg-white lysozyme. A three-dimensional Fourier synthesis at 2 Ångström resolution. *Nature* 206:757–761.

Boehr, D. D., D. McElheny, H. J. Dyson, and P. E. Wright. 2006. The dynamic energy land-scape of dihydrofolate reductase catalysis. *Science* 313:1638–1642.

Boekelheide, N., R. Salomon-Ferrer, and T. F. Miller. 2011. Dynamics and dissipation in enzyme catalysis. *Proc. Natl. Acad. Sci. U. S. A.* 108:16159–16163.

Bondar, A. N., S. Fischer, J. C. Smith, M. Elstner, and S. Suhai. 2004. Key role of electrostatic interactions in bacteriorhodopsin proton transfer. *J. Am. Chem. Soc.* 126:14668–14677.

Bosco, D. A., E. Z. Eisenmesser, M. W. Clarkson, M. Wolf-Watz, W. Labeikovsky, O. Millet, and D. Kern. 2010. Dissecting the microscopic steps of the cyclophilin A enzymatic cycle on the biological HIV-1 capsid substrate by NMR. *J. Mol. Biol.* 403:723–738.

Bosco, D. A., E. Z. Eisenmesser, S. Pochapsky, W. I. Sundquist, and D. Kern. 2002. Catalysis of cis/trans isomerization in native HIV-1 capsid by human cyclophilin A. *Proc. Natl. Acad. Sci. U. S. A.* 99:5247–5252.

Bosco, D. A., and D. Kern. 2004. Catalysis and binding of cyclophilin a with different HIV-1 capsid constructs. *Biochemistry* 43:6110–6119.

Braun-Sand, S., P. K. Sharma, Z. T. Chu, A. V. Pisliakov, and A. Warshel. 2008. The energet-ics of the primary proton transfer in bacteriorhodopsin revisited: It is a sequential light induced charge separation after all. *Biochim. Biophys. Acta* 1777:441–452.

Brown, R. S. 2011. Biomimetic and nonbiological dinuclear Mx+ complex-catalyzed alcoholysis reactions of phosphoryl transfer reactions. In *Progress in Inorganic Chemistry*, K. D. Karlin, ed., Vol. 57, 55–117. Hoboken, NJ: John Wiley & Sons.

Bu, Z., R. Biehl, M. Monkenbusch, D. Richter, and D. J. Callaway. 2005. Coupled protein domain motion in Taq polymerase revealed by neutron spin-echo spectroscopy. *Proc. Natl. Acad. Sci. U. S. A.* 102:17646–17651.

Cannon, W. R., S. F. Singleton, and S. J. Benkovic. 1996. A perspective on biological cataly-sis. *Nat. Struct. Biol.* 3:821–833.

Careri, G., P. Fasella, and E. Gratton. 1979. Enzyme dynamics: The statistical physics approach. *Annu. Rev. Biophys. Bioeng.* 8:69–97.

Castillo, R., J. Andres, and V. Moliner. 1999. Catalytic mechanism of dihydrofolate reduc-tase enzyme. A combined quantum-mechanical/molecular-mechanical character-ization of transition state structure for the hydride transfer step. *J. Am. Chem. Soc.* 121:12140–12147.

Castillo, R., M. Roca, A. Soriano, V. Moliner, and I. Tuñón. 2008. Using Grote-Hynes theory to quantify dynamical effects on the reaction rate of enzymatic processes. The case of methyltransferases. *J. Phys. Chem. B* 112:529–534.

Cheung, M. S., L. L. Chavez, and J. N. Onuchic. 2004. The energy landscape for protein fold-ing and possible connections to function. *Polymer* 45:547–555.

Coan, K. E. D., M. J. Swann, and J. Ottl. 2012. Measurement and differentiation of ligand-induced calmodulin conformations by dual polarization interferometry. *Anal. Chem.* 84:1586–1591.

Craig, D. B., E. A. Arriaga, J. C. Y. Wong, H. Lu, and N. J. Dovichi. 1996. Studies on single alkaline phosphatase molecules: Reaction rate and activation energy of a reaction catalyzed by a single molecule and the effect of thermal denaturation—The death of an enzyme. *J. Am. Chem. Soc.* 118:5245–5253.

Craig, I. R., and D. E. Manolopoulos. 2004. Quantum statistics and classical mechanics: Real time correlation functions from ring polymer molecular dynamics. *J. Chem. Phys.* 121:3368–3373.

Craig, I. R., and D. E. Manolopoulos. 2005. A refined ring polymer molecular dynamics theory of chemical reaction rates. *J. Chem. Phys.* 123:034102.

Cremer, D. 2011. Møller-Plesset perturbation theory: From small molecule methods to methods for thousands of atoms. *WIREs: Comput. Mol. Sci.* 1:509–530.

Dams, T., and R. Jaenicke. 1999. Stability and folding of dihydrofolate reductase from the hyperthermophilic bacterium *Thermotoga maritima*. *Biochemistry* 38:9169–9178.

Daniel, R. M., R. V. Dunn, J. L. Finney, and J. C. Smith. 2003. The role of dynamics in enzyme activity. *Annu. Rev. Biophys. Biomol. Struct.* 32:69–92.

Davies, J. F., T. J. Delcamp, N. J. Prendergast, V. A. Ashford, J. H. Freisheim, and J. Kraut. 1990. Crystal-structures of recombinant human dihydrofolate-reductase complexed with folate and 5-deazafolate. *Biochemistry* 29:9467–9479.

Davis, T. L., J. R. Walker, V. Campagna-Slater, P. J. Finerty, R. Paramanathan, G. Bernstein, F. MacKenzie, W. Tempel, H. Ouyang, W. H. Lee, E. Z. Eisenmesser, and S. Dhe-Paganon. 2010. Structural and biochemical characterization of the human cyclophilin family of peptidyl-prolyl isomerases. *PLoS Biol.* 8:e1000439.

de Jong, W. A., E. Bylaska, N. Govind, C. L. Janssen, K. Kowalski, T. Muller, I. M. B. Nielsen, H. J. J. van Dam, V. Veryazov, and R. Lindh. 2010. Utilizing high performance computing for chemistry: Parallel computational chemistry. *Phys. Chem. Chem. Phys.* 12:6896–6920.

Del Ben, M., J. Hutter, and J. VandeVondele. 2012. Second-order Møller-Plesset perturbation theory in the condensed phase: An efficient and massively parallel Gaussian and plane waves approach. *J. Chem. Theory Comput.* 8:4177–4188.

Diederichs, K., and G. E. Schulz. 1991. The refined structure of the complex between adenylate kinase from beef-heart mitochondrial matrix and its substrate AMP at 1.85 Å resolution. *J. Mol. Biol.* 217:541–549.

Doshi, U., L. C. McGowan, S. T. Ladani, and D. Hamelberg. 2012. Resolving the complex role of enzyme conformational dynamics in catalytic function. *Proc. Natl. Acad. Sci. U. S. A.* 109:5699–5704.

Doublie, S., M. R. Sawaya, and T. Ellenberger. 1999. An open and closed case for all polymerases. *Structure.* 7:R31–R35.

Dreusicke, D., P. A. Karplus, and G. E. Schulz. 1988. Refined structure of porcine cytosolic adenylate kinase at 2.1-A resolution. *J. Mol. Biol.* 199:359–371.

Dror, R. O., M. O. Jensen, D. W. Borhani, and D. E. Shaw. 2010. Exploring atomic resolution physiology on a femtosecond to millisecond timescale using molecular dynamics simulations. *J. Gen. Phys.* 135:555–562.

Dryga, A., S. Chakrabarty, S. Vicatos, and A. Warshel. 2012. Realistic simulation of the activation of voltage-gated ion channels. *Proc. Natl. Acad. Sci. U. S. A.* 109:3335–3340.

Dryga, A., and A. Warshel. 2010. Renormalizing SMD: The renormalization approach and its use in long timescale simulations and accelerated PMF calculations of macromolecules. *J. Phys. Chem. B* 114:12720–12728.

Ehrlich, S., J. Moellmann, and S. Grimme. 2013. Dispersion-corrected density functional theory for aromatic interactions in complex systems. *Acc. Chem. Res.* 46:916–926.

Eisenmesser, E. Z., D. A. Bosco, M. Akke, and D. Kern. 2002. Enzyme dynamics during catalysis. *Science* 295:1520–1523.

Eisenmesser, E. Z., O. Millet, W. Labeikovsky, D. M. Korzhnev, M. Wolf-Watz, D. A. Bosco, J. J. Skalicky, L. E. Kay, and D. Kern. 2005. Intrinsic dynamics of an enzyme underlies catalysis. *Nature* 438:117–121.

Engelkamp, H., N. S. Hatzakis, J. Hofkens, F. C. De Schryver, R. J. M. Nolte, and A. E. Rowan. 2006. Do enzymes sleep and work? *Chem. Commun.* 9:935–940.

English, B. P., W. Min, A. M. van Oijen, K. T. Lee, G. B. Luo, H. Y. Sun, B. J. Cherayil, S. C. Kou, and S. N. Xie. 2006. Erratum. Ever-fluctuating single enzyme molecules: Michaelis-Menten equation revisited. *Nat. Chem. Biol.* 2:168 (first published in *Nat. Chem. Biol.* 2:87–94, 2006).

Ess, D. H., S. E. Wheeler, R. G. Iafe, L. Xu, N. Celebi-Olcum, and K. N. Houk. 2008. Bifurcations on potential energy surfaces of organic reactions. *Angew. Chem. Int. Ed.* 47:7592–7601.

Evans, M. G., and M. Polanyi. 1935. Some applications of the transition state method to the calculation of reaction velocities, especially in solution. *Trans. Faraday Soc.* 31:875–893.

Fan, Y., A. Cembran, S. Ma, and J. Gao. 2013. Connecting protein conformational dynamics with catalytic function as illustrated in dihydrofolate reductase. *Biochemistry* 52:2036–2049.

Farago, B., J. Li, G. Cornilescu, D. J. Callaway, and Z. Bu. 2010. Activation of nanoscale allosteric protein domain motion revealed by neutron spin echo spectroscopy. *Biophys. J.* 99:3473–3482.

Fiaux, J., E. B. Bertelsen, A. L. Horwich, and K. Wuthrich. 2002. NMR analysis of a 900K GroEL-GroES complex. *Nature* 418:207–211.

Fierke, C. A., K. A. Johnson, and S. J. Benkovic. 1987. Construction and evaluation of the kinetic scheme associated with dihydrofolate-reductase from *Escherichia coli*. *Biochemistry* 26:4085–4092.

Florián, J., M. F. Goodman, and A. Warshel. 2005. Computer simulations of protein functions: Searching for the molecular origin of the replication fidelity of DNA polymerases. *Proc. Natl. Acad. Sci. U. S. A.* 102:6819–6824.

Fraser, J. S., M. W. Clarkson, S. C. Degnan, R. Erion, D. Kern, and T. Alber. 2009. Hidden alternative structures of proline isomerase essential for catalysis. *Nature* 462:669–673.

Frauenfelder, H., S. G. Sligar, and P. G. Wolynes. 1991. The energy landscapes and motions of proteins. *Science* 254:1958–1603.

Frey, P. A., and A. D. Hegeman. 2006. *Enzymatic Reaction Mechanisms*. New York: Oxford University Press.

Gao, J. L., S. H. Ma, D. T. Major, K. Nam, J. Z. Pu, and D. G. Truhlar. 2006. Mechanisms and free energies of enzymatic reactions. *Chem. Rev.* 106:3188–3209.

Gao, J., and D. G. Truhlar. 2002. Quantum mechanical methods for enzyme kinetics. *Annu. Rev. Phys. Chem.* 53:467–505.

Garcia-Viloca, M., D. G. Truhlar, and J. L. Gao. 2003. Reaction-path energetics and kinetics of the hydride transfer reaction catalyzed by dihydrofolate reductase. *Biochemistry* 42:13558–13575.

Gavish, B., and M. M. Werber. 1979. Viscosity-dependent structural fluctuations in enzyme catalysis. *Biochemistry* 18:1269–1275.

Glowacki, D. R., J. N. Harvey, and A. J. Mulholland. 2012a. Protein dynamics and enzyme catalysis: The ghost in the machine? *Biochem. Soc. Trans.* 40:515–521.

Glowacki, D. R., J. N. Harvey, and A. J. Mulholland. 2012b. Taking Ockham's razor to enzyme dynamics and catalysis. *Nat. Chem.* 4:169–176.

Grant, B. J., A. A. Gorfe, and J. A. McCammon. 2010. Large conformational changes in proteins: Signaling and other functions. *Curr. Opin. Struct. Biol.* 20:142–147.

Grimmelmann, E. K., J. C. Tully, and E. Helfand. 1981. Molecular dynamics of infrequent events: Thermal desorption of xenon from a platinum surface. *J. Chem. Phys.* 74:5300–5310.

Grote, R. F., and J. T. Hynes. 1980. The stable states picture of chemical reactions. 2. Rate constants for condensed and gas-phase reaction models. *J. Chem. Phys.* 73:2715–2732.

Grunwald, E. 1985. Reaction-mechanism from structure energy relations. 2. Acid-catalyzed addition of alcohols to formaldehyde. *J. Am. Chem. Soc.* 107:4715–4720.

Guthrie, J. P. 1990. Concertedness and E2 elimination reactions: Prediction of transition state position and reaction-rates using 2-dimensional reaction surfaces based on quadratic and quartic approximations. *Can. J. Chem.* 68:1643–1652.

Hamelberg, D., L. Mongan, and J. A. McCammon. 2004. Accelerated molecular dynamics: A promising and efficient simulation method for biomolecules. *J. Chem. Phys.* 120:11919–11929.

Hammes, G. G. 2002. Multiple conformational changes in enzyme catalysis. *Biochemistry* 41:8221–8228.

Hammes, G. G., S. J. Benkovic, and S. Hammes-Schiffer. 2011. Flexibility, diversity and cooperativity: Pillars of enzyme catalysis. *Biochemistry* 50:10422–10430.

Hammes-Schiffer, S. 2013. Catalytic efficiency of enzymes: A theoretical analysis. *Biochemistry* 52:2012–2020.

Hammes-Schiffer, S., and S. J. Benkovic. 2006. Relating protein motion to catalysis. *Annu. Rev. Biochem.* 75:519–541.

Hanggi, P., P. Talkner, and M. Borkovec. 1990. Reaction-rate theory: Fifty years after Kramers. *Rev. Mod. Phys.* 62:251–341.

Hanson, J. A., K. Duderstadt, L. P. Watkins, S. Bhattacharyya, J. Brokaw, J. W. Chu, and H. Yang. 2007. Illuminating the mechanistic roles of enzyme conformational dynamics. *Proc. Natl. Acad. Sci. U. S. A.* 104:18055–18060.

Hatzakis, N. S., L. Wei, S. K. Jorgensen, A. H. Kunding, P. Y. Bolinger, N. Ehrlich, I. Makarov, M. Skjot, A. Svendsen, P. Hedegård, and D. Stamou. 2012. Single enzyme studies reveal the existence of discrete functional states for monomeric enzymes and how they are selected upon allosteric regulation. *J. Am. Chem. Soc.* 134:9296–9302.

Heath, A. P., L. E. Kavraki, and C. Clementi. 2007. From coarse-grain to all-atom: Toward multiscale analysis of protein landscapes. *Proteins: Struct. Funct. Bioinformat.* 68:646–661.

Henderson, R., J. M. Baldwin, T. A. Ceska, F. Zemlin, E. Beckmann, and K. H. Downing. 1990. Model for the structure of bacteriorhodopsin based on high-resolution electron cryo-microscopy. *J. Mol. Biol.* 213:899–929.

Henzler-Wildman, K., and D. Kern. 2007. Dynamic personalities of proteins. *Nature* 450:964–972.

Henzler-Wildman, K. A., M. Lei, V. Thai, S. J. Kerns, M. Karplus, and D. Kern. 2007. A hierarchy of timescales in protein dynamics is linked to enzyme catalysis. *Nature* 450:913–916.

Henzler-Wildman, K. A., V. Thai, M. Lei, M. Ott, M. Wolf-Watz, T. Fenn, E. Pozharski, M. A. Wilson, G. A. Petsko, M. Karplus, C. G. Hubner, and D. Kern. 2007. Intrinsic motions along an enzymatic reaction trajectory. *Nature* 450:838–844.

Homme, M. B., C. Carter, and S. Scarlata. 2005. The cysteine residues of HIV-1 capsid regulate oligomerization and cyclophilin A-induced changes. *Biophys. J.* 88:2078–2088.

· Hong, G., E. Rosta, and A. Warshel. 2006. Using the constrained DFT approach in generating diabatic surfaces and off diagonal empirical valence bond terms for modeling reactions in condensed phases. *J. Phys. Chem. B* 110:19570–19574.

Hong, Y. J., and D. J. Tantillo. 2010. Quantum chemical dissection of the classic terpinyl/pinyl/bornyl/camphyl cation conundrum-the role of pyrophosphate in manipulating pathways to monoterpenes. *Org. Biomol. Chem.* 8:4589–4600.

Hubbard, B. P., A. P. Gomes, H. Dai, J. Li, A. W. Case, T. Considine, T. V. Riera, J. E. Lee, E. S. Yen, D. W. Lamming, B. L. Pentelute, E. R. Schuman, L. A. Stevens, A. J. Y. Ling, S. M. Armour, S. Michan, H. Z. Zhao, Y. Jiang, S. M. Sweitzer, C. A. Blum, J. S. Disch, P. Y. Ng, K. T. Howitz, A. P. Rolo, Y. Hamuro, J. Moss, R. B. Perni, J. L. Ellis, G. P. Vlasuk, and D. A. Sinclair. 2013. Evidence for a common mechanism of SIRT1 regulation by allosteric activators. *Science* 339:1216–1219.

Hwang, J. K., G. King, S. Creighton, and A. Warshel. 1988. Simulation of free energy relationships and dynamics of S_N2 reactions in aqueous solution. *J. Am. Chem. Soc.* 110:5297–5311.

Hwang, J. K., and A. Warshel. 1987. Microscopic examination of free-energy relationships for electron-transfer in polar-solvents. *J. Am. Chem. Soc.* 109:715–720.

Hwang, J. K., and A. Warshel. 1996. How important are quantum mechanical nuclear motions in enzyme catalysis? *J. Am. Chem. Soc.* 118:11745–11751.

Ionescu, R. M., V. F. Smith, J. C. O'Neill, and C. R. Matthews. 2000. Multistate equilibrium unfolding of *Escherichia coli* dihydrofolate reductase: Thermodynamic and spectroscopic description of the native, intermediate, and unfolded ensembles. *Biochemistry* 39:9540–9550.

Isralewitz, B., M. Gao, and K. Schulten. 2001. Steered molecular dynamics and mechanical functions of proteins. *Curr. Opin. Struct. Biol.* 11:224–230.

Jencks, W. P. 1972. General acid-base catalysis of complex reactions in water. *Chem. Rev.* 72:705–718.

Jensen, F. 2007. *Introduction to Computational Chemistry*, 2nd ed. Chichester, England: John Wiley & Sons.

Johnson, K. A. 1993. Conformational coupling in DNA-polymerase fidelity. *Annu. Rev. Biochem.* 62:685–713.

Johnson, L. N. 1998. The early history of lysozyme. *Nat. Struct. Mol. Biol.* 5:942–944.

Joyce, C. M., and S. J. Benkovic. 2004. DNA polymerase fidelity: Kinetics, structure, and checkpoints. *Biochemistry* 43:14317–14324.

Kale, S., G. Ulas, J. Song, G. W. Brudvig, W. Furey, and F. Jordan. 2008. Efficient coupling of catalysis and dynamics in the E1 component of *Escherichia coli* pyruvate dehydrogenase multienzyme complex. *Proc. Natl. Acad. Sci. U. S. A.* 105:1158–1163.

Kamerlin, S. C. L., Z. T. Chu, and A. Warshel. 2010. On catalytic preorganization in oxyanion holes: Highlighting the problems with gas-phase modeling of oxyanion holes and illustrating the need for complete enzyme models. *J. Org. Chem.* 75:6391–6401.

Kamerlin, S. C., M. Haranczyk, and A. Warshel. 2009. Progress in ab initio QM/MM free-energy simulations of electrostatic energies in proteins: Accelerated QM/MM studies of pKa, redox reactions and solvation free energies. *J. Phys. Chem. B* 113:1253–1272.

Kamerlin, S. C. L., J. Mavri, and A. Warshel. 2010. Examining the case for the effect of barrier compression on tunneling, vibrationally enhanced catalysis, catalytic entropy and related issues. *FEBS Lett.* 584:2759–2766.

Kamerlin, S. C. L., S. Vicatos, A. Dryga, and A. Warshel. 2011. Coarse-grained (multiscale) simulations in studies of biophysical and chemical systems. *Annu. Rev. Phys. Chem.* 62:41–64.

Kamerlin, S. C. L., and A. Warshel. 2010a. At the dawn of the 21st century: Is dynamics the missing link for understanding enzyme catalysis? *Proteins: Struct. Func. Bioinformat.* 78:1339–1375.

Kamerlin, S. C. L., and A. Warshel. 2010b. The EVB as a quantitative tool for formulating simulations and analyzing biological and chemical reactions. *Faraday Discuss.* 145:71–106.

Kamerlin, S. C. L., and A. Warshel. 2011a. The empirical valence bond model: Theory and applications. *WIREs: Comp. Mol. Sci.* 1:30–45.

Kamerlin, S. C. L., and A. Warshel. 2011b. Multiscale modeling of biological functions. *Phys. Chem. Phys.* 13:10401–10411.

Karabencheva-Christova, T. G., U. Carlsson, K. Balali-Mood, G. W. Black, and C. Z. Christov. 2013. Conformational effects on the circular dichroism of human carbonic anhydrase II: A multilevel computational study. *PLoS One* 8:e56874.

Karplus, M., and J. A. McCammon. 1983. Dynamics of proteins: Elements and function. *Annu. Rev. Biochem.* 53:263–300.

Kato, M., and A. Warshel. 2005. Through the channel and around the channel: Validating and comparing microscopic approaches for the evaluation of free energy profiles for ion penetration through ion channels. *J. Phys. Chem. B.* 109:19516–19522.

Keck, J. C. 1966. Variational theory of reaction rates. *Adv. Chem. Phys.* 13:85–121.

Kern, D., E. Z. Eisenmesser, and M. Wolf-Watz. 2005. Enzyme dynamics during catalysis measured by NMR spectroscopy. *Methods Enzymol.* 394:507–524.

Kern, D., G. Kern, G. Scherer, G. Fischer, and T. Drakenberg. 1995. Kinetic-analysis of cyclophilin-catalyzed prolyl cis/trans isomerization by dynamic NMR-spectroscopy. *Biochemistry* 34:13594–13602.

Kern, D., M. Schutkowski, and T. Drakenberg. 1997. Rotational barriers of cis/trans isomerization of proline analogues and their catalysis by cyclophilin. *J. Am. Chem. Soc.* 119:8403–8408.

Kern, G., D. Kern, R. Jaenicke, and R. Seckler. 1993. Kinetics of folding and association of differently glycosylated variants of invertase from saccharomyces-cerevisiae. *Protein Sci.* 2:1862–1868.

Kern, G., D. Kern, F. X. Schmid, and G. Fischer. 1994. Reassessment of the putative chaperone function of prolyl-cis/trans-isomerases. *FEBS Lett.* 348:145–148.

Kern, G., D. Kern, F. X. Schmid, and G. Fischer. 1995. A kinetic-analysis of the folding of human carbonic-anhydrase-II and its catalysis by cyclophilin. *J. Biol. Chem.* 270:740–745.

Kimura, Y., D. G. Vassylyev, A. Miyazawa, A. Kidera, M. Matsushima, K. Mitsuoka, K. Murata, T. Hirai, and Y. Fujiyoshi. 1997. Surface of bacteriorhodopsin revealed by high-resolution electron crystallography. *Nature* 389:206–211.

Klähn, M., S. Braun-Sand, E. Rosta, and A. Warshel. 2005. On possible pitfalls in ab initio quantum mechanics/molecular mechanics minimization approaches for studies of enzymatic reactions. *J. Phys. Chem. B* 109:15645–15650.

Kline, C. D., M. Mayfield, and N. J. Blackburn. 2013. HHM motif at the CuH-site of peptidyl-glycine monooxygenase is a pH-dependent conformational switch. *Biochemistry* 52:2586–2596.

Klinman, J. P. 2013. Importance of protein dynamics during enzymatic C-H bond cleavage catalysis. *Biochemistry* 52:2068–2077.

Knapp, M. J., K. Rickert, and J. P. Klinman. 2002. Temperature-dependent isotope effects in soybean lipoxygenase-1: Correlating hydrogen tunneling with protein dynamics. *J. Am. Chem. Soc.* 124:3865–3874.

Kohen, A., R. Cannio, S. Bartolucci, and J. P. Klinman. 1999. Enzyme dynamics and hydrogen tunnelling in a thermophilic alcohol dehydrogenase. *Nature* 399:496–499.

Kramers, H. A. 1940. Brownian motion in a field of force and the diffusion model of chemical reactions. *Physica* 7:284–304.

Kuhlman, B., and D. Baker. 2004. Exploring folding free energy landscapes using computational protein design. *Curr. Opin. Struct. Biol.* 14:89–95.

Kumar, S., B. Y. Ma, C. J. Tsai, N. Sinha, and R. Nussinov. 2000. Folding and binding cascades: Dynamic landscapes and population shifts. *Protein Sci.* 9:10–19.

Kurkcouglu, Z., A. Bakan, D. Kocaman, I. Bahar, and P. Doruker. 2012. Coupling between catalytic loop motions and enzyme global dynamics. *PLoS Comput. Biol.* 8:e1002705.

Labas, A., E. Szabo, L. Mones, and M. Fuxreiter. 2013. Optimization of reorganization energy drives evolution of the designed Kemp eliminase KE07. *Biochim. Biophys. Acta* 1834:908–917.

Lane, T. J., D. Shukla, K. A. Beauchamp, and V. S. Pande. 2013. To milliseconds and beyond: Challenges in the simulation of protein folding. *Curr. Opin. Struc. Biol.* 23:58–65.

Lanyi, J. K. 2004. Bacteriorhodopsin. *Annu. Rev. Physiol.* 66:665–688.

Lanyi, J. K., and B. Schobert. 2003. Mechanism of proton transport in bacteriorhodopsin from crystallographic structures of the K, L, M1, M2, and M2' intermediates of the photocycle. *J. Mol. Biol.* 328:439–450.

Lee, F. S., Z. T. Chu, and A. Warshel. 1993. Microscopic and semimicroscopic calculations of electrostatic energies in proteins by the polaris and enzymix programs. *J. Comput. Chem.* 14:161–185.

Levitt, M., and A. Warshel. 1975. Computer-simulation of protein folding. *Nature* 253:694–698.

Liao, R. Z., and W. Thiel. 2012. Comparison of QM-only and QM/MM models for the mechanism of tungsten-dependent acetylene hydratase. *J. Chem. Theory Comput.* 8:3793–3803.

Liu, H., Y. Shi, X. S. Chen, and A. Warshel. 2009. Simulating the electrostatic guidance of the vectorial translocations in hexameric helicases and translocases. *Proc. Natl. Acad. Sci. U. S. A.* 106:7449–7454.

Liu, H. B., and A. Warshel. 2007a. Origin of the temperature dependence of isotope effects in enzymatic reactions: The case of dihydrofolate reductase. *J. Phys. Chem. B* 111:7852–7861.

Liu, H., and A. Warshel. 2007b. The catalytic effect of dihydrofolate reductase and its mutants is determined by reorganization energies. *Biochemistry* 46:6011–6025.

Liu, W., E. Chun, A. A. Thompson, P. Chubukov, F. Xu, V. Katritch, G. W. Han, C. B. Roth, L. H. Heitman, A. P. Ijzerman, V. Cherezov, and R. C. Stevens. 2012. Structural basis for allosteric regulation of GPCRs by sodium ions. *Science* 337:232–236.

Lodola, A., J. Sirirak, N. Fey, S. Rivara, M. Mor, and A. J. Mulholland. 2010. Structural fluctuations in enzyme-catalyzed reactions: Determinants of reactivity in fatty acid amide hydrolase from multivariate statistical analysis of quantum mechanics/molecular mechanics paths. *J. Chem. Theory Comput.* 6:2948–2960.

Loveridge, E. J., E. M. Behiry, J. N. Guo, and R. K. Allemann. 2012. Evidence that a "dynamic knockout" in *Escherichia coli* dihydrofolate reductase does not affect the chemical step of catalysis. *Nat. Chem.* 4:292–297.

Loveridge, E. J., L. H. Tey, and R. K. Allemann. 2010. Solvent effects on catalysis by *Escherichia coli* dihydrofolate reductase. *J. Am. Chem. Soc.* 132:1137–1143.

Loveridge, E. J., L. H. Tey, E. M. Behiry, W. M. Dawson, R. M. Evans, S. B. M. Whittaker, U. L. Gunther, C. Williams, M. P. Crump, and R. K. Allemann. 2011. The role of large-scale motions in catalysis by dihydrofolate reductase. *J. Am. Chem. Soc.* 133:20561–20570.

Lu, H. P., L. Y. Xun, and X. S. Xie. 1998. Single-molecule enzymatic dynamics. *Science* 282:1877–1882.

Lu, H. P., L. Y. Xun, and X. S. Xie. 1999. Single-molecule spectroscopic study of protein conformational and enzymatic dynamics. *Biophys. J.* 76:A136.

Lu, Q., and J. Wang. 2008. Single molecule conformational dynamics of adenylate kinase: Energy landscape, structural correlations and transition state ensembles. *J. Am. Chem. Soc.* 130:4772–4783.

Luecke, H., B. Schobert, H. T. Richter, J. P. Cartailler, and J. K. Lanyi. 1999a. Structural changes in bacteriorhodopsin during ion transport at 2 Ångström resolution. *Science* 286:255–261.

Luecke, H., B. Schobert, H. T. Richter, J. P. Cartailler, and J. K. Lanyi. 1999b. Structure of bacteriorhodopsin at 1.55 Å resolution. *J. Mol. Biol.* 291:899–911.

Luo, J, B. van Loo, and S. C. L. Kamerlin. 2012. Catalytic promiscuity in Pseudomonas aeruginosa arylsulfatase as an example of chemistry-driven protein evolution. *FEBS Lett.* 586:1622–1630.

Maglia, G., M. H. Javed, and R. K. Allemann. 2003. Hydride transfer during catalysis by dihydrofolate reductase from *Thermotoga maritima*. *Biochem. J.* 374:529–535.

Major, D. T., and M. Weitman. 2012. Electrostatically guided dynamics—The root of fidelity in a promiscuous terpene synthase? *J. Am. Chem. Soc.* 134:19454–19462.

Maragakis, P., and M. Karplus. 2005. Large amplitude conformational change in proteins explored with a plastic network model: Adenylate kinase. *J. Mol. Biol.* 352:807–822.

Marcus, R. A., and N. Sutin. 1985. Electron transfers in chemistry and biology. *Biochim. Biophys. Acta* 811:265–322.

Masgrau, L., A. Roujeinikova, L. O. Johannissen, P. Hothi, J. Basran, K. E. Ranaghan, A. J. Mulholland, M. J. Sutcliffe, N. S. Scrutton, and D. Leys. 2006. Atomic description of an enzyme reaction dominated by proton tunneling. *Science* 312:237–241.

Mathies, R. A., C. H. Brito Cruz, W. T. Pollard, and C. V. Shank. 1988. Direct observation of the femtosecond excited-state cis-trans isomerization in bacteriorhodopsin. *Science* 240:777–779.

Mauldin, R. V., and R. T. Sauer. 2013. Allosteric regulation of DegS protease subunits through a shared energy landscape. *Nat. Chem. Biol.* 9:90–96.

Mayor, U., N. R. Guydosh, C. M. Johnson, J. G. Grossmann, S. Sato, G. S. Jas, S. M. V. Freund, D. O. V. Alonso, V. Daggett, and A. R. Fersht. 2003. The complete folding pathway of a protein from nanoseconds to microseconds. *Nature* 421:863–867.

McCammon, J. A., B. R. Gelin, and M. Karplus. 1977. Dynamics of folded proteins. *Nature* 267:585–590.

McCammon, J. A., P. G. Wolynes, and M. Karplus. 1979. Picosecond dynamics of tyrosine side chains in proteins. *Biochemistry* 18:927–942.

McElheny, D., J. R. Schnell, J. C. Lansing, H. J. Dyson, and P. E. Wright. 2005. Defining the role of active-site loop fluctuations in dihydrofolate reductase catalysis. *Proc. Natl. Acad. Sci. U. S. A.* 102:5032–5037.

McGeagh, J. D., K. E. Ranaghan, and A. J. Mulholland. 2011. Protein dynamics and enzyme catalysis: Insights from simulations. *BBA-Proteins Proteom.* 1814:1077–1092.

McGowan, L. C., and D. Hamelberg. 2013. Conformational plasticity of an enzyme during catalysis: Intricate coupling between Cyclophilin A dynamics and substrate turnover. *Biophys. J.* 104:216–226.

Messer, B. M., M. Roca, Z. T. Chu, S. Vicatos, A. Kilshtain, and A. Warshel. 2010. Multiscale simulations of protein landscapes: Using coarse grained models as reference potentials to full explicit models. *Proteins: Struct. Funct. Bioinformat.* 78:1212–1227.

Michaelis, L., M. L. Menten, K. A. Johnson, and R. S. Goody. 2011. The original Michaelis constant: Translation of the 1913 Michaelis-Menten paper. *Biochemistry* 50:8264–8269.

Miller, G. P., and S. J. Benkovic. 1998a. Deletion of a highly motional residue affects formation of the Michaelis complex for *Escherichia coli* dihydrofolate reductase. *Biochemistry* 37:6327–6335.

Miller, G. P., and S. J. Benkovic. 1998b. Strength of an interloop hydrogen bond determines the kinetic pathway in catalysis by *Escherichia coli* dihydrofolate reductase. *Biochemistry* 37:6336–6342.

Miller, G. P., and S. J. Benkovic. 1998c. Stretching exercises–flexibility in dihydrofolate reductase catalysis. *Chem. Biol.* 5:R105–R113.

Miller, G. P., D. C. Wahnon, and S. J. Benkovic. 2001. Interloop contacts modulate ligand cycling during catalysis by *Escherichia coli* dihydrofolate reductase. *Biochemistry* 40:867–875.

Min, W., S. Xie, and B. Bagchi. 2008. Two-dimensional reaction free energy surfaces of catalytic reaction: Effects of protein conformational dynamics on enzyme catalysis. *J. Phys. Chem. B* 112:454–466.

Mones, L., P. Kulhanek, I. Simon, A. Laio, and M. Fuxreiter. 2009. The energy gap as a universal reaction coordinate for the simulation of chemical reactions. *J. Phys. Chem. B* 113:7867–7873.

Muegge, I., P. X. Qi, A. J. Wand, Z. T. Chu, and A. Warshel. 1997. The reorganization energy of cytochrome *c* revisited. *J. Phys. Chem. B.* 101:825–836.

Mukherjee, S., and A. Warshel. 2012. Realistic simulations of the coupling between the protomotive force and the mechanical rotation of the F0-ATPase. *Proc. Natl. Acad. Sci. U. S. A.* 109:14876–14881.

Mulholland, A. J., A. Roitberg, and I. Tuñón. 2012. Enzyme dynamics and catalysis in the mechanism of DNA polymerase. *Theor. Chem. Acc.* 131:1286–1288.

Muller, C. W., G. J. Schlauderer, J. Reinstein, and G. E. Schulz. 1996. Adenylate kinase motions during catalysis: An energetic counterweight balancing substrate binding. *Structure* 4:147–156.

Muller, C. W., and G. E. Schulz. 1992. Structure of the complex between adenylate kinase from Escherichia-coli and the inhibitor Ap5a refined at 1.9 Å resolution: A model for a catalytic transition-state. *J. Mol. Biol.* 224:159–177.

Muller, C. W., and G. E. Schulz. 1993. Crystal structures of two mutants of adenylate kinase from Escherichia coli that modify the Gly-loop. *Proteins* 15:42–49.

Muller-Dieckmann, H. J., and G. E. Schulz. 1994. The structure of uridylate kinase with its substrates, showing the transition-state geometry. *J. Mol. Biol.* 236:361–367.

Muller-Dieckmann, H. J., and G. E. Schulz. 1995. Substrate specificity and assembly of the catalytic center derived from two structures of ligated uridylate kinase. *J. Mol. Biol.* 246:522–530.

Nagel, Z. D., and J. P. Klinman. 2009. A 21st century revisionist's view at a turning point in enzymology. *Nat. Chem. Biol.* 5:543–550.

Nam, K., X. Prat-Resina, M. Garcia-Viloca, L. S. Devi-Kesavan, and J. L. Gao. 2004. Dynamics of an enzymatic substitution reaction in haloalkane dehalogenase. *J. Am. Chem. Soc.* 126:1369–1376.

Nelson, D. L., A. L. Lehninger, and M. M. Cox. 2008. *Lehninger Principles of Biochemistry*, 5th ed. New York: W.H. Freeman.

Neria, E., and M. Karplus. 1997. Molecular dynamics of an enzyme reaction: Proton transfer in TIM. *Chem. Phys. Lett.* 267:23–30.

Northrup, S. H., M. R. Pear, C. Y. Lee, J. A. Mccammon, and M. Karplus. 1982. Dynamical theory of activated processes in globular-proteins. *Proc. Natl. Acad. Sci. U. S. A.* 79:4035–4039.

Nymeyer, H., N. D. Socci, and J. N. Onuchic. 2000. Landscape approaches for determining the ensemble of folding transition states: Success and failure hinge on the degree of frustration. *Proc. Natl. Acad. Sci. U. S. A.* 97:634–639.

O'Ferrall, R. A. 1970. Relationships between E2 and ElcB mechanism of β-elimination. *J. Chem. Soc. B.* 1970:274–277.

Okazaki, K. I., N. Koga, S. Takada, J. N. Onuchic, and P. G. Wolynes. 2006. Multiple-basin energy landscapes for large-amplitude conformational motions of proteins: Structure-based molecular dynamics simulations. *Proc. Natl. Acad. Sci. U. S. A.* 103:11844–11849.

Olsson, M. H. M., W. W. Parson, and A. Warshel. 2006. Dynamical contributions to enzyme catalysis: Critical tests of a popular hypothesis. *Chem. Rev.* 106:1737–1756.

Olsson, U., and M. Wolf-Watz. 2010. Overlap between folding and functional energy landscapes for adenylate kinase conformational change. *Nature Commun.* 1:111.

Onuchic, J. N., Z. Luthey-Schulten, and P. G. Wolynes. 1997. Theory of protein folding: The energy landscape perspective. *Annu. Rev. Phys. Chem.* 48:545–600.

Palczewski, K., T. Kumasaka, T. Hori, C. A. Behnke, H. Motoshima, B. A. Fox, I. Le Trong, D. C. Teller, T. Okada, R. E. Stenkamp, M. Yamamoto, and M. Miyano. 2000. Crystal structure of rhodopsin: A G protein-coupled receptor. *Science* 289:739–745.

Pande, V. S., K. Beauchamp, and G. R. Bowman. 2010. Everything you wanted to know about Markov state models but were afraid to ask. *Methods* 52:99–105.

Patel, S. S., I. Wong, and K. A. Johnson. 1991. Pre-steady-state kinetic analysis of processive DNA replication including complete characterization of an exonuclease-deficient mutant. *Biochemistry* 30:511–525.

Pervushin, K., K. Vamcava, B. Vögeli, and D. Hilvert. 2007. Structure and dynamics of a molten globular enzyme. *Nat. Struct. Mol. Biol.* 14:1202–1206.

Pisliakov, A. V., J. Cao, S. C. L. Kamerlin, and A. Warshel. 2009. Enzyme millisecond conformational dynamics do not catalyze the chemical step. *Proc. Natl. Acad. Sci. U. S. A.* 106:17359–17364.

Prasad, B. R., S. C. L. Kamerlin, J. Florián, and A. Warshel. 2012. Prechemistry barriers and checkpoints do not contribute to fidelity and catalysis as long as they are not rate limiting. *Theor. Chem. Acc.* 131:1288–1302.

Prasad, R. B., and A. Warshel. 2011. Prechemistry versus preorganization in DNA replication fidelity. *Proteins* 79:2900–2919.

Pronk, S., P. Larsson, I. Pouya, G. R. Browman, I. S. Haque, K. A. Beauchamp, B. Hess, V. S. Pande, P. M. Kasson, and E. Lindahl. 2011. Copernicus: A new paradigm for parallel adaptive molecular dynamics. Paper read at Proceedings of the 2011 ACM/IEEE International Conference for High Performance Computing, Networking, Storage and Analysis.

Radhakrishnan, R., K. Arora, Y. L. Wang, W. A. Beard, S. H. Wilson, and T. Schlick. 2006. Regulation of DNA repair fidelity by molecular checkpoints: Gates in DNA polymerase β's substrate selected. *Biochemistry* 45:15142–15156.

Radhakrishnan, R., and T. Schlick. 2004. Orchestration of cooperative events in DNA synthesis and repair mechanism unraveled by transition path sampling of DNA polymerase β's closing. *Proc. Natl. Acad. Sci. U. S. A.* 101:5970–5975.

Radhakrishnan, R., and T. Schlick. 2005. Fidelity discrimination in DNA polymerase β: Differing closing profiles for a mismatched (G: A) versus matched (G: C) base pair. *J. Am. Chem. Soc.* 127:13245–13252.

Radhakrishnan, R., and T. Schlick. 2006. Correct and incorrect nucleotide incorporation pathways in DNA polymerase β. *Biochem. Biophys. Res. Commun.* 350:521–529.

Radkiewicz, J. L., and C. L. Brooks. 2000. Protein dynamics in enzymatic catalysis: Exploration of dihydrofolate reductase. *J. Am. Chem. Soc.* 122:225–231.

Rajagopalan, P. T., and S. J. Benkovic. 2002. Preorganization and protein dynamics in enzyme catalysis. *Chem. Rec.* 2:24–36.

Razvi, A., and J. M. Scholtz. 2006. Lessons in stability from thermophilic proteins. *Protein Sci.* 15:1569–1578.

Reuner, C., M. Hable, M. Wilmanns, E. Kiefer, E. Schiltz, and G. E. Schulz. 1988. Amino acid sequence and three-dimensional structure of cytosolic adenylate kinase from carp muscle. *Protein Seq. Data Anal.* 1:335–343.

Roca, M., H. Liu, B. Messer, and A. Warshel. 2007. On the relationship between thermal stability and catalytic power of enzymes. *Biochemistry* 46:15076–15088.

Roca, M., B. Messer, D. Hilvert, and A. Warshel. 2008. On the relationship between folding and chemical landscapes in enzyme catalysis. *Proc. Natl. Acad. Sci. U. S. A.* 105:13877–13882.

Roca, M., V. Moliner, I. Tuñón, and J. T. Hynes. 2006. Coupling between protein and reaction dynamics in enzymatic processes: Application of Grote-Hynes theory to catechol O-methyltransferase. *J. Am. Chem. Soc.* 128:6186–6193.

Roca, M., M. Oliva, R. Castillo, V. Moliner, and I. Tuñón. 2010. Do dynamic effects play a significant role in enzymatic catalysis? A theoretical analysis of formate dehydrogenase. *Chem. Eur. J.* 16:11399–11411.

Rod, T. H., J. L. Radkiewicz, and C. L. Brooks, 3rd. 2003. Correlated motion and the effect of distal mutations in dihydrofolate reductase. *Proc. Natl. Acad. Sci. U. S. A.* 100:6980–6985.

Rohrdanz, M. A., W. Zheng, and C. Clementi. 2013. Discovering mountain passes via torchlight: Methods for the definition of reaction coordinates and pathways in complex macromolecular reactions. *Annu. Rev. Phys. Chem.* 64:295–316.

Rohrdanz, M. A., W. W. Zheng, M. Maggioni, and C. Clementi. 2011. Determination of reaction coordinates via locally scaled diffusion map. *J. Chem. Phys.* 134:124116.

Rousseau, R., V. Kleinschmidt, U. W. Schmitt, and D. Marx. 2004. Modeling protonated water networks in bacteriorhodopsin. *Phys. Chem. Phys.* 6:1848–1859.

Ruiz-Pernia, J. J., I. Tuñón, V. Moliner, J. T. Hynes, and M. Roca. 2008. Dynamic effects on reaction rates in a Michael addition catalyzed by chalcone isomerase. Beyond the frozen environment approach. *J. Am. Chem. Soc.* 130:7477–7488.

Rychkova, A., S. Vicatos, and A. Warshel. 2010. On the energetics of translocon-assisted insertion of charged transmembrane helices into membranes. *Proc. Natl. Acad. Sci. U. S. A.* 107:17598–17603.

Rychkova, A., and A. Warshel. 2013. Exploring the nature of the translocon-assisted protein insertion. *Proc. Natl. Acad. Sci. U. S. A.* 110:495–500.

Saen-Oon, S., M. Ghanem, V. L. Schramm, and S. D. Schwart. 2008. Remote mutations and active site dynamics correlate with catalytic properties of purine nucleoside phosphorylase. *Biophys. J.* 94:4078–4088.

Schenter, G. K., H. P. Lu, and X. S. Xie. 1999. Statistical analyses and theoretical models of single-molecule enzymatic dynamics. *J. Phys. Chem. A* 103:10477–10488.

Schlick, T. 2009. Molecular dynamics-based approaches for enhanced sampling of long-time, large-scale conformational changes in biomolecules. *F1000 Biol. Rep.* 1:51.

Schlick, T., R. Collepardo-Guevara, L. A. Halvorsen, S. Jung, and X. Xiao. 2011. Biomolecular modeling and simulation: A field coming of age. *Q. Rev. Biophys.* 44:191–228.

Schlitter, J., M. Engels, P. Kruger, E. Jacoby, and A. Wollmer. 1993. Targeted molecular dynamics simulation of conformational change-application to the T↔R transition in insulin. *Mol. Sim.* 10:291–308.

Schnell, J. R., H. J. Dyson, and P. E. Wright. 2004. Structure, dynamics, and catalytic function of dihydrofolate reductase. *Annu. Rev. Biophys. Biomol. Struct.* 33:119–140.

Schoenlein, R. W., L. A. Peteanu, R. A. Mathies, and C. V. Shank. 1991. The first step in vision: Femtosecond isomerization of rhodopsin. *Science* 254:412–415.

Schulten, K., and P. Tavan. 1978. Mechanism for light-driven proton pump of halobacterium-halobium. *Nature* 272:85–86.

Scudder, P. H. 1990. Use of reaction cubes for generation and display of multiple mechanistic pathways. *J. Org. Chem.* 55:4238–4240.

Seco, J., C. Ferrer-Costa, J. M. Campanera, R. Soliva, and X. Barril. 2012. Allosteric regulation of PKC theta: Understanding multistep phosphorylation and priming by ligands in AGC kinases. *Proteins: Struct. Func. Bioinformat.* 80:269–280.

Sena, A. M. P., T. Miyazaki, and D. R. Bowler. 2011. Linear scaling constrained density functional theory in CONQUEST. *J. Chem. Theory Comput.* 7:884–889.

Shaw, D. E. 2013. Millisecond-long molecular dynamics simulations of proteins on a special-purpose machine. *Biophys. J.* 104:45a.

Siegbahn, P. E., and F. Himo. 2009. Recent developments of the quantum chemical cluster approach for modeling enzyme reactions. *J. Biol. Inorg. Chem.* 14:643–651.

Soskine, M., and D. S. Tawfik. 2010. Mutational effects and the evolution of new protein functions. *Nat. Rev. Gen.* 11:572–582.

Sousa, S. F., P. A. Fernandes, and M. J. Ramos. 2012. Computational enzymatic catalysis–clarifying enzymatic mechanisms with the help of computers. *Phys. Chem. Phys.* 14:12431–12441.

Stadtman, E. R. 2006. Allosteric regulation of enzyme activity. In *Advances in Enzymology and Related Areas of Molecular Biology*, E. J. Toone, ed., 41–154. Hoboken, NJ: John Wiley & Sons.

Stehle, T., and G. E. Schulz. 1992. Refined structure of the complex between guanylate kinase and its substrate GMP at 2·0 Å resolution. *J. Mol. Biol.* 224:1127–1141.

Strajbl, M., A. Shurki, and A. Warshel. 2003. Converting conformational changes to electrostatic energy in molecular motors: The energetics of ATP synthase. *Proc. Natl. Acad. Sci. U. S. A.* 100:14834–14839.

Subramaniam, S., I. Lindahl, P. Bullough, A. R. Faruqi, J. Tittor, D. Oesterhelt, L. Brown, J. Lanyi, and R. Henderson. 1999. Protein conformational changes in the bacteriorhodopsin photocycle. *J. Mol. Biol.* 287:145–161.

Sulkowska, J. I., J. K. Noel, and J. N. Onuchic. 2012. Energy landscape of knotted protein folding. *Proc. Natl. Acad. Sci. U. S. A.* 109:17783–17788.

Tan, W. H., and E. S. Yeung. 1997. Monitoring the reactions of single enzyme molecules and single metal ions. *Anal. Chem.* 69:4242–4248.

te Velde, G., F. M. Bickelhaupt, E. J. Baerends, C. F. Guerra, S. J. A. Van Gisbergen, J. G. Snijders, and T. Ziegler. 2001. Chemistry with ADF. *J. Comp. Chem.* 22:931–967.

Thai, V., P. Renesto, C. A. Fowler, D. J. Browni, T. Davis, W. J. Gu, D. D. Pollock, D. Kern, D. Raoult, and E. Z. Eisenmosser. 2008. Structural, biochemical, and in vivo characterization of the first virally encoded cyclophilin from the Mimivirus. *J. Mol. Biol.* 378:71–86.

Torres, T., and M. Levitus. 2007. Measuring conformational dynamics: A new FCS-FRET approach. *J. Phys. Chem. B.* 111:7392–7400.

Torrie, G. M., and J. P. Valleau. 1977. Non-physical sampling distributions in monte-carlo free-energy estimation-umbrella sampling. *J. Comput. Phys.* 23:187–199.

Truong, K., and M. Ikura. 2001. The use of FRET imaging microscopy to detect protein-protein interactions and protein conformational changes in vivo. *Curr. Opin. Struct. Biol.* 11:573–578.

Trushkov, I. V., V. V. Zhdankin, A. S. Kozmin, and N. S. Zefirov. 1990. Cubic reaction coordinate diagram in the nucleophilic-substitution process. *Tetrahedron Lett.* 31:3199–3200.

Tsai, M. D., X. J. Zhong, J. W. Arndt, W. M. Gong, Z. Lin, C. Paxson, and M. R. Chan. 1999. Structure of DNA polymerase beta/DNA template-primer/Cr(III)dTMPPCP complex: Insight into structural bases of conformational changes and fidelity. *FASEB J.* 13:A1364.

Valiyaveetil, F. I., and R. H. Fillingame. 1997. On the role of Arg-210 and Glu-219 of subunit a in proton translocation by the *Escherichia coli* F_0F_1-ATP synthase. *J. Biol. Chem.* 272:32635–32641.

Vamvaca, K., B. Vögeli, P. Kast, K. Pervushin, and D. Hilvert. 2004. An enzymatic molten globule: Efficient coupling of folding and catalysis. *Proc. Natl. Acad. Sci. U. S. A.* 101:12860–12864.

van der Kamp, M. W., R. Chaudret, and A. J. Mulholland. 2013. QM/MM modelling of keto-steroid isomerase reactivity indicates that active site closure is integral to catalysis. *FEBS J.* 13:3120–3131.

Vendruscolo, M., and C. M. Dobson. 2006. Dynamic visions of enzymatic reactions. *Science* 313:1586–1587.

Venkitakrishnan, R. P., E. Zaborowski, D. McElheny, S. J. Benkovic, H. J. Dyson, and P. E. Wright. 2004. Conformational changes in the active site loops of dihydrofolate reductase during the catalytic cycle. *Biochemistry* 43:16046–16055.

Villali, J., and D. Kern. 2010. Choreographing an enzyme's dance. *Curr. Opin. Chem. Biol.* 14:636–643.

Vonrhein, C., G. J. Schlauderer, and G. E. Schulz. 1995. Movie of the structural changes during a catalytic cycle of nucleoside monophosphate kinases. *Structure* 3:483–490.

Wang, Y., and T. Schlick. 2008. Quantum mechanics/molecular mechanics investigation of the chemical reaction in Dpo4 reveals water-dependent pathways and requirements for active site reorganization. *J. Am. Chem. Soc.* 130:13240–13250.

Warshel, A. 1976. Bicycle-pedal model for the first step in the vision process. *Nature* 260:679–683.

Warshel, A. 1978. Energetics of enzyme catalysis. *Proc. Natl. Acad. Sci. U. S. A.* 75:5250–5254.

Warshel, A. 1982. Dynamics of reactions in polar solvents. Semiclassical trajectory studies of electron-transfer and proton-transfer reactions. *J. Phys. Chem.* 86:2218–2224.

Warshel, A. 1991. *Computer Modeling of Chemical Reactions in Enzymes and Solutions.* New York: Wiley.

Warshel, A., and N. Barboy. 1982. Energy storage and reaction pathways in the first step of the vision process. *J. Am. Chem. Soc.* 104:1469–1476.

Warshel, A., Z. T. Chu, and W. W. Parson. 1989. Dispersed polaron simulations of electron transfer in photosynthetic reaction centers. *Science* 246:112–116.

Warshel, A., J. Florian, M. Strajbl, and J. Villa. 2001. Circe effect versus enzyme preorganization: What can be learned from the structure of the most proficient enzyme? *Chem. Bio. Chem.* 2:109–111.

Warshel, A., and M. Levitt. 1976. Theoretical studies of enzymic reactions: Dielectric, electrostatic and steric stabilization of carbonium-ion in reaction of lysozyme. *J. Mol. Biol.* 103:227–249.

Warshel, A., P. K. Sharma, M. Kato, and W. W. Parson. 2006a. Modeling electrostatic effects in proteins. *Biochim. Biophys. Acta-Proteins Proteom.* 1764:1647–1676.

Warshel, A., P. K. Sharma, M. Kato, Y. Xiang, H. B. Liu, and M. H. M. Olsson. 2006b. Electrostatic basis for enzyme catalysis. *Chem. Rev.* 106:3210–3235.

Watney, J. B., P. K. Agarwal, and S. Hammes-Schiffer. 2003. Effect of mutation on enzyme motion in dihydrofolate reductase. *J. Am. Chem. Soc.* 125:3745–3750.

Weinan, E., and E. Vanden-Eijnden. 2010. Transition-path theory and path-finding algorithms for the study of rare events. *Annu. Rev. Phys. Chem.* 61:391–420.

Weitman, M., and D. T. Major. 2010. Challenges posed to bornyl diphosphate synthase: Diverging reaction mechanisms in monoterpenes. *J. Am. Chem. Soc.* 132:6349–6360.

Whitford, P. C., O. Miyashita, Y. Levy, and J. N. Onuchic. 2007. Conformational transitions of adenylate kinase: Switching by cracking. *J. Mol. Biol.* 366:1661–1671.

Wieland, B., A. G. Tomasselli, L. H. Noda, R. Frank, and G. E. Schulz. 1984. The amino-acid-sequence of GTP:AMP phosphotransferase from beef-heart mitochondria— Extensive homology with cytosolic adenylate kinase. *Eur. J. Biochem.* 143:331–339.

Williams, G. 2010. Elastic network model of allosteric regulation in protein kinase PDK1. *BMC Struct. Biol.* 10:11.

Wise, M. L., T. J. Savage, E. Katahira, and R. Croteau. 1998. Monoterpene synthases from common sage (Salvia officinalis). cDNA isolation, characterization, and functional expression of (+)-sabinene synthase, 1,8-cineole synthase, and (+)-bornyl diphosphate synthase. *J. Biol. Chem.* 273:14891–14899.

Wold, S., K. Esbensen, and P. Geladi. 1987. Principal component analysis. *Chemometr. Intell. Lab.* 2:37–52.

Wolfenden, R., and M. J. Snider. 2001. The depth of chemical time and the power of enzymes as catalysts. *Acc. Chem. Res.* 34:938–945.

Wolf-Watz, M., V. Thai, K. Henzler-Wildman, G. Hadjipavlou, E. Z. Eisenmesser, and D. Kern. 2004. Linkage between dynamics and catalysis in a thermophilic-mesophilic enzyme pair. *Nat. Struct. Mol. Biol.* 11:945–949.

Wong, K. F., T. Selzer, S. J. Benkovic, and S. Hammes-Schiffer. 2005. Impact of distal mutations on the network of coupled motions correlated to hydride transfer in dihydrofolate reductase. *Proc. Natl. Acad. Sci. U. S. A.* 102:6807–6812.

Wrba, A., A. Schweiger, V. Schultes, R. Jaenicke, and P. Zavodszky. 1990. Extremely thermostable D-glyceraldehyde-3-phosphate dehydrogenase from the eubacterium *Thermotoga maritima. Biochemistry* 29:7584–7592.

Xiang, Y., M. F. Goodman, W. A. Beard, S. H. Wilson, and A. Warshel. 2008. Exploring the role of large conformational changes in the fidelity of DNA polymerase β. *Proteins.* 70:231–247.

Xie, X. S. 2002. Single-molecule approach to dispersed kinetics and dynamic disorder: Probing conformational fluctuation and enzymatic dynamics. *J. Chem. Phys.* 117:11024–11032.

Xue, Q., and E. S. Yeung. 1995. Differences in the chemical reactivity of individual molecules of an enzyme. *Nature.* 373:681–683.

Yu, H., A. N. Gupta, X. Liu, K. Neupane, A. M. Brigley, I. Sosova, and M. T. Woodside. 2012. Energy landscape analysis of native folding of the prion protein yields the diffusion constant, transition path time, and rates. *Proc. Natl. Acad. Sci. U. S. A.* 109:14452–14457.

Zalatan, J. G., and D. Herschlag. 2009. The far reaches of enzymology. *Nat. Chem. Biol.* 5:516–520.

Zavodszky, P., J. Kardos, A. Svingor, and G. A. Petsko. 1998. Adjustment of conformational flexibility is a key event in the thermal adaptation of proteins. *Proc. Natl. Acad. Sci. U. S. A.* 95:7406–7411.

Zhuravlev, P. I., C. K. Materese, and G. A. Papoian. 2009. Deconstructing the native state: Energy landscapes, function, and dynamics of globular proteins. *J. Phys. Chem. B.* 113:8800–8812.

Chapter 4

Interplay between Enzyme Function and Protein Dynamics

A Multiscale Approach to the Study of the NAG Kinase Family and Two Class II Aldolases

Enrique Marcos, Melchor Sanchez-Martinez, and Ramon Crehuet

CONTENTS

INTRODUCTION

Proteins under native state conditions exhibit a wide range of motions enabling the performance of their biological function. Fast protein motions (pico–nanosecond timescale) are local and involve conformational fluctuations of a few angstroms, including side chain rotations and small backbone movements. By contrast, at slow timescales (micro-milliseconds and beyond), large collective motions of protein domains and entire subunits in oligomeric assemblies allow changes in conformation of tenths of angstroms that are typically associated to active site opening/closing in ligand binding and vast structural rearrangements in allosteric events. It is the balanced interplay of this hierarchy of motions that allows proteins to adopt conformations outstandingly complementary to their binding partners. Toward unveiling the relationship between protein function and dynamics, computational methods, such as those covered in this book, in combination with biophysics experiments have been successful in describing relevant dynamic properties, but still many fundamental questions remain to be answered. For instance, why should enzymes move? Of course, they move because temperature, at the molecular level, shakes everything. But evolution can regulate the rigidity of proteins in compliance with stability, as observed in thermophilic-mesophilic pairs of enzymes (Kumar and Nussinov 2001; Razvi and Scholtz 2006; Sterpone and Melchionna 2012; Vieille and Zeikus 2001; Wolf-Watz et al. 2004). So it seems that enzymes could be much more rigid than what they are; suggesting that flexibility is related to their function and likely to their capacity to evolve toward new functions (Dellus-Gur et al. 2013). While there are many experimental (Bhabha et al. 2011; Cameron and Benkovic 1997; Eisenmesser et al. 2005; Henzler-Wildman et al. 2007a,b; Rajagopalan et al. 2002; Wolf-Watz et al. 2004) and computational (Agarwal 2005, 2006; Agarwal et al. 2002, 2004; Hammes-Schiffer and Benkovic 2006; Ma et al. 2000; Yang and Bahar 2005) results that confirm a linkage between enzyme flexibility and function, some computational studies indicate the effects of dynamics are, if any, too small (García-Meseguer et al. 2013; Glowacki et al. 2012; Kamerlin and Warshel 2010a,b; Martí et al. 2003; Olsson et al. 2006; Pisliakov et al. 2009). Indeed, there are different conceptions of what "dynamical effects" are and this has been the core of controversy in recent years. On one hand there are arguments based on transition state theory (TST). This is an equilibrium theory based on stationary properties, and therefore deviations from this theory are called dynamical effects. These deviations are different in the enzyme and in water, but their effect on catalysis is rather small (García-Meseguer et al. 2013; Kamerlin and Warshel 2010a,b; Kanaan et al. 2010; Olsson et al. 2006; Roca et al. 2010). On the other hand, many experimentalists agree that there is ample evidence that the dynamics of enzymes is a prerequisite for their function and that altering the dynamics affects their rates (Agarwal et al. 2002; Benkovic and Hammes-Schiffer 2003; Bhabha et al. 2011; Eisenmesser et al. 2002, 2005; Engelkamp et al. 2006; English et al. 2006; Hammes-Schiffer and Benkovic 2006; Min et al. 2005; Osborne et al. 2001; Rajagopalan et al. 2002). This, of course, is not in contradiction with an explanation based on TST, which would describe these effects of dynamics as nondynamical effects! As long as the chemical step remains the rate-limiting step, TST remains valid. But even when that step is not rate-limiting, TST could be applied if one can define the transition state for a protein motion. On the basis of the large number of computational studies that reproduced enzyme rate constants by only modeling the chemical step with hybrid quantum mechanics/molecular mechanics (QM/MM) methods, this step seems to be rate-limiting for several enzymes, but that also remains controversial (Benkovic and Hammes-Schiffer 2003; Bhabha et al. 2011; Cameron and Benkovic 1997; Eisenmesser et al. 2002, 2005; Hammes-Schiffer and Benkovic 2006; Wolf-Watz et al. 2004). Natural selection only puts evolutionary pressure to the rate-limiting step: once the chemical step is no longer the rate-limiting step, it will not decrease much further, and thus will remain relatively close the new rate-limiting step (e.g., large-scale enzyme motions). This turns the identification and modeling of the rate-limiting step into a computational challenge. We believe part of the seemingly contradictory facts in this field arise from this evolutionary conundrum.

Even if we accept the importance of motions along the catalytic cycle, we still need to understand their function. Here we suggest a simple explanation and link it to related proposals. If an enzyme moves within a configuration or alternates configurations, the energy barrier for different instantaneous conformations will vary. Is a fluctuating barrier advantageous?

Imagine a conformation of an enzyme with an energy barrier E_b, and imagine that fluctuations around this conformation produce a Gaussian distribution of energy barriers $p(E)$ with mean E_b and dispersion ΔE. The average rate will proportional to $p(E)\exp(-E/RT)$ (this is true provided that the rate-limiting step is not the fluctuation among configurations). The Boltzmann average or effective energy barrier $\langle E \rangle$ can be defined as the barrier that would give the same rate constant (Lodola et al. 2007), which is the average of all rates for the different energy barriers. This gives:

$$\langle k \rangle = C \int p(E)\exp\left(\frac{-E_b}{RT}\right)dE.$$

$$\langle E \rangle = -RT\log\frac{\langle k \rangle}{C} = -RT\log\int p(E)\exp\left(\frac{-E_b}{RT}\right)dE = E_b - \frac{\Delta E^2}{4RT}$$

where C is a proportionality constant, R is the gas constant, and T the temperature. The decrease in the barrier arises from lower barriers having exponentially larger contributions than higher ones. This effect is significant. For a barrier of 30 kcal/mol and a standard deviation of 3 kcal/mol, the resulting apparent barrier at 300K is 22 kcal/mol, a remarkable reduction. Figure 4.1 plots the results for 5000 different conformations.

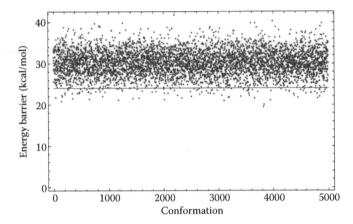

Figure 4.1 Plot of the energy barriers of 5000 conformations for a hypothetical system with an average energy barrier of 30 kcal/mol and a standard deviation of 3 kcal/mol. The solid line corresponds to the effective energy barrier that at 300K. Note that it is far below the mean energy barrier.

However, this argument is misleading because we are comparing a barrier E_b with fluctuations that can reduce it. But if the most stable configuration has the lowest barrier and any fluctuation increases it, the average barrier will always increase, and then the most proficient enzyme will be the most rigid with fewest fluctuations (Kamerlin and Warshel 2010a,b; Olsson et al. 2006; Pisliakov et al. 2009; Warshel et al. 2006).

Flexibility plays still another role. Enzymes not only catalyze chemical reactions; they also need to bind reactants and release products. For that purpose, they need an active site open to the solvent. The more solvent-accessible the active site, the less protein surface the enzyme the more solvent-accessible the active site, the less protein surface is available to stabilize the transition state. If it closes around the substrates, their release will be slower. By fluctuating between different conformations, the enzyme is able to visit a conformation where the substrates can get in and out. This conformation is, presumably, less suited to catalyze the reaction. Then it can fluctuate to another conformation more proficient for catalysis. This is even more relevant for multistep reactions—less computationally studied but ubiquitous in nature—where the enzyme needs to stabilize chemically different transition state structures.

An inspiring work by Xie and coworkers (Min et al. 2008) proposed a two-dimensional method to describe enzyme catalysis. One dimension corresponds to the reaction coordinate and the other to the collective motions of the enzyme. This model gives the mathematical description that we have tried to convey. Figure 4.2 represents its extension to a multistep reaction. As a model, it fits several experimental data, but it remains a challenge to calculate the actual shape of the free energy for a real enzyme along the conformational coordinate. The model shows that a fixed active site in the right conformation cannot be catalytically as efficient as a moving enzyme. The right conformation does not exist, however, because the best conformation for catalysis may be too occluded to let the substrates enter or leave the active site, as happens in the *N*-acetyl-glutamate (NAG) kinase described in the following, or the need to catalyse different chemical steps in a single active site may need different interactions to stabilize different transition states, as in the aldolases we also describe. A flexible active site may be able to achieve this with more proficiency than a rigid one. The examples of NAG kinase and aldolases presented in the following nicely illustrate the diversity of dynamic events that enable a proficient execution of the enzymatic function and allosteric regulation. We will show particular examples of how these dynamic features are indeed encoded in the protein architecture, thus bridging the gap between structure and function.

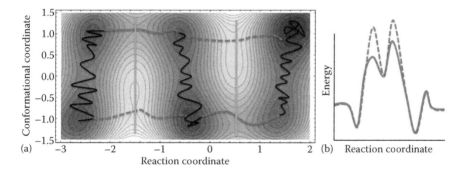

Figure 4.2   (a) A model surface depicting the combination of a reaction coordinate for the chemical step and a conformational coordinate. The enzyme needs to adapt to different conformations to optimally catalyze the reaction (gray solid lines), otherwise the energy barriers are higher (gray dashed lines). If the sampling of the chemical transition state (vertical shaded lines) is sufficient we will capture the necessary conformational reorganization (black wiggled lines). (b) A one-dimensional energy profile of the same model, where only the chemical step is considered. In such a case, the energy barrier seems to change with time as the enzyme samples different conformations. Without considering this sampling the enzyme would have to surmount, in one of the two steps, a higher-energy barrier (dashed curve).

COMPUTATIONAL APPROACHES TO DESCRIBE SLOW CONFORMATIONAL DYNAMICS

From the computational perspective, molecular dynamics (MD) simulations routinely allow the description of fast dynamic events and only in some exceptional studies of small proteins have they been used for describing slow dynamics and even folding processes (Dror et al. 2012; Lane et al. 2013). But when it comes to studying large-amplitude motions as featured in oligomeric assemblies, the large conformational space to be sampled makes MD simulations unaffordable, so that more approximate and less expensive methods become necessary (Zhuravlev and Papoian 2010). Of increasing importance are coarse-grained models, which vastly reduce the number of degrees of freedom and interaction sites by replacing sets of atoms by beads losing atomic detail. In particular, elastic network models (ENMs) use a coarse-grained representation of the protein that in combination with normal mode analysis (NMA) allow the global motion of the protein to be decomposed into a set of modes of motion that give insight into conformational changes of functional relevance at very low computational expense. In this section we will illustrate how ENMs find wide use in describing a variety of functional slow motions in oligomeric proteins.

ENM

ENMs represent the protein as a network, in which C-alpha atoms of amino acid residues define N nodes, and each pair of nodes (within a cutoff distance) interacts via a harmonic potential. Different types of ENMs have been developed, mostly differing in the definition of the force constant and the degree of coarse-graining, but all provide a consistent description of large-amplitude motions in proteins (Kondrashov et al. 2007). Among them, here we will make reference to the anisotropic network model (ANM) by Bahar and coworkers (Atilgan et al. 2001) and the similar C-alpha force field developed by Hinsen et al. (2000) and implemented in the molecular modelling toolkit (MMTK) (Hinsen 2000). A NMA is carried out from the $3N \times 3N$ Hessian matrix $\mathbf{H}$ of the ENM potential. The modes of motion and its frequencies are extracted thus from diagonalization of $\mathbf{H}$ yielding $3N-6$ nonzero modes. A wealth of studies showed that the low-frequency modes (large-amplitude motions) obtained under this simplification fit nicely with the directionality of functional conformational changes observed experimentally (Bahar et al. 2010; Bakan and Bahar 2009; Cui and Bahar 2006; Tobi and Bahar 2005). This is consistent with the notion that large-amplitude motions depend more on the fold and shape of the protein than on atomic details (Tama and Brooks 2006). In this regard, ENM is well suited for studying large protein structures.

With the normal modes of motion in hand, one can extract useful information with a set of algebraic operations. For instance, to ascertain which normal modes are more relevant to protein function, one can determine the overlap between each mode and an experimentally observed conformational change (e.g., deformation between apo and holo protein states). Moreover, the degree of similarity between the conformational spaces accessible by two systems can be quantified by the overlap between the subspaces described by a given subset of modes (subspace overlap). To identify rigid-body motions of parts of the system, the use of distance variation maps is particularly helpful (Marcos et al. 2011). In addition, to understand how the dynamics of a protein region is affected by the rest of the system, the use of an effective Hessian for the subsystem of interest (Zheng and Brooks 2005) in the NMA is very useful in a variety of applications (e.g., oligomerization effects in the dynamics of subunits or comparing the dynamics of proteins with different sequences but similar in structure). A variety of computational studies have used these approaches to analyze the information contained in normal modes for different purposes, but here we will focus on how these were applied for analyzing functional motions in the amino acid kinase family and dihydroxyacetone phosphate (DHAP)-dependent aldolases.

Slow Dynamic Fingerprints of the Amino Acid Kinase Family: NAGK Dynamic Paradigm

The amino acid kinase (AAK) family includes a series of enzymes similar in terms of sequence and fold that undergo large conformational changes for substrate binding and allosteric regulation (Figure 4.3). AAK members present different oligomeric states, so that this family represents a suitable case for exploring the role of oligomeric architecture in determining functional motions. Among them, N-acetyl-glutamate kinase from *Escherichia coli* (*E. coli*) (EcNAGK) has been the most exhaustively studied family member by crystallography and is regarded as the structural paradigm of this family (Fernández-Murga and Rubio 2008; Gil-Ortiz et al. 2003, 2010; Ramón-Maiques et al. 2002a,b, 2006). In this section, we first show to which extent the dynamic patterns found in EcNAGK are shared by other family members and then show how these dynamic features are indeed modulated by the oligomeric assembly.

EcNAGK is a homodimer that uses adenosine triphosphate (ATP) to phosphorylate N-acetyl-glutamate in the metabolic route of arginine biosynthesis. It undergoes a large conformational change opening/closing the active site that allows substrate binding/release (Gil-Ortiz et al. 2003, 2010; Ramón-Maiques et al. 2002b). It involves the C-domain, where ATP is bound, and the hairpin of the N-domain, which behaves as a lid of the NAG binding site in a hinge-bending motion. Our recent study (Marcos et al. 2010) based on the ANM model showed that the lowest frequency modes of motion indeed describe the conformational change between the apo (open) and holo (closed) forms of the enzyme very well, pointing to the importance of the overall protein fold in determining the

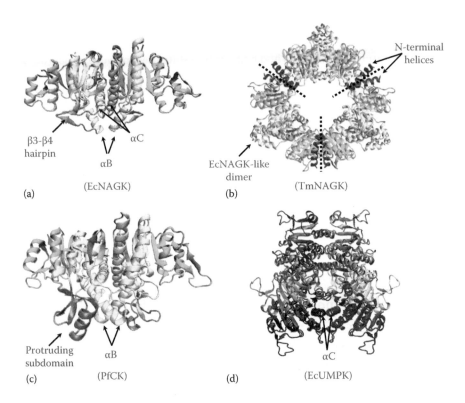

N-terminal
helices

β3-β4
hairpin   αC

αB

(EcNAGK)   EcNAGK-like
dimer
(a)   (TmNAGK)
(b)

Protruding
subdomain   αB   αC

(c)   (PfCK)   (d)   (EcUMPK)

Figure 4.3 AAK family members studied. (a) Dimeric NAGK from *E. coli*, (b) NAGK from *Thermotoga maritima* (TmNAGK), (c) carbamate kinase from *Pyrococcus furiosus* (PfCK), (d) UMPK from *E. coli*. Panels a, b, and c color the ATP-binding domain in green and the aminoacid-binding domain in orange. These are the two most flexible domains in these topologies. Critical secondary structure elements for the dynamics interface are denoted with arrows. Note in (d) the different assembly of the subunits with respect to other family members. Here, colors indicate the different monomeric chains.

intrinsic motions that enable both substrates enter the active site. Additionally, molecular dynamics simulations also captured this conformational transition from the closed to the open state in tenths of nanoseconds, pointing to the feasibility of this protein motion as well as the consistency of the two computational approaches (Sanchez-Martinez et al. 2013).

Therefore, if the monomeric fold along with the assembly architecture determines the slow motions of EcNAGK—the family paradigm—the rest of the family members should exhibit similar dynamic features. To assess the similarity of the slow motions between different family members with different sequences it is very convenient to use an effective Hessian of the structurally aligned

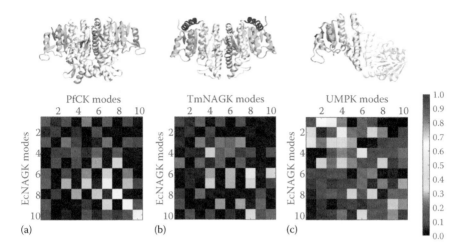

Figure 4.4   Comparison of lowest frequency modes among AAK family members. Overlap of the 10 lowest frequency modes of the EcNAGK with (a) PfCK, (b) TmNAGK, and (c) UMPK. Normal modes used for comparison were computed from the effective Hessian, where subsystems are the structurally aligned residues of a given protein pair.

residues of a given pair of proteins (Zheng and Brooks 2005). The structurally aligned residues define the *subsystem* of interest and the rest of the protein defines the *environment*. Then the Hessian is partitioned in terms of the subsystem and the environment allowing the construction of an effective Hessian. An NMA on this effective Hessian provides a set of modes that are now directly comparable. One can calculate the cumulative overlap between subsets of these modes for a given protein pair, which quantifies the degree of similarity of the conformational spaces accessible to the two proteins. It is instructive to consider that this approach is readily applicable to other families of proteins and indeed permits us to identify dynamic fingerprints shared among family members. For the AAK family, all protein members were found to exhibit well-defined dynamic patterns that are encoded in their shared architecture pointing to similar mechanisms of function (see Figure 4.4).

Catalytic Proficiency of Different Conformations

The motions described so far allow the opening of the active site for substrate entrance and product release. Do they also affect the chemical step of the catalytic cycle—the phosphoryl transfer from ATP to NAG?

We performed MD simulations of the reactants from four different crystal structures of EcNAGK complexed with different substrate and transition state analogs representing different stages of the catalytic cycle. We selected five snapshots of each trajectory and calculated the energy profile for each of these 20 snapshots with the nudged elastic band method (Henkelman and Jónsson 2000; Jónsson et al. 1998) in combination with a density functional theory (DFT) QM/MM potential. We found a remarkable heterogeneity in the computed reaction profiles even among those from the same crystal structure, but all shared a common associative and exothermic mechanism. We characterized the conformational diversity of the 20 snapshots by means of a statistical analysis and found that different geometrical parameters play a primary role in the relative stabilization between reactants and transition states and thus in the energetic barrier for the phosphoryl (Sanchez-Martinez et al. 2013).

The wide range of energy barriers could arise from different flexible reactant conformations having to reach a rigid preorganized transition state structure. This is not what the energy profiles tell us. The O-O distance presents a standard deviation (σ), over PDB structures 1OHA and 1OH9 structures of 0.16 Å at the reactants and 0.20 Å at the transition state, while for the O-P-O angle the trend reverses and σ is 4.69° for the reactants and 1.94° for the transition state. Each reactant conformation proceeds along its reaction valley with a different energy barrier.

Conformational Compression in the Chemical Step

Rubio and coworkers noticed that the crystal structures corresponding to transition state analogs had shorter substrate distances than those corresponding to reactants (Gil-Ortiz et al. 2003). From this geometrical observation they inferred that the compression of substrates by the enzyme pushes the reaction forward. Our energy profiles for different crystal structures and different trajectory snapshots for each structure provided us with data to check if the conformational compression hypothesis was tenable. Among all crystal structures considered, we found that two of them (1OHA and 1OH9) were particularly useful to address this hypothesis for reasons described in (Sanchez-Martinez et al. 2013).

The variety of energy profiles (see Figure 4.5) indicate that the energy barrier is not determined by how much the distance (between nucleophile and leaving group, i.e., O-O distance) changes along the reaction process; that is, it does not depend on how much compression is needed from reactants to transition state. Instead, there is a striking correlation between the substrate distance in the reactants and the energy barrier. The lower the distance between nucleophile and leaving group (O-O distance), the lower the energy barrier. In other words,

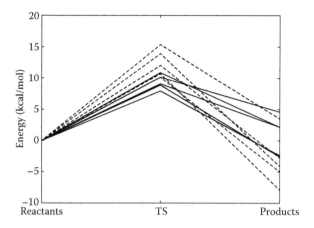

Figure 4.5   Reactants, transition state, and product energies of the five 1OH9 (dashed) and 1OHA (solid) structures. The dispersion of energy values is large even for snapshots coming from the same crystal structure.

the more compressed the active site is, the more reactive it is. If there was a rigid transition state geometry that different reactants conformations had to reach, the difference in O-O distance between reactants and the transition state would determine the barrier because the barrier would depend on how far they are from the transition state. However, the statistical analysis reveals that this difference does not determine the barrier and that each reactant conformation proceeds through its reactant valley with a transition state that will be more unstable (determining a higher barrier) when the reactants are far apart.

Additionally, a linear angle of the transferring phosphoryl with the nucleophile and leaving groups was also found to lower the energy barrier. With only two parameters, the O-O distance and the O-P-O angle in reactants, the energy barrier can be predicted with accuracy for a wide range of values spanning the 10 snapshots (see Figure 4.6).

Overall, the structure of the prereactive complex contains relevant predictive information on the energy barrier. In essence, our results indicate that the catalytic proficiency of the enzyme lies in collective motions accessing properly oriented and highly compressed active site conformations, thus supporting the conformational compression hypothesis inferred from x-ray crystallography. If the energy barrier is so sensitive on the reactants conformation, it might be that the chemical step is not rate-limiting, and instead, these motions leading to catalytic compressed conformations turn out to be the actual bottleneck.

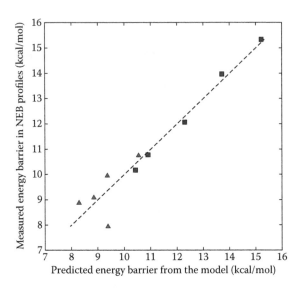

Figure 4.6   A linear model with two dependent variables, the O-O distance and the O-P-O angle in reactants can predict the energy barrier obtained in the energy profiles of the 10 snapshots coming from 1OHA (triangles) and 1OH9 (squares) crystal structures. The Pearson coefficient $r^2 = 0.93$ indicates the fit goodness.

Is Enzyme Dynamics the Rate-Limiting Step?

The calculated energy barrier for the chemical step for all our conformations is significantly lower than the apparent energy barrier. The measured energy barrier is obtained from an application of transition state theory, $\Delta E = 15.8$ kcal/mol and it corresponds to a free energy that implicitly incorporates dynamical and tunneling corrections to the transition state. These corrections, however, depend logarithmically on the energy barrier, and thus their effect is significantly smaller. It is also true that our calculations contain several approximations—from the choice of the QM region and its treatment to the limited sampling of conformations—that introduce uncertainties in the calculation of the energy barrier. However, the experimental energy barrier almost doubles the calculated energy barrier for the chemical step (15.8 vs. 9.3 kcal/mol). This reinforces the suggestion that the chemical step is not the rate-limiting step for this enzyme and that conformational motions associated with the lid opening and closing can be slower than the chemical reaction. This hypothesis has also been put forward for other enzymes, such as adenylate kinase (Wolf-Watz et al. 2004), cyclophilin A (Eisenmesser et al. 2005), or dihydrofolate reductase (Bhabha et al. 2011). In line with what was stated in the Introduction, the rate-limiting step is likely to be dictated by motions leading to active site conformations compressed to the extent that the resultant chemical barrier is close to the conformational barrier.

ROLE OF OLIGOMERIC ASSEMBLY IN FUNCTIONAL MOTIONS

Our previous NMA across the AAK family points to the importance of large collective motions in the monomeric subunits for substrate binding. But to what extent does the overall architecture of the oligomeric assembly determine these subunit collective motions? Is the intrinsic dynamics of the subunit modified or conserved by the interface? Does the assembly structure play an active or passive role in building the slow dynamics of the subunits relevant for binding? We have addressed these questions in three different AAK family members by using a similar approach to that shown above for comparing low-frequency modes of different proteins (i.e., the effective Hessian of the ENM). Here, we aim to compare the dynamics of a component subunit isolated from the rest of the oligomer with the subunit in the context of the assembly. Thus, for the calculation of the effective Hessian, the subunit is the subsystem and the rest of the oligomer is the environment. For instance, in the case of a hexamer, not only monomers but dimers can be considered component subunits of the oligomer and thus this can provide clues for understanding the role of different interfaces in the overall dynamics of the assembly.

This approach proved to be useful to analyze how the oligomerization confers new cooperative modes of motion that exploit the intrinsic dynamics of the subunits (Marcos et al. 2011; Perica et al. 2012). By comparing the low-frequency modes of the component subunits of EcNAGK, carbamate kinase, and hexameric NAGK with those of the oligomer, we found that large-amplitude motions intrinsically accessible by the subunits are preserved in the oligomeric state in a high extent. This indicates that oligomeric interfaces stabilize monomeric folds with intrinsic motions suitable for function. Beyond stabilizing the fold of monomeric subunits, the oligomeric assembly plays a second role in determining functional dynamics. It provides relative rigid body motions between subunits with functional significance. Now we examine two different examples of the dynamic role of the interface.

Hexameric NAGK

Acquired cooperative modes of motion in the assembly are found to be associated to the allosteric regulation of hexameric NAGK. The hexamer is regarded as a trimer of EcNAGK-like dimers and in contrast to the dimeric form, it is allosterically inhibited by arginine (Ramón-Maiques et al. 2006). What is interesting from the dynamics perspective is that inhibitor binding at the interface between EcNAGK-like dimers (The interface between chains A and F as denoted in Figure 4.7a) triggers a conformational change of the assembly that changes the size of the hexameric ring and opens the active sites in each subunit disabling substrate binding. Indeed, it involves rigid-body motions of the dimeric subunits. To elucidate the role of interfaces in this functional conformational change it is very convenient to contrast the dynamics of different component subunits. We calculated the subspace overlap of the 20 lowest frequency modes of the structural component of interest in the presence and absence of the rest of the hexamer (Figure 4.7b and c). The stronger correlation between both subspaces found for the newly formed interface (The interface between chains A and F with respect to the dimer interface between chains A and B, as shown in Figure 4.7) indicates that the intrinsic dynamics of the isolated AF dimer is more preserved in the hexamer pointing to the primary role

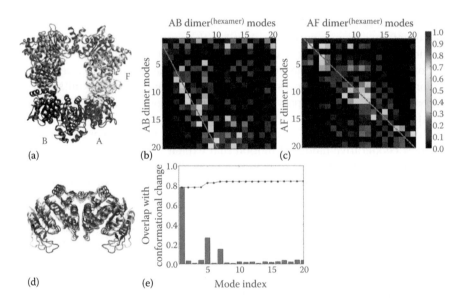

(a)   (b)   (c)
(d)   (e)   Mode index

Figure 4.7 NMA of hexameric TmNAGK and EcUMPK. (a) Representation of TmNAGK, which has two types of interfaces indicated by AB and AF dimers. (b, c) Overlap between the 20 lowest modes of the AB and AF dimers, respectively, in the presence and absence of the rest of hexamer. (d) Representation of different conformations of EcUMPK dimer component along the lowest-frequency mode. (e) Comparison of EcUMPK modes with the allosteric conformational change. Note the striking similarity between the motion shown in (d) and the allosteric structural change.

of this interface in providing the hexamer with new slow dynamic properties. It is indeed the AF interface, and not the AB interface, that enables rigid-body motions in the hexamer that are linked to the allosteric regulation.

Hexameric UMP Kinase

Another case of interest was that of hexameric uridine monophosphate kinase (UMPK) (Briozzo et al. 2005; Meyer et al. 2008), which exhibits an assembly of the dimeric subunits that is strikingly different from that present in the rest of the AAK family. By studying the low-frequency modes of the dimeric subunit we found that the unique interface displayed by UMPK (the disposition of two interface helices is not found in other AAK members) allows rigid-body motions of the monomeric subunits that are not allowed by the EcNAGK-like dimeric architecture (Figure 4.7d). Interestingly, crystallographic structures of UMPK (Briozzo et al. 2005; Meyer et al. 2008) with and without the allosteric effector guanosine triphosphate (GTP) showed that these rigid-body motions are necessary for allosteric regulation and are well predicted by the lowest frequency modes (Figure 4.7e). Overall, the examples of hexameric NAGK and UMPK nicely illustrate how different interface architectures mediate different dynamic mechanisms of function.

MOTIONS IN RHAMNULOSE-1-PHOSPHATE AND FUCULOSE-1-PHOSPHATE ALDOLASES

In this second example we will show how the combination of protein normal modes with a coarse-grained method and high-level DFT calculations reveal the role of motions in the rhamnulose-1-phosphate aldolase (RhuA) and fuculose-1-phosphate aldolase (FucA).

RhuA and FucA catalyze the aldol addition of dihydroxyacetone phosphate (DHAP) to lactaldehyde (LA) to give rhamnulose-1-phosphate (Kroemer et al. 2003) and FucA catalyses the stereocomplementary reaction (Dreyer and Schulz 1996a,b). Both enzymes are homotetramers, where each monomer is composed of the N- and C-domains. Close to the domain interface lies the active site, where a zinc ion binds DHAP. The main role of this ion is to increase the acidity of DHAP, which contributes to the catalysis of the aldol addition.

Studies based on anisotropic temperature factors by Schulz and coworkers (Grueninger and Schulz 2008; Kroemer et al. 2003) showed a rotational displacement of the N-domain with respect to the C-domain. Interestingly, mutations that altered motions between these domains—but not the active site—were found to decrease the enzymatic activity. In addition, an apparently unrelated observation is that the DHAP substrate in RhuA is bidentate (i.e., binding the zinc ion through two oxygens) whereas it is a monodentate ligand in FucA (Dreyer and Schulz 1996b; Kroemer et al. 2003).

Normal mode calculations of the RhuA monomer showed that the lowest modes significantly affected the shape of the active site (Jimenez et al. 2008). Applying these distortions to a small model of the active site resulted in a change of coordination number of the zinc ion (Jimenez et al. 2008). That prompted us to relate the observed rotational displacement of the RhuA domains with the coordination of the metal in the active site and its different coordination number in FucA and RhuA crystals.

A more detailed analysis of the largest amplitude normal modes of the RhuA tetramer showed that these modes represent rotational motions of the N-domain with respect to the C-domain. The agreement with the conclusions from the temperature factors analysis (Grueninger and Schulz 2008) reinforced the idea that these motions must contribute to the catalytic turnover of the enzyme.

To test this hypothesis, we then built a larger model that included 11 residues of the active site, the metal ion, DHAP, and a water molecule that bind the zinc ion in the absence of the other substrate, LA (Figure 4.8). First, we wanted to understand the different coordination number of the active site metal in FucA and RhuA. It could be that the different chemical nature of the residues caused the change in coordination number (Dudev and Lim 2000, 2003; Rulisek and Vondrasek 1998). However, when we modified the geometry of the RhuA active site so that its residues had the same geometry as the FucA residues, the zinc metal released the water molecule. Not surprisingly, the FucA active site could then bind a water molecule when distorted to adopt the RhuA geometry. Consequently, it is their different conformations, and not the nature of the residues, that determines the metal coordination. Metals having coordinations different from their most stable are called entactic. Entactic metals are found in several active sites (Williams 1995) and serve to modulate the intrinsic metal chemistry. Most entactic metals are buried in rigid regions of the proteins, but in FucA and RhuA the active site is located at the interface between the two domains in a highly deformable region.

The domain motions of these aldolases therefore cause conformational changes in the active site that in turn change the metal coordination. Because the protein is flexible for these types of deformations, this would explain why FucA has a pentacoordinated zinc (Zn) ion whereas in RhuA it is hexacoordinated: the crystals have captured different conformational states of the natural motion of these enzymes, but in principle both enzymes can visit conformations with different coordination preferences. This also explains why these enzymes chose Zn as the active site metal: zinc is a very labile metal, adopting coordination numbers from 4 to 6 depending on the environment (Dudev et al. 2006; Rulisek and Vondrasek 1998). Other metals, such as magnesium—which

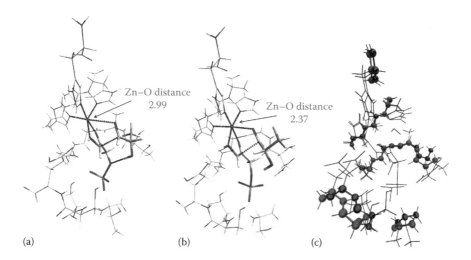

Figure 4.8   Models of the active site for FucA (a) and RhuA (b) depicting the substrates DHAP and LA in liquorice and the active site residues in lines. The geometry corresponds to the optimized transition state for each model where the difference in the Zn–O distance determines their different energy. Panel c shows both models superimposed (RhuA in red and FucA in blue). The atoms depicted in balls were displaced from the red positions to the blue ones to generate the deformed RhuA model with the FucA geometry and vice versa. The model was treated with a two-layer ONIOM methodology. (From Maseras, F., and Morokuma, K., *J. Comput. Chem.* 16 (9):1170–1179, 1995.) This image was generated with VMD. (From Humphrey, W. et al., *J. Mol. Graph.* 14 (1):33–38, 1996.)

would give a similar acidic character to DHAP—are much more comfortable with a constant hexacoordination.

But how is all this related to catalysis? The catalytic cycle involves the following steps: (1) deprotonation of DHAP, (2) C–C bond formation with LA, and (3) reprotonation of the products. We focused on the high-energy region of the deprotonated intermediate and the new C–C bond formation. We located the structures of the reactants, products, and transition state for this step in the active site models of RhuA and FucA. The energy barriers in RhuA and FucA are 1.0 and 7.2 kcal/mol, respectively. The source of this difference is that the aldehyde group, which bears a nascent negative charge at the transition state and can bind the Zn ion in RhuA but not in FucA because of the coordination preferences of the ion. To check this we can use the deformed models previously explained. Indeed, the RhuA model deformed in the FucA geometry increases its energy barrier to 4 kcal/mol, and the FucA model with the RhuA geometry reduces its barrier to 2.6 kcal/mol. Why does Zn coordination affect the barrier (i.e., affect differentially the stability of the reactants and the transition state)? In all four models (crystal and deformed conformations) Zn is pentacoordinated in the reactants and hexacoordinated in the products, but they differ in the transition state. The Zn–O distances are short in the RhuA geometries but much longer in the FucA geometries (Figure 4.8). This short distance stabilizes the negative charge at the transition state and thus reduces the barrier. Experiments in other enzymes have shown that Zn changes its coordination number along the catalytic cycle (Kleifeld et al. 2000, 2003, 2004).

If there are such differences between FucA and RhuA, how do they have such similar rate constants k_{cat} of 19.3 (Joerger et al. 2000) and 9.1 s^{-1} (Grueninger and Schulz 2008; Kroemer et al. 2003), respectively? To explain this, it is worth remembering the domain motions shared by these enzymes. The FucA and RhuA models that we used indeed correspond to two frozen conformation of these enzymes, but their intrinsic dynamics allow both enzymes to visit these two conformations. Therefore, both should have similar rate constants in accord with the experimental observation. The linkage of these motions to the catalytic step also explains why mutations that perturb this motion affect the rate constant (Grueninger and Schulz 2008).

Despite analyzing the step corresponding to the reactive intermediate, it could be that this is not the rate-limiting step for the catalytic cycle. However, because domain motions affect the geometry of the active site, they probably affect all the steps of the reaction, thus reassuring the role of these motions for catalysis.

CONCLUDING REMARKS

Certainly our results are qualitative. Present-day QM/MM methods are able to give an accurate energy barrier for the chemical step. Calculations of free energies are also possible—at a lower computational level—but we still lack quantitative methods to evaluate the rate of the slow large-amplitude motions. ENMs are only good at describing directions of motion, not their rate. As we have seen in the lid opening of NAGK, this rate can vary a lot depending on the presence of ligands or substrate that interact in specific ways beyond the description of coarse-grained methods. Some novel methods can give the free-energy barrier along these large-scale motions (Batista et al. 2011) and integrating this information in the chemical step is the next goal toward a more realistic modeling of enzyme catalysis. However, motions along these coordinates—with low barriers and large amplitudes—will have an important diffusive character (Hayward et al. 1993; Hinsen et al. 2000; Ma 2005), so that the friction along these modes will also be a necessary ingredient in our quantitative simulation of the complete turnover of enzymes.

ACKNOWLEDGMENTS

We acknowledge financial support from the Ministerio de Innovación y Competitividad (CTQ2009-08223 and CTQ2012-33324) and the Generalitat de Catalunya (2009SGR01472). M.S-M. thanks the Ministerio de Economia y Competitividad for a predoctoral fellowship. E.M. acknowledges a postdoctoral fellowship from the European Union (FP7-PEOPLE-2011-IOF 298976).

REFERENCES

Agarwal, P. K. 2005. Role of protein dynamics in reaction rate enhancement by enzymes. *J. Am. Chem. Soc.* 127 (43):15248–15256.

Agarwal, P. K. 2006. Enzymes: An integrated view of structure, dynamics and function. *Microb. Cell Fact.* 5 (1):2.

Agarwal, P. K., Billeter, S. R., Rajagopalan, P. T. R., Benkovic, S. J., and Hammes-Schiffer, S. 2002. Network of coupled promoting motions in enzyme catalysis. *Proc. Natl. Acad. Sci. U. S. A.* 99 (5):2794–2799.

Agarwal, P. K., Geist, A., and Gorin, A. 2004. Protein dynamics and enzymatic catalysis: Investigating the peptidyl-prolyl cis-trans isomerization activity of cyclophilin A. *Biochemistry* 43:10605–10618.

Atilgan, A. R., Durell, S. R., Jernigan, R. L., Demirel, M. C., Keskin, O., and Bahar, I. 2001. Anisotropy of fluctuation dynamics of proteins with an elastic network model. *Biophys. J.* 80 (1):505–515.

Bahar, I., Lezon, T. R., Yang, L.-W., and Eyal, E. 2010. Global dynamics of proteins: Bridging between structure and function. *Annu. Rev. Biophys.* 39:23–42.

Bakan, A., and Bahar, I. 2009. The intrinsic dynamics of enzymes plays a dominant role in determining the structural changes induced upon inhibitor binding. *Proc. Natl. Acad. Sci. U. S. A.* 106 (34):14349–14354.

Batista, P. R., Pandey, G., Pascutti, P. G., Bisch, P. M., Perahia, D., and Robert, C. H. 2011. Free energy profiles along consensus normal modes provide insight into HIV-1 protease flap opening. *J. Chem. Theory Comput.* 7 (8):2348–2352.

Benkovic, S. J., and Hammes-Schiffer, S. 2003. A perspective on enzyme catalysis. *Science* 301:1196–1202.

Bhabha, G., Lee, J., Ekiert, D. C., Gam, J., Wilson, I. A., Dyson, H. J., Benkovic, S. J., and Wright, P. E. 2011. A dynamic knockout reveals that conformational fluctuations influence the chemical step of enzyme catalysis. *Science* 332:234–238.

Briozzo, P., Evrin, C., Meyer, P., Assairi, L., Joly, N., Barzu, O., and Gilles, A. M. 2005. Structure of *Escherichia coli* UMP kinase differs from that of other nucleoside monophosphate kinases and sheds new light on enzyme regulation. *J. Biol. Chem.* 280 (27):25533–25540.

Cameron, C. E., and Benkovic, S. J. 1997. Evidence for a functional role of the dynamics of glycine-121 of *Escherichia coli* dihydrofolate reductase obtained from kinetic analysis of a site-directed mutant. *Biochemistry* 36:15792–15800.

Cui, Q., and Bahar, I. (eds.). 2006. *Normal Mode Analysis. Theory and Applications to Biological and Chemical Systems*, Mathematical and Computational Biology Series. Boca Raton, FL: Chapman & Hall/CRC.

Dellus-Gur, E., Toth-Petroczy, A., Elias, M., and Tawfik, D. S. 2013. What makes a protein fold amenable to functional innovation? Fold polarity and stability trade-offs. *J. Mol. Biol.* 425 (14):2609–2621.

Dreyer, M. K., and Schulz, G. E. 1996a. Catalytic mechanism of the metal-dependent fuculose aldolase from *Escherichia coli* as derived from the structure. *J. Mol. Biol.* 259 (3):458–466.

Dreyer, M. K., and Schulz, G. E. 1996b. Refined high-resolution structure of the metal-ion dependent L-fuculose-1-phosphate aldolase (class II) from *Escherichia coli. Acta Crystallogr. D Biol. Crystallogr.* 52 (Pt. 6):1082–1091.

Dror, R. O., Dirks, R. M., Grossman, J. P., Xu, H., and Shaw, D. E. 2012. Biomolecular simulation: A computational microscope for molecular biology. *Annu. Rev. Biophys.* 41:429–452.

Dudev, M., Wang, J., Dudev, T., and Lim, C. 2006. Factors governing the metal coordination number in metal complexes from Cambridge Structural Database analyses. *J. Phys. Chem. B* 110 (4):1889–1895.

Dudev, T., and Lim, C. 2000. Metal binding in proteins: The effect of the dielectric medium. *J. Phys. Chem. B* 104 (15):3692–3694.

Dudev, T., and Lim, C. 2003. Principles governing Mg, Ca, and Zn binding and selectivity in proteins. *Chem. Rev.* 103:773–787.

Eisenmesser, E. Z., Bosco, D. A., Akke, M., and Kern, D. 2002. Enzyme dynamics during catalysis. *Science* 295 (5559):1520–1523.

Eisenmesser, E. Z., Millet, O., Labeikovsky, W., Korzhnev, D. M., Wolf-Watz, M., Bosco, D. A., Skalicky, J. J., Kay, L. E., and Kern, D. 2005. Intrinsic dynamics of an enzyme underlies catalysis. *Nature* 438 (7064):117–121.

Engelkamp, H., Hatzakis, N. S., Hofkens, J., De Schryver, F. C., Nolte, R. J., and Rowan, A. E. 2006. Do enzymes sleep and work? *Chem. Commun.* (9):935–940.

English, B. P., Min, W., van Oijen, A. M., Lee, K. T., Luo, G., Sun, H., Cherayil, B. J., Kou, S. C., and Xie, X. S. 2006. Ever-fluctuating single enzyme molecules: Michaelis-Menten equation revisited. *Nat. Chem. Biol.* 2 (2):87–94.

Fernández-Murga, M. L., and Rubio, V. 2008. Basis of arginine sensitivity of microbial N-acetyl-l-glutamate kinases: Mutagenesis and protein engineering study with the *Pseudomonas aeruginosa* and *Escherichia coli* enzymes. *J. Bacteriol.* 190 (8):3018–3025.

García-Meseguer, R., Martí, S., Ruiz-Pernía, J. J., Moliner, V., and Tuñón, I. 2013. Studying the role of protein dynamics in an SN2 enzyme reaction using free-energy surfaces and solvent coordinates. *Nat. Chem.* 5 (7):566–571.

Gil-Ortiz, F., Ramón-Maiques, S., Fernández-Murga, M. L., Fita, I., and Rubio, V. 2010. Two crystal structures of *Escherichia coli* N-acetyl-L-glutamate kinase demonstrate the cycling between open and closed conformations. *J. Mol. Biol.* 399:476–490.

Gil-Ortiz, F., Ramón-Maiques, S., Fita, I., and Rubio, V. 2003. The course of phosphorus in the reaction of N-acetyl-l-glutamate kinase, determined from the structures of crystalline complexes, including a complex with an AlF4–transition state mimic. *J. Mol. Biol.* 331:231–244.

Glowacki, D. R., Harvey, J. N., and Mulholland, A. J. 2012. Taking Ockham's razor to enzyme dynamics and catalysis. *Nat. Chem.* 4 (3):169–176.

Grueninger, D., and Schulz, G. E. 2008. Antenna domain mobility and enzymatic reaction of L-rhamnulose-1-phosphate aldolase. *Biochemistry* 47 (2):607–614.

Hammes-Schiffer, S., and Benkovic, S. J. 2006. Relating protein motion to catalysis. *Annu. Rev. Biochem.* 75 (1):519–541.

Hayward, S., Kitao, A., Hirata, F., and Gō, N. 1993. Effect of solvent on collective motions in globular protein. *J. Mol. Biol.* 234 (4):1207–1217.

Henkelman, G., and Jónsson, H. 2000. Improved tangent estimate in the nudged elastic band method for finding minimum energy paths and saddle points. *J. Chem. Phys.* 113 (22):9978–9985.

Henzler-Wildman, K. A., Lei, M., Thai, V., Kerns, S. J., Karplus, M., and Kern, D. 2007a. A hierarchy of timescales in protein dynamics is linked to enzyme catalysis. *Nature* 450 (7171):913–916.

Henzler-Wildman, K. A., Thai, V., Lei, M., Ott, M., Wolf-Watz, M., Fenn, T., Pozharski, E., Wilson, M. A., Petsko, G. A., Karplus, M., Hubner, C. G., and Kern, D. 2007b. Intrinsic motions along an enzymatic reaction trajectory. *Nature* 450 (7171): 838–844.

Hinsen, K. 2000. The molecular modeling toolkit: A new approach to molecular simulations. *J. Comput. Chem.* 21 (2):79–85.

Hinsen, K., Petrescu, A. J., Dellerue, S., Bellissent-Funel, M. C., and Kneller, G. R. 2000. Harmonicity in slow protein dynamics. *Chem. Phys.* 261 (1–2):25–37.

Humphrey, W., Dalke, A., and Schulten, K. 1996. VMD: Visual molecular dynamics. *J. Mol. Graph.* 14 (1):33–38.

Jimenez, A., Clapes, P., and Crehuet, R. 2008. A dynamic view of enzyme catalysis. *J. Mol. Model.* 14 (8):735–746.

Joerger, A. C., Gosse, C., Fessner, W. D., and Schulz, G. E. 2000. Catalytic action of fuculose 1-phosphate aldolase (class II) as derived from structure-directed mutagenesis. *Biochemistry* 39 (20):6033–6041.

Jónsson, H., Mills, G., and Jacobsen, K. W. 1998. Nudged elastic band method for finding minimum energy paths of transitions, in *Classical and Quantum Dynamics in Condensed Phase Simulations*, B. J. Berne, G. Ciccotti, and D. F. Coker (eds.). Singapore: World Scientific.

Kamerlin, S. C. L., and Warshel, A. 2010a. At the dawn of the 21st century: Is dynamics the missing link for understanding enzyme catalysis? *Proteins* 78 (6):1339–1375.

Kamerlin, S. C. L., and Warshel, A. 2010b. Reply to Karplus: Conformational dynamics have no role in the chemical step. *Proc. Natl. Acad. Sci. U. S. A.* 107:E72.

Kanaan, N., Roca, M., Tuñón, I. A., Martí, S., and Moliner, V. 2010. Application of Grote–Hynes theory to the reaction catalyzed by thymidylate synthase. *J. Phys. Chem. B* 114 (42):13593–13600.

Kleifeld, O., Frenkel, A., Bogin, O., Eisenstein, M., Brumfeld, V., Burstein, Y., and Sagi, I. 2000. Spectroscopic studies of inhibited alcohol dehydrogenase from *Thermoanaerobacter brockii*: Proposed structure for the catalytic intermediate state. *Biochemistry* 39 (26):7702–7711.

Kleifeld, O., Frenkel, A., Martin, J., and Sagi, I. 2003. Active site electronic structure and dynamics during metalloenzyme catalysis. *Nat. Struct. Biol.* 10 (2):98–103.

Kleifeld, O., Rulisek, L., Bogin, O., Frenkel, A., Havlas, Z., Burstein, Y., and Sagi, I. 2004. Higher metal-ligand coordination in the catalytic site of cobalt-substituted *Thermoanaerobacter brockii* alcohol dehydrogenase lowers the barrier for enzyme catalysis. *Biochemistry* 43 (22):7151–7161.

Kondrashov, D. A., Van Wynsberghe, A. W., Bannen, R. M., Cui, Q., and Phillips, G. N. 2007. Protein structural variation in computational models and crystallographic data. *Structure* 15 (2):169–177.

Kroemer, M., Merkel, I., and Schulz, G. E. 2003. Structure and catalytic mechanism of L-rhamnulose-1-phosphate aldolase. *Biochemistry* 42 (36):10560–10568.

Kumar, S., and Nussinov, R. 2001. How do thermophilic proteins deal with heat? *Cell. Mol. Life Sci.* 58 (9):1216–1233.

Lane, T. J., Shukla, D., Beauchamp, K. A., and Pande, V. S. 2013. To milliseconds and beyond: Challenges in the simulation of protein folding. *Curr. Opin. Struct. Biol.* 23 (1):58–65.

Lodola, A., Mor, M., Zurek, J., Tarzia, G., Piomelli, D., Harvey, J. N., and Mulholland, A. J. 2007. Conformational effects in enzyme catalysis: Reaction via a high energy conformation in fatty acid amide hydrolase. *Biophys. J.* 92 (2):L20–L22.

Ma, B., Kumar, S., Tsai, C.-J., Hu, Z., and Nussinov, R. 2000. Transition-state ensemble in enzyme catalysis: Possibility, reality, or necessity? *J. Theor. Biol.* 203:383–397.

Ma, J. 2005. Usefulness and limitations of normal mode analysis in modeling dynamics of biomolecular complexes. *Structure* 13 (3):373–380.

Marcos, E., Crehuet, R., and Bahar, I. 2010. On the conservation of the slow conformational dynamics within the amino acid kinase family: NAGK the paradigm. *PLoS Comput. Biol.* 6 (4):e1000738.

Marcos, E., Crehuet, R., and Bahar, I. 2011. Changes in dynamics upon oligomerization regulate substrate binding and allostery in amino acid kinase family members. *PLoS Comput. Biol.* 7:e1002201.

Martí, S., Roca, M., Andrés, J., Moliner, V., Silla, E., Tuñón, I., and Bertrán, J. 2003. Theoretical insights in enzyme catalysis *Chem. Soc. Rev.* 33:98–107.

Maseras, F., and Morokuma, K. 1995. IMOMM—A new integrated ab-initio plus molecular mechanics geometry optimization scheme of equilibrium structures and transition-states. *J. Comput. Chem.* 16 (9):1170–1179.

Meyer, P., Evrin, C., Briozzo, P., Joly, N., Barzu, O., and Gilles, A. M. 2008. Structural and functional characterization of *Escherichia coli* UMP kinase in complex with its allosteric regulator GTP. *J. Biol. Chem.* 283 (51):36011–36018.

Min, W., English, B. P., Luo, G., Cherayil, B. J., Kou, S. C., and Xie, X. S. 2005. Fluctuating enzymes: Lessons from single-molecule studies. *Acc. Chem. Res.* 38:923–931.

Min, W., Xie, X. S., and Bagchi, B. 2008. Two-dimensional reaction free energy surfaces of catalytic reaction: Effects of protein conformational dynamics on enzyme catalysis. *J. Phys. Chem. B* 112 (2):454–466.

Olsson, M. H. M., Parson, W. W., and Warshel, A. 2006. Dynamical contributions to enzyme catalysis: Critical tests of a popular hypothesis. *Chem. Rev.* 106 (5):1737–1756.

Osborne, M. J., Schnell, J., Benkovic, S. J., Dyson, H. J., and Wright, P. E. 2001. Backbone dynamics in dihydrofolate reductase complexes: Role of loop flexibility in the catalytic mechanism. *Biochemistry* 40:9846–9859.

Perica, T., Marsh, J. A., Sousa, F. L., Natan, E., Colwell, L. J., Ahnert, S. E., and Teichmann, S. A. 2012. The emergence of protein complexes: Quaternary structure, dynamics and allostery. Colworth Medal Lecture. *Biochem. Soc. Trans.* 40 (3):475–491.

Pisliakov, A. V., Cao, J., Kamerlin, S. C. L., and Warshel, A. 2009. Enzyme millisecond conformational dynamics do not catalyze the chemical step. *Proc. Natl. Acad. Sci. U. S. A.* 106:17359–17364.

Rajagopalan, P. T. R., Lutz, S., and Benkovic, S. J. 2002. Coupling interactions of distal residues enhance dihydrofolate reductase catalysis: Mutational effects on hydride transfer rates. *Biochemistry* 41:12618–12628.

Ramón-Maiques, S., Britton, H. G., and Rubio, V. 2002a. Molecular physiology of phosphoryl group transfer from carbamoyl phosphate by a hyperthermophilic enzyme at low temperature. *Biochemistry* 41:3916–3924.

Ramón-Maiques, S., Fernández-Murga, M. L., Gil-Ortiz, F., Vagin, A., Fita, I., and Rubio, V. 2006. Structural bases of feed-back control of arginine biosynthesis, revealed by the structures of two hexameric N-acetylglutamate kinases, from *Thermotoga maritima* and *Pseudomonas aeruginosa. J. Mol. Biol.* 356 (3):695–713.

Ramón-Maiques, S., Marina, A., Gil-Ortiz, F., Fita, I., and Rubio, V. 2002b. Structure of acetylglutamate kinase, a key enzyme for arginine biosynthesis and a prototype for the amino acid kinase enzyme family, during catalysis. *Structure* 10:329–342.

Razvi, A., and Scholtz, J. M. 2006. Lessons in stability from thermophilic proteins. *Protein Sci.* 15 (7):1569–1578.

Roca, M., Oliva, M., Castillo, R., Moliner, V., and Tuñón, I. 2010. Do dynamic effects play a significant role in enzymatic catalysis? A theoretical analysis of formate dehydrogenase. *Chem. Eur. J.* 16 (37):11399–11411.

Rulisek, L., and Vondrasek, J. 1998. Coordination geometries of selected transition metal ions (Co2+, Ni2+, Cu2+, Zn2+, Cd2+, and Hg2+) in metalloproteins. *J. Inorg. Biochem.* 71 (3–4):115–127.

Sanchez-Martinez, M., Marcos, E., Tauler, R., Field, M. J., and Crehuet, R. 2013. Conformational compression and barrier height heterogeneity in the N-acetylglutamate kinase. *J. Phys. Chem. B* 117 (46):14261–14272.

Sterpone, F., and Melchionna, S. 2012. Thermophilic proteins: Insight and perspective from in silico experiments. *Chem. Soc. Rev.* 41 (5):1665–1676.

Tama, F., and Brooks, C. L. 2006. Symmetry, form, and shape: Guiding principles for robustness in macromolecular machines. *Annu. Rev. Biophys. Biomol. Struct.* 35:115–133.

Tobi, D., and Bahar, I. 2005. Structural changes involved in protein binding correlate with intrinsic motions of proteins in the unbound state. *Proc. Natl. Acad. Sci. U. S. A.* 102 (52):18908–18913.

Vieille, C., and Zeikus, G. J. 2001. Hyperthermophilic enzymes: Sources, uses, and molecular mechanisms for thermostability. *Microbiol. Mol. Biol. Rev.* 65 (1):1–43.

Warshel, A., Sharma, P. K., Kato, M., Xiang, Y., Liu, H., and Olsson, M. H. M. 2006. Electrostatic basis for enzyme catalysis. *Chem. Rev.* 106 (8):3210–3235.

Williams, R. J. 1995. Energised (entatic) states of groups and of secondary structures in proteins and metalloproteins. *Eur. J. Biochem.* 234 (2):363–381.

Wolf-Watz, M., Thai, V., Henzler-Wildman, K., Hadjipavlou, G., Eisenmesser, E. Z., and Kern, D. 2004. Linkage between dynamics and catalysis in a thermophilic-mesophilic enzyme pair. *Nat. Struct. Mol. Biol.* 11:945–949.

Yang, L.-W., and Bahar, I. 2005. Coupling between catalytic site and collective dynamics: A requirement for mechanochemical activity of enzymes. *Structure* 13 (6):893–904.

Zheng, W., and Brooks, B. R. 2005. Probing the local dynamics of nucleotide-binding pocket coupled to the global dynamics: Myosin versus kinesin. *Biophys. J.* 89 (1):167–178.

Zhuravlev, P. I., and Papoian, G. A. 2010. Protein functional landscapes, dynamics, allostery: A tortuous path towards a universal theoretical framework. *Q. Rev. Biophys.* 43 (3):295–332.

Protein Motions: Flexibility Analysis

Chapter 5

Simplified Flexibility
Analysis of Proteins

Yves-Henri Sanejouand

CONTENTS

INTRODUCTION

In order to understand the function of a protein, the knowledge of its structure is of utmost importance. However, it is becoming more and more clear that in most cases this is not enough and that the relevant information needed is its *structural ensemble*; that is, a fair sample of the set of conformations a protein can visit at room temperature.

A straightforward way to obtain this information is to perform a *long enough* molecular dynamics (MD) simulation in explicit solvent. In practice, noteworthy in the case of enzymes, the timescale that has to be reached is at least of the order of the microsecond since the most efficient enzymes, like catalase, have turnovers of this magnitude. Today, for standard-size proteins, MD simulations that long are routinely performed on supercomputers. Actually, using a dedicated application-specific integrated circuit (ASIC) based supercomputer, the millisecond timescale has recently been reached, in the case of the small bovine pancreatic trypsin inhibitor (BPTI) model system [1]. Moreover, starting from random initial configurations, accurate *ab initio* folding of several fast-folding peptides has been obtained [1,2], revealing all the potential of the brute force approach for the years to come.

However, even with a supercomputer, obtaining the relevant structural ensemble through this approach often takes months. Moreover, this can hardly be done for very large systems such as the ribosome, whole virus capsids, and so forth. So, simplified methods are still welcome and are noteworthy because, as a consequence of their low computational cost, they can be implemented on web servers where anyone can give them a try.

NORMAL MODE ANALYSIS

Background

A simple way to study the flexibility of a molecule is to perform a normal mode analysis (NMA). This method rests on the fact that the classical equations of motion for a set of N atoms can be solved analytically when the displacements of the atoms in the vicinity of their equilibrium positions are small enough. As a consequence, V, the potential energy of the studied system, can be approximated by the first terms of a Taylor series. Moreover, because the system is expected to be at equilibrium (or on a saddle point), the term with the first derivatives of the energy (the forces) can be dropped, so that

$$V = \frac{1}{2} \sum_{i=1}^{3N} \sum_{j=1}^{3N} \left(\frac{\partial^2 V}{\partial r_i \partial r_j} \right)_0 \left(r_i - r_i^0 \right) \left(r_j - r_j^0 \right) \tag{5.1}$$

where r_i is the ith coordinate of the system, r_i^0 being its equilibrium value.

In other words, within the frame of this approximation, the potential energy of a system can be written as a quadratic form. In such a case, it is quite straightforward to show that the equations of atomic motion have the following solutions [3,4]:

$$r_i(t) = r_i^0 + \frac{1}{\sqrt{m_i}} \sum_{k=1}^{3N} C_k a_{ik} \cos\left(2\pi v_k t + \Phi_k\right)$$

(5.2)

where m_i is the atomic mass and where C_k and Φ_k, the amplitude and phases of the so-called normal mode of vibration k, depend on the initial conditions; that is, on atomic positions and velocities at $t = 0$.

In practice, v_k, the frequency of mode k, is obtained through the kth eigenvalue of the mass-weighted Hessian of the potential energy; that is, the matrix whose elements are the mass-weighted second derivatives of the energy, the a_{ik}'s being the coordinates of the corresponding eigenvector.

In Figure 5.1, the frequency spectrum thus obtained for the protease of human immunodeficiency virus (HIV) is shown when a standard empirical energy function is used; namely, as available in the software CHARMM* [5]. The modes with the highest frequencies correspond to motions of couples of atoms that are chemically bonded and the reason why these frequencies are much higher than the others is because a hydrogen (i.e., a light particle) is involved

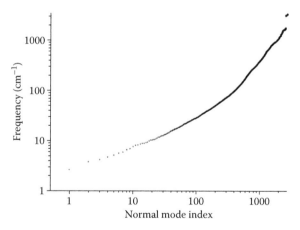

Figure 5.1   Spectrum of the HIV protease as obtained using standard normal mode analysis. The structure considered is the monomer found in PDB 1HHP after energy minimization. High-frequency modes correspond to localized motions of chemically bonded atoms. The highest frequency ones correspond to motions in which a hydrogen atom is involved. There are only a few (164) because an extended-atom model was considered [6].

---

* Chemistry at Harvard Macromolecular Mechanics.

in these bonds. Generally speaking, modes with high frequencies are localized; that is, only a few atoms are moving significantly at these frequencies.

On the contrary, on the low-frequency part of the spectrum, modes tend to involve large parts of the structure. This is probably why their comparison with protein functional motions was undertaken, the latter being often described as relative motions of structural domains [7]. As a matter of fact, in the first NMA study of a protein with well-defined domains; namely, the human lysozyme, the lowest-frequency mode ($\nu_1 = 3.6$ cm^{-1}) was found to correspond to a hinge-bending motion [8]. Later on, the lowest frequency modes of citrate synthase [9] and hemoglobin [10] were indeed found able to provide a fair description of the rather complex and large-amplitude motion these proteins experience upon ligand binding.

Technical Issue

Getting the eigenvalues and eigenvectors of a large matrix can prove challenging, in particular because methods available in mathematical libraries usually require the storage of the whole matrix in the computer memory. For a small protein like monomeric HIV protease (99 residues) this is not an issue since, with an extended-atom model, the storage of its Hessian takes ≈ 10 Mo. For dimeric citrate synthase (2 × 450 residues), it takes 3 Go. Twenty years ago, this used to be an issue [9]. Of course, it is not any more. However, for not-so-uncommon multimeric systems like aspartate transcarbamylase (nearly 3000 residues), using standard approaches would still require access, like at the time [11], either to use smarter numerical methods and/or supercomputers, as the storage of the whole Hessian would take 30 Go of computer memory.

Useful Methods and Approximations

To overcome this practical limitation, several methods have been proposed. Note that within the frame of most of them it is not possible to get the $3N$ normal modes of the system (i.e., its full spectrum).

When a good enough guess of a required eigenvector can be made, for instance when it is expected to be a relative motion of well-defined structural domains, a method of choice is the Lanczos one, especially because it only involves calculations of matrix-vector products. This method was applied to the case of human lysozyme [8] as well as, later on, using a more sophisticated version of the algorithm to the case of dimeric citrate synthase [9].

Other methods usually rely on the splitting of the Hessian into blocks [12,13] and/or on the choice of new coordinates that allow for the building of a smaller Hessian whose eigensolutions are as close as possible to the original ones [14]. For instance, the principle of the RTB approximation [15] is to use the six

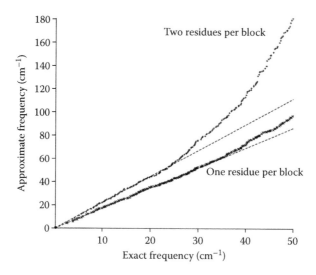

Figure 5.2 Approximate frequencies of the HIV protease as obtained through the RTB approximation, as a function of exact ones, when one or two amino acid residues are put in each block. Dotted lines are linear fits of the low-frequency part of the spectra.

rotation (R) and translation (T) vectors of blocks (B) of atoms as the new set of coordinates.* When each block contains a whole amino-acid residue, since there is, on average, ≈ 16 atoms (48 coordinates) per residue, the order of the Hessian is reduced by a factor of ≈ 8. Interestingly, if the frequencies calculated with this approximation are found to be, as expected, higher than those of the full Hessian, on the low-frequency part of the spectrum a proportionality is observed between approximate and exact values [19] (Figure 5.2). Moreover, the corresponding approximate eigenvectors are found to be remarkably similar to exact ones [15,19]. Of course, several amino acid residues can be put in a given block [15], and the way atoms are grouped into blocks can be done on smarter grounds [20,21].

Note that reducing the size of the Hessian is a coarse-graining process, which has the advantage of preserving the interactions between blocks observed in the all-atom model. On the other hand, the fact that low-frequency modes are little perturbed by the process suggest that they are robust in the sense that the corresponding pattern of displacements does not depend significantly on the details of the description of the system.

---

* This approximation has been implemented in CHARMM under the block normal modes (BNMs) acronym [16]. It is used by the ElNémo web server [17] as well as by software such as PHASER [18] and DIAGRTB. The latter is in the public domain and can be downloaded at http://ecole .modelisation.free.fr/modes.html.

NETWORK MODELS

Tirion's Model

The fact that an approximation like Equation 5.1 can prove useful for the study of protein functional motions is far from being obvious. First, because NMA is usually performed for a single minimum of the potential energy surface, while it is well known that at room temperature a protein explores a huge number of different ones [22]. More generally, Equation 5.1 means that the system is studied as if it were a solid, while it is quite clear that the liquidlike character of the dynamical behavior of a protein, noteworthy because of its amino-acid side chains [23], is required for its function [24].

Moreover, prior to the development of simple implicit solvent models like EEF1 [6], NMA was usually performed *in vacuo* even though it was well known that water is essential for protein function,* not to mention the significant distortion a system experiences during the required preliminary energy minimization when solvent effects are only taken into account through an effective dielectric constant [25].

So, from the very beginning, the usefulness of NMA in the field of protein dynamics had to mean that functionally relevant normal modes are highly robust ones, similar not only when they are obtained for different minima of the energy function [26,27], but also when they are obtained with different force fields. As an extreme check of the latter point, Tirion proposed to replace electrostatics and Lennard-Jones interactions in a protein by an harmonic term such that

$$V_{nb} = \frac{1}{2} k_{enm} \sum_{d_{ij}^0 < R_c} \left(d_{ij} - d_{ij}^0 \right)^2 \tag{5.3}$$

where d_{ij} is the actual distance between atoms i and j with d_{ij}^0 being their distance in the studied structure [28]. In other words, in Tirion's model, all pairs of atoms less than R_c Å away from each other are linked by Hookean springs. Note that when, as herein, k_{enm}, the force constant, is the same for all atom pairs, it only plays the role of a scaling factor for the frequencies of the system. As a corollary, the only parameter of the model is R_c. To evaluate how similar modes obtained with such a description of the non-bonded interactions in a protein are to those obtained with a standard, empirical, one, a useful quantity is the overlap:

$$O_{ij} = \left(\sum_k^{3N} a_{ki} a_{kj} \right)^2 \tag{5.4}$$

---

* For its proper folding, in the first place.

that is, the square of the scalar product of the eigenvectors obtained in both cases. Note that because eigenvectors are normalized, for each of them

$$\sum_{j}^{3N} O_{ij} = 1.$$

As shown in Figure 5.3, the overlap between an eigenvector obtained with Equation 5.3 and an eigenvector obtained with the energy function available in CHARMM [6] can be as high as 0.8, as in the case of the lowest frequency mode of the HIV protease. So, obviously, this mode is a very robust one in the sense that it does not depend much on the nature of the atomic interactions in the structure. Note in particular that in the simplified version of Tirion's model considered above, at variance with Tirion's original work [28], chemically bonded atoms are treated on the same footing as other pairs of close atoms; that is, by linking them with a spring of same force constant (k_{enm}).

In the HIV protease case, only three low-frequency modes have an overlap with a mode obtained using Tirion's model that is larger than 0.4. However, this does not necessary mean that HIV protease has only three highly robust modes, in particular because a given subset of modes can prove robust as a whole even when each of them is not. To quantify this, D_i^{eff}, the effective dimension of standard mode i (i.e., the effective number of Tirion's modes involved in its description), can be calculated as follows [29,30]:

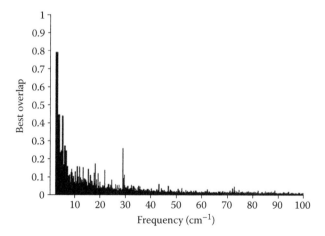

Figure 5.3 For each eigenvector of the HIV protease, as obtained using a standard empirical energy function, the best overlap with an eigenvector obtained using Tirion's model is given. The structure considered is the monomer found in PDB 1HHP after energy minimization. Tirion's model was built using $R_c = 5$ Å.

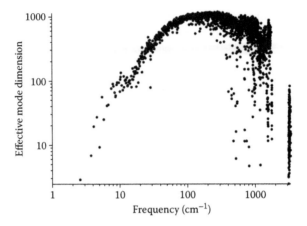

Figure 5.4 For each eigenvector of the HIV protease as obtained using a standard empirical energy function, the effective number of Tirion's eigenvectors involved in its description is given. The structure considered is the monomer found in PDB 1HHP after energy minimization. Tirion's model was built using $R_c = 5$ Å.

$$D_i^{eff} = \exp\left(-\sum_{j}^{3N} O_{ij} \log O_{ij}\right) \qquad (5.5)$$

For the lowest frequency mode of the HIV protease, $D_i^{eff} = 2.8$ and indeed, as shown in Figure 5.4, on the low-frequency side of the spectrum only three standard modes of the HIV protease can be described accurately with less than 10 of Tirion's modes; however, three other ones can be described with less than 30 of Tirion's modes. Note also that some high-frequency modes can also be described with a handful of Tirion's modes. This is because atomic masses are taken into account in Tirion's model. As a consequence, also within the frame of this model, high-frequency modes are well localized, and likewise, they correspond for the most part to localized motions in which hydrogen atoms are involved.

Proteins as Undirected Graphs

Describing interactions between amino-acid residues in terms of short-range harmonic springs (e.g., with $R_c = 5$ Å) as proposed by Tirion, and ending with a subset of low-frequency modes that are highly similar to those obtained with a much more complex description, suggest that such modes are due to some generic property.

A possibility is that what determines the nature of these modes is the pattern of interactions between residues. As a matter of fact, simple methods have been proposed for evaluating protein flexibility through the study of the graph corresponding to the set of interactions observed in a given structure (Figure 5.5 shows such a graph, in the case of the HIV protease). For instance, it has been shown that hinges and flexible loops can be identified in proteins like the

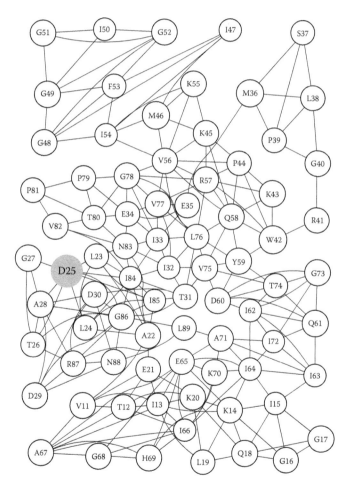

Figure 5.5   Graph of the interactions between HIV-protease residues. The catalytic residue (Asp 25) is indicated in gray. For the sake of clarity, only one monomer is considered and the 10 first and last residues were omitted. Drawn with Graphviz-Neato.

HIV protease and adenylate kinase [31] using the highly efficient pebble game algorithm [32]. On the other hand, diagonalizing directly the corresponding so-called adjacency matrix provides fair estimates of the amplitude of atomic fluctuations, as observed experimentally, noteworthy through crystallographic Debye-Wäller factors [33,34] (see the section titled "Crystallographic B-Factors").

Interestingly, while the small eigenvalues of the adjacency matrix are enough for providing such estimates, large eigenvalues also seem able to provide useful information. Indeed, because eigenvectors corresponding to large eigenvalues pinpoint spots in the structure where residue density (in a coarse-grained sense) is the highest, it has been suggested that they may correspond to protein-folding cores [35].

For instance, in the case of the HIV protease [35], the eigenvector corresponding to the largest eigenvalue is dominated by the motion of Gly 86 (Figure 5.6) a residue close to Asp 25, the catalytic residue (see Figure 5.5). As a matter of fact, there is a hydrogen bond between the backbone carbonyl oxygen of the previous residue, Ile 85, and the backbone amide nitrogen of Asp 25. The fact that the folding core could prove that close to the enzyme active site in such a small protein makes sense, since enzymatic activity usually requires a precise positioning of the residues involved in the catalytic mechanism; that is, a rather rigid local environment.

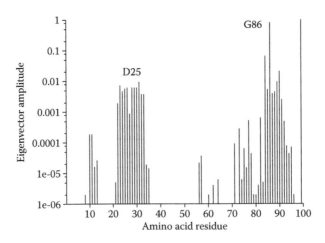

Figure 5.6 Eigenvector of the adjacency matrix of the HIV protease (PDB 1HHP) corresponding to the largest eigenvalue. Here, two residues are defined as interacting ones (i.e., the corresponding matrix element is –1) when the distance between their C_α is less than 6.5 Å.

Elastic Networks

However, by considering a protein as a graph, important information is lost; namely, the directionality of the motions. This is probably why, among the family of simple protein models, three-dimensional elastic network models (3-D ENMs) seem to be the most popular ones. Such models are just one simplification step further from Tirion's model: here, residues are described as beads, often a single one, as initially proposed [36], although beads can also represent groups of residues [37] as well as whole protein monomers [38].

Using as beads the C_α atoms (C_α-ENM), it was shown that a few low-frequency modes of the coarse-grained elastic networks thus obtained are enough for describing accurately the motion a protein experiences upon ligand binding [30,39,40] as long as a significant portion of the protein is involved [30] (e.g., whole domains), at least when the amplitude of the motion is large enough [41] (typically more than ≈ 2 Å of C_α-rmsd*).

To measure how well normal mode i describes a given motion, a useful quantity is I_i, its involvement coefficient [9,25]:

$$I_i = \left(\sum_j^{3N} a_{ji} \frac{\overrightarrow{\Delta r_i}}{|\Delta_r|} \right)^2 \tag{5.6}$$

where $\overrightarrow{\Delta r} = \overrightarrow{r_b} - \overrightarrow{r_a}$, $\overrightarrow{r_a}$ and $\overrightarrow{r_b}$ are the atomic positions observed in conformations a and b, respectively. Note that in order to have meaningful involvement coefficients, conformation b needs to be fitted onto conformation a if the normal modes were obtained for the latter. Note also that Equation 5.6 is the scalar product between $\overrightarrow{\Delta r}$ and normal mode i. So, as a consequence of normal modes orthogonality: $\sum_i^{3N} I_i = 1$. In other words, I_i gives the fraction of the protein motion that can be described just by considering the displacement of the system along mode i.

$\sum_i^n I_i$, the quality of the description of the closure motion of guanylate kinase (see Figure 5.7) as a function of n, the number of modes taken into account, is shown in Figure 5.8, for the modes of the C_α-ENM built with an open form (PDB 1EX6; $R_c = 10$ Å). In this case, which is far from being an exceptional one [30,40], the lowest frequency mode is enough for describing 72% of the functional motion ($I_1 = 0.72$). Note that such a high value of the involvement coefficient means that both patterns of atomic displacements are remarkably

---

* rmsd: root-mean-square deviation.

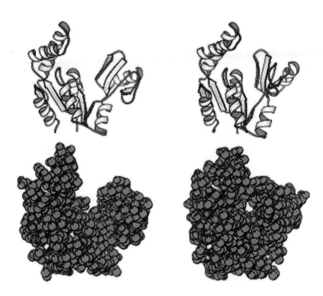

Figure 5.7 Conformational change of *Saccharomyces cerevisiae* guanylate kinase. Left: open form (PDB 1EX6). Right: closed form (PDB 1EX7). Top: standard sketch. Bottom: van-der-Waals spheres. Drawn with Molscript [42].

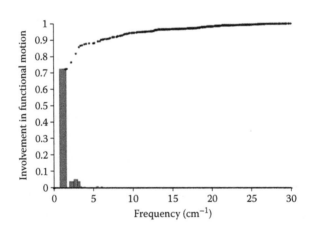

Figure 5.8 Accuracy of the description of the conformational change of *Saccharomyces cerevisiae* guanylate kinase with low-frequency modes as obtained using an elastic network model (PDB 1EX6, R_c = 10 Å). Boxes: involvement coefficients. Black disks: cumulative sum. The force constant of the springs was chosen so as to have a lowest frequency of 1.5 cm^{-1}.

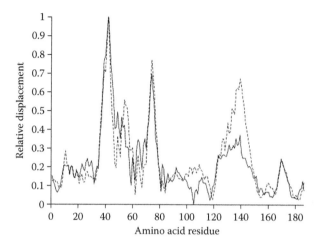

Figure 5.9 Comparison of the conformational change of *Saccharomyces cerevisiae* guanylate kinase (plain line) with its lowest frequency mode (dotted line) as obtained using an ENM (PDB 1EX6, R_c = 10 Å). Both sets of C_α displacements are normalized with respect to the maximum value.

similar (Figure 5.9). Indeed, the main difference concerns the relative amplitude of the motion of helices 125–135 and 141–157. On the other hand, together modes three to five are able to describe 13% of the motion (I_3 = 0.04, I_4 = 0.05, I_5 = 0.04). So, four modes *only* are enough for describing 85% of the closure motion of guanylate kinase.

Robust Modes

However, a way to identify *a priori* the modes the most involved in the functional motion is required to turn such a qualitative description into a possibly useful prediction. To do so, low-resolution experimental data can prove enough; for example, those obtained by cryoelectromicroscopy [43–45]. A more general possibility is to build upon the robustness of these modes; namely, to look for modes that are not, or very little, sensitive to the protein model used [41]. For instance, instead of setting springs between pairs of C_α atoms less than R_c Å away from each other, as above, the springs can be established so that each C_α atom is linked to $\approx n_c$ of its closest neighbors [41]. Then, overlaps between both sets of modes can be obtained (Equation 5.4), with those that can be described accurately with a small number of modes (Equation 5.5) of the other set being indeed robust ones. In the case of the open form of guanylate kinase, with R_c = 10 Å and n_c = 10, the first four modes calculated with R_c = 10 Å can be described with less than three modes calculated with n_c = 10, while all others need more than eight. Note that the four robust modes thus identified are enough for describing 81% of the closure motion of guanylate kinase (although mode two does not contribute significantly; see Figure 5.8).

The case of guanylate kinase may look too simple since the functional motion can be correctly guessed just by looking at the structures (see Figure 5.7). Interestingly, results obtained with this model system seem to have a general character [41]. For instance, in the case of citrate synthase, it is possible to guess where the active site is, and as a consequence, where the closure motion should occur. However, the structure is more complex (Figure 5.10) and it is hardly feasible to decide where are the limits of each structural domain. Nevertheless, results obtained through a normal mode analysis of a C_α-ENM of dimeric citrate synthase are almost as impressive as those obtained for guanylate kinase. Note that with R_c = 10 Å, I_2 = 0.29 and I_3 = 0.48, which means that 77% of the conformational change of citrate synthase can be described with these

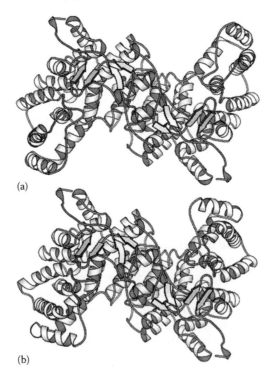

(a)

(b)

Figure 5.10 Conformational change of *Thermus thermophilus* citrate synthase. (a) Open form (PDB 1IOM). (b) Closed form (PDB 1IXE). Drawn with Molscript [42].

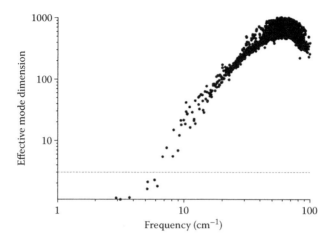

Figure 5.11 Effective dimension of each eigenvector of dimeric citrate synthase as obtained with a C_α-ENM (PDB 1IOM; $R_c = 10$). Here, the effective dimension of a mode corresponds to the effective number of eigenvectors involved in its description when the latter are obtained with $n_c = 10$. The force constant of the springs was chosen so as to have a lowest frequency of 2.9 cm^{-1}, with $R_c = 10$. The dotted line indicates an effective dimension of three.

two modes. On the other hand, with $n_c = 10$, $I_2 = 0.27$ and $I_3 = 0.48$. As expected, modes two and three are among the most robust ones. However, in the case of dimeric citrate synthase, there are more than two obviously robust modes since the first seven modes obtained with $R_c = 10$ can be described accurately with less than three modes obtained with $n_c = 10$ (Figure 5.11).

When compared with modes obtained with $n_c = 10$, high-frequency modes obtained with $R_c = 10$ can certainly not be considered as being robust (their effective dimension is 200 or more; see Figure 5.11). This is because, as a consequence of the cutoff criterion, such modes correspond to motions of residues belonging to parts of the structure where density (in a coarse-grained sense) is the highest. This is not the case with $n_c = 10$. As a matter of fact, the latter kind of ENM was designed so as to show that low-frequency modes do not result from density patterns inside a structure but from mass distribution in space (i.e., from the overall shape of the structure) [41].

Crystallographic B-Factors

Having access to an analytical solution of the equations of atomic motion (Equation 5.2) allows for the calculation of many quantities [46] such as the fluctuations of atomic coordinates around their equilibrium values. For instance, in the case of the x coordinate of atom i:

$$\left\langle \Delta x_i^2 \right\rangle = k_B T \sum_{k=1}^{n} \frac{a_{ik}^2}{4\pi^2 m_i v_k^2} \tag{5.7}$$

where T is the temperature, k_B, the Boltzmann constant, n, the number of modes taken into account ($n = 3N - 6$, unless specified otherwise) and where <> denotes a time average. Note that the six rigid-body modes (translation and rotation ones) are usually excluded from the summation as a consequence of their null frequencies.

Interestingly, crystallographic B-factors are expected to derive from the fluctuations of atomic positions within the crystal cell, namely:

$$B_i = \frac{8\pi^2}{3} \left\langle \Delta x_i^2 + \Delta y_i^2 + \Delta z_i^2 \right\rangle$$

where B_i is the isotropic B-factor of atom i. Although other factors contribute to the experimental values, such as crystal disorder or phonons, fair correlations have been obtained with values calculated using ENMs [47], especially when effects of neighboring molecules in the crystal are included [48,49]. Figure 5.12

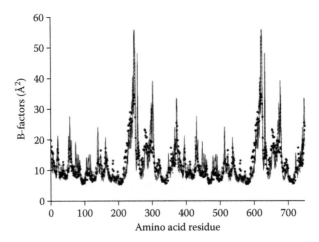

Figure 5.12 B-factors of dimeric citrate synthase. Filled circles: experimental values (PDB 1IOM). Plain and dotted lines, respectively: values calculated either with a standard ENM ($R_c = 10$) or with $n_c = 10$. Calculated values are shifted and scaled so that minimum and average values are the same as experimental ones. A justification for such a treatment is that experimental B-factors include various physical effects other than thermal intraprotein fluctuations.

illustrates how accurate B-factors calculated with ENMs can be, even without taking such subtleties into account. Here, the correlation between experimental and calculated values is 0.80 when a cutoff-based ENM is used and 0.85 when a constant-number-of-neighbors ENM is used. Note that predictions made with both kinds of ENMs are hardly distinguishable. This is a mere consequence of the weight of the low-frequency modes in Equation 5.7. Indeed, taking into account only the seven robust modes identified previously ($n = 7$) yields a correlation of 0.82; that is, a value as high as when all modes are used ($n = 3N - 6$).

Flexibility versus Rigidity

As briefly discussed in the section titled "Proteins as Undirected Graphs," while low-frequency modes can provide information on the overall flexibility of a structure, when a cutoff-based C_α-ENM is considered, high-frequency modes pinpoint parts of the structure where amino acid density is the highest. Though it is tempting to view such parts as being the most rigid ones, the relationship between protein rigidity and density is not expected to be that straightforward. Also, because high-frequency modes are usually localized, several of them need to be considered in order to get a consistent picture of the overall rigidity inside a structure. Unfortunately, selecting the corresponding subset of high-frequency modes in a rigorous way is not that obvious.

Several alternatives have been proposed. For instance, k_i, an effective local force constant associated to atom i, can be defined as follows [50,51]:

$$k_i = \frac{3k_B T}{\left\langle \left(\bar{d}_i - \langle \bar{d}_i \rangle \right)^2 \right\rangle}$$

where $\bar{d}_i$ is the average distance of atom i from all other atoms in the structure. Note that it is when this average distance fluctuates little that the force constant (the rigidity) is high. Of course, when the considered protein is a multidomain one, the average is meaningful only if it is calculated for atoms belonging to the same domain.

Interestingly, this measure involves an ensemble averaging that can be performed using any protein model (e.g., all-atom as well as coarse-grained ones). For instance, a C_α-ENM can be used together with Equation 5.2. Figure 5.13 shows the result in the case of the HIV protease. As with the top eigenvector of the adjacency matrix (see Figure 5.6), the peptidic stretch near G86 is identified as being rigid, but the site now identified as being the most rigid, V75, was not pinpointed with the previous approach. However, V75 is found to be significantly involved in the eigenvector corresponding to the fourth

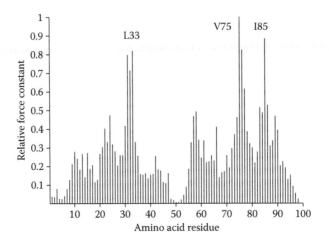

Figure 5.13   Rigidity of a monomer of the HIV protease (PDB 1HHP, R_c = 10). Effective force constants have been normalized so that the maximum value is one.

highest eigenvalue. This suggests that when several eigenvectors are taken into account, both methods may provide similar informatios about the rigidity of protein structures.

Figure 5.14 shows what such an analysis yields in the case of a larger protein; namely, myosin (747 amino acid residues). Interestingly, T178 and I455, the sites identified as being the most rigid, are both quite close to the enzyme active site, their carbonyl oxygens being 6 and 4 Å away, respectively, from

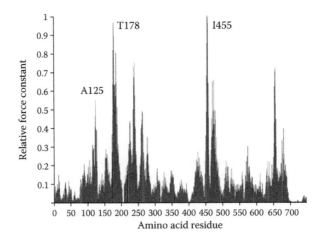

Figure 5.14   Rigidity of *Dictyostelium discoideum* myosin (PDB 1VOM, R_c = 10).

phosphate oxygens belonging to the ATP analog observed in the studied structure (PDB 1VOM).

Nonlinear Network Models

ENMs have proved useful, in particular because they are very simple. As a consequence, analyses performed with such models can be very quick, allowing for an almost instantaneous check of a hypothesis as well as for large-scale studies. However, for many specific applications, ENMs are expected to prove too naive. Therefore, it is of interest to develop more complex models. However, when complexity increases, the time it takes to perform an analysis usually also increases (either for a human being or for a computer), and therefore the most useful are expected to be models that are only a bit more complex than ENMs. In practice, since ENMs are, in essence, single-parameter models, increasing their complexity means increasing their number of parameters. So, in order not to increase their complexity too much, a natural choice is to keep the number of parameters as small as possible.

Along this line of thought, in order to recover one of the major property of the potential energy surface of a protein; namely, that it is a multiminima one, network models with several (usually two) energy minima have been proposed [52,53]. However, a more basic property of such a surface is that it is highly anharmonic [54,55] and so it seems worth starting by adding explicit nonlinearity* into an elastic network model (e.g., [56]):

$$V = \sum_{d_{ij}^0 < R_c} \frac{k_2}{2}\left(d_{ij} - d_{ij}^0\right)^2 + \frac{k_4}{4}\left(d_{ij} - d_{ij}^0\right)^4 \qquad (5.8)$$

the choice of an additional term with a power of four, instead of three, being mostly for symmetry reasons but also because previous works on one-dimensional (1-D) and two-dimensional (2-D) systems had shown that the dynamical properties of systems with such an energy function can be quite spectacular [57,58]. Note that when $k_4 = 0$, Equation 5.8 corresponds to Equation 5.3; that is, to a standard ENM (up to now, only protein models with $k_4/k_2 = 1\text{Å}^2$ have been studied in depth [59]).

One of the dynamical nonlinear phenomena that can occur in this context is the birth and the (rather) long-time survival of a discrete breather (DB) [57]; that is, a localized mode whose frequency is high enough so that energy exchange with the rest of the system can prove extremely slow (thousands of

---

* Anharmonicity and nonlinearity may sound more familiar, respectively, to biophysicists and physicists.

periods of the DB). A way to observe such a phenomenon without making any *a priori* assumption on its nature (localized or not, etc.) is to perform a molecular dynamics simulation, starting with a high initial temperature and cooling the system through friction on its surface atoms [56]. When a DB sets up, the (kinetic) energy remaining in the system becomes (quite suddenly) localized, with the few atoms involved in the DB being the only ones with a significant motion whereas all other ones are almost frozen.

Interestingly, DBs tend to appear in the most rigid parts of a structure [56]. This is due to the fact that most are related to one of the high-frequency modes of the elastic network. As a matter of fact, the more energetic a DB is, the higher its frequency [56], and the more localized [56,60] and different it is from the high-frequency (normal) mode it comes from [60]. As a consequence, a way to obtain DBs is to provide energy to one of the high-frequency modes of the network [61]. Because energetic DBs are localized, such a protocol allows us to pinpoint a few specific residues (a single one per high-frequency mode).

For instance, in the case of the HIV protease, exciting the highest frequency mode highlights residues I85 and T31 (see Figure 5.15 and compare to Figures 5.6 and 5.13) while in the case of myosin, such an approach highlights residue A125 (see Figure 5.16 and compare to Figure 5.14). Note that in both cases, more than 50% of the kinetic energy initially given to the highest frequency mode of the system is observed during the simulation on a single residue.

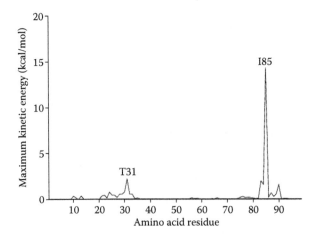

Figure 5.15 Discrete breather arising from the highest frequency mode of the HIV protease (PDB 1HHP, $R_c = 10$). Maximum kinetic energies observed during a 200-ps MD simulation are given when the highest frequency mode of the elastic network is excited with 20 kcal/mol.

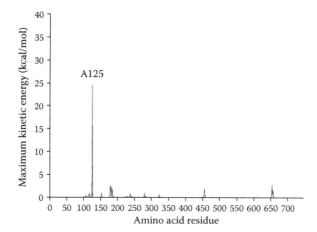

Figure 5.16 Discrete breather arising from the highest frequency mode of *Dictyostelium discoideum* myosin (PDB 1VOM, R_c = 10). Maximum kinetic energies observed during a 200-ps MD simulation are given when the highest frequency mode of the elastic network is excited with 40 kcal/mol.

CONCLUSION

The study of simple models where a protein is described as a set of Hookean springs linking neighboring amino acid residues often provide useful information about its flexibility. Note that the correlation between calculated residue fluctuations and experimental B-factors can be quite high (see Figure 5.12). Moreover, in many cases, a few low-frequency modes are found able to provide a fair description of the functional motion of a protein (see Figures 5.8 and 5.9). This is far from being an obvious result, since, for instance, the energy function of a set of Hookean springs has a single minimum while a conformational change is expected to involve (at least) two significantly different minima of the potential energy surface.

Interestingly, modes involved in functional motions were found to be robust [41]; that is, very little sensitivity to changes in the model used to describe the protein. Actually, the robustness of a small subset of the lowest frequency modes of a protein explains why coarse-grained models can be used on the same footing as highly detailed ones as far as low-frequency (and large amplitude) motions are concerned. Reciprocally, seeking for robust modes allows us to get a small set of coordinates (eigenvectors) that is able to provide a fair description of the functional motion a given protein can perform.

More surprisingly, high-frequency modes of cutoff-based elastic networks may also prove useful. Indeed, they pinpoint parts of the structure where the residue density (in a coarse-grained sense) is the highest, and it seems that such parts can be important for the proper folding [35] and/or for the function [37,51,56] of many proteins.

REFERENCES

1. Shaw, D. et al. Atomic-level characterization of the structural dynamics of proteins. *Science* **330**, 341–346 (2010).
2. Lindorff-Larsen, K., Piana, S., Dror, R. O. & Shaw, D. E. How fast-folding proteins fold. *Science* **334**, 517–520 (2011).
3. Goldstein, H. *Classical Mechanics*. Addison-Wesley, Reading, MA (1950).
4. Wilson, E., Decius, J. & Cross, P. *Molecular Vibrations*. McGraw-Hill, New York (1955).
5. Brooks, B. R. et al. CHARMM: A program for macromolecular energy, minimization, and dynamics calculations. *J. Comput. Chem.* **4**, 187–217 (1983).
6. Lazaridis, T. & Karplus, M. Effective energy function for proteins in solution. *Proteins* **35**, 133–152 (1999).
7. Gerstein, M. & Krebs, W. A database of macromolecular motions. *Nucleic Acids Res.* **26**, 4280–4290 (1998).
8. Brooks, B. R. & Karplus, M. Normal modes for specific motions of macromolecules: Application to the hinge-bending mode of lysozyme. *Proc. Natl. Acad. Sci. U. S. A.* **82**, 4995–4999 (1985).
9. Marques, O. & Sanejouand, Y.-H. Hinge-bending motion in citrate synthase arising from normal mode calculations. *Proteins* **23**, 557–560 (1995).
10. Guilbert, C., Perahia, D. & Mouawad, L. A method to explore transition paths in macromolecules. Applications to hemoglobin and phosphoglycerate kinase. *Comput. Phys. Commun.* **91**, 263–273 (1995).
11. Thomas, A., Field, M. J., Mouawad, L. & Perahia, D. Analysis of the low frequency normal modes of the T-state of aspartate transcarbamylase. *J. Mol. Biol.* **257**, 1070–1087 (1996).
12. Harrison, R. Variational calculation of the normal modes of a large macromolecules: Methods and some initial results. *Biopolymers* **23**, 2943–2949 (1984).
13. Mouawad, L. & Perahia, D. DIMB: Diagonalization in a mixed basis. A method to compute low-frequency normal modes for large macromolecules. *Biopolymers* **33**, 569–611 (1993).
14. Ghysels, A., Miller, B. T., Pickard, F. C. & Brooks, B. R. Comparing normal modes across different models and scales: Hessian reduction versus coarse-graining. *J. Comput. Chem.* **33**, 2250–2275 (2012).
15. Durand, P., Trinquier, G. & Sanejouand, Y. H. A new approach for determining low-frequency normal modes in macromolecules. *Biopolymers* **34**, 759–771 (1994).
16. Li, G. & Cui, Q. A coarse-grained normal mode approach for macromolecules: An efficient implementation and application to Ca(2+)-ATPase. *Biophys. J.* **83**, 2457–2474 (2002).
17. Suhre, K. & Sanejouand, Y.-H. ElNémo: A normal mode server for protein movement analysis and the generation of templates for molecular replacement. *Nucleic Acids Res* **32**, W610–W614 (2004).

18. McCoy, A. J. et al. Phaser crystallographic software. *J. Appl. Cryst.* **40**, 658–674 (2007).

19. Tama, F., Gadea, F.-X., Marques, O. & Sanejouand, Y.-H. Building-block approach for determining low-frequency normal modes of macromolecules. *Proteins* **41**, 1–7 (2000).

20. Gohlke, H. & Thorpe, M. A natural coarse graining for simulating large biomolecular motion. *Biophys. J.* **91**, 2115–2120 (2006).

21. Ahmed, A. & Gohlke, H. Multiscale modeling of macromolecular conformational changes combining concepts from rigidity and elastic network theory. *Proteins* **63**, 1038–1051 (2006).

22. Elber, R. & Karplus, M. Multiple conformational states of proteins: A molecular dynamics analysis of myoglobin. *Science* **235**, 318–321 (1987).

23. Kneller, G. R. & Smith, J. C. Liquid-like side-chain dynamics in myoglobin. *J. Mol. Biol.* **242**, 181–185 (1994).

24. Teeter, M., Yamano, A., Stec, B. & Mohanty, U. On the nature of a glassy state of matter in a hydrated protein: Relation to protein function. *Proc. Natl. Acad. Sci. U. S. A.* **98**, 11242–11247 (2001).

25. Ma, J. & Karplus, M. Ligand-induced conformational changes in ras p21: A normal mode and energy minimization analysis. *J. Mol. Biol.* **274**, 114–131 (1997).

26. Lamy, A., Souaille, M. & Smith, J. Simulation evidence for experimentally detectable low-temperature vibrational inhomogeneity in a globular protein. *Biopolymers* **39**, 471–478 (1996).

27. Batista, P. R. et al. Consensus modes, a robust description of protein collective motions from multiple-minima normal mode analysis application to the HIV-1 protease. *Phys. Chem. Chem. Phys.* **12**, 2850–2859 (2010).

28. Tirion, M. Low-amplitude elastic motions in proteins from a single-parameter atomic analysis. *Phys. Rev. Lett.* **77**, 1905–1908 (1996).

29. Bruschweiler, R. Collective protein dynamics and nuclear spin relaxation. *J. Chem. Phys.* **102**, 3396–3403 (1995).

30. Tama, F. & Sanejouand, Y. H. Conformational change of proteins arising from normal mode calculations. *Protein Eng.* **14**, 1–6 (2001).

31. Jacobs, D. J., Rader, A. J., Kuhn, L. A. & Thorpe, M. F. Protein flexibility predictions using graph theory. *Proteins* **44**, 150–165 (2001).

32. Jacobs, D. J. & Thorpe, M. F. Generic rigidity percolation: The pebble game. *Phys. Rev. Lett.* **75**, 4051–4054 (1995).

33. Haliloglu, T., Bahar, I. & Erman, B. Gaussian dynamics of folded proteins. *Phys. Rev. Lett.* **79**, 3090–3093 (1997).

34. Kondrashov, D., Cui, Q. & Phillips, G. Optimization and evaluation of a coarse-grained model of protein motion using x-ray crystal data. *Biophys. J.* **91**, 2760–2767 (2006).

35. Bahar, I., Atilgan, A. R., Demirel, M. C. & Erman, B. Vibrational dynamics of folded proteins: Significance of slow and fast motions in relation to function and stability. *Phys. Rev. Lett.* **80**, 2733–2736 (1998).

36. Hinsen, K. Analysis of domain motions by approximate normal mode calculations. *Proteins* **33**, 417–429 (1998).

37. Yang, L. & Bahar, I. Coupling between catalytic site and collective dynamics: A requirement for mechanochemical activity of enzymes. *Structure* **13**, 893–904 (2005).

38. Tama, F. & Brooks III, C. The mechanism and pathway of pH induced swelling in cowpea chlorotic mottle virus. *J. Mol. Biol.* **318**, 733–747 (2002).
39. Delarue, M. & Sanejouand, Y.-H. Simplified normal modes analysis of conformational transitions in DNA-dependant polymerases: The elastic network model. *J. Mol. Biol.* **320**, 1011–1024 (2002).
40. Krebs, W. G. et al. Normal mode analysis of macromolecular motions in a database framework: Developing mode concentration as a useful classifying statistic. *Proteins* **48**, 682–695 (2002).
41. Nicolay, S. & Sanejouand, Y.-H. Functional modes of proteins are among the most robust. *Phys. Rev. Lett.* **96**, 078104 (2006).
42. Kraulis, P. Molscript: A program to produce both detailed and schematic plots of protein structures. *J. Appl. Cryst.* **24**, 946–950 (1991).
43. Tama, F., Valle, M., Frank, J. & Brooks III, C. L. Dynamic reorganization of the functionally active ribosome explored by normal mode analysis and cryo-electron microscopy. *Proc. Natl. Acad. Sci. U. S. A.* **100**, 9319–9323 (2003).
44. Delarue, M. & Dumas, P. On the use of low-frequency normal modes to enforce collective movements in refining macromolecular structural models. *Proc. Natl. Acad. Sci. U. S. A.* **101**, 6957–6962 (2004).
45. Suhre, K., Navaza, J. & Sanejouand, Y.-H. NORMA: A tool for flexible fitting of high resolution protein structures into low resolution electron microscopy derived density maps. *Acta Crystallogr. D Biol. Crystallogr.* **62**, 1098–1100 (2006).
46. Bahar, I. & Cui, Q. (eds.). *Normal Mode Analysis: Theory and Applications to Biological and Chemical Systems. C&H/CRC Mathematical & Computational Biology Series*, Vol. 9. CRC Press, Boca Raton, FL (2005).
47. Bahar, I., Atilgan, A. R. & Erman, B. Direct evaluation of thermal fluctuations in proteins using a single-parameter harmonic potential. *Fold. Des.* **2**, 173–181 (1997).
48. Kundu, S., Melton, J., Sorensen, D. & Phillips Jr., G. Dynamics of proteins in crystals: Comparison of experiment with simple models. *Biophys. J.* **83**, 723–732 (2002).
49. Hinsen, K. Structural flexibility in proteins: Impact of the crystal environment. *Bioinformatics* **24**, 521 (2008).
50. Sacquin-Mora, S. & Lavery, R. Investigating the local flexibility of functional residues in hemoproteins. *Biophys. J.* **90**, 2706–2717 (2006).
51. Sacquin-Mora, S., Laforet, E. & Lavery, R. Locating the active sites of enzymes using mechanical properties. *Proteins* **67**, 350–359 (2007).
52. Maragakis, P. & Karplus, M. Large amplitude conformational change in proteins explored with a plastic network model: Adenylate kinase. *J. Mol. Biol.* **352**, 807–822 (2005).
53. Chu, J.-W. & Voth, G. A. Coarse-grained free energy functions for studying protein conformational changes: A double-well network model. *Biophys. J.* **93**, 3860–3871 (2007).
54. Levy, R., Perahia, D. & Karplus, M. Molecular dynamics of an alpha-helical polypeptide: Temperature dependence and deviation from harmonic behavior. *Proc. Natl. Acad. Sci. U. S. A.* **79**, 1346–1350 (1982).
55. Hayward, S. & Go, N. Collective variable description of native protein dynamics. *Annu. Rev. Phys. Chem.* **46**, 223–250 (1995).
56. Juanico, B., Sanejouand, Y.-H., Piazza, F. & De Los Rios, P. Discrete breathers in nonlinear network models of proteins. *Phys. Rev. Lett.* **99**, 238104 (2007).

57. Flach, S., Kladko, K. & Willis, C. R. Localized excitations in two-dimensional Hamiltonian lattices. *Phys. Rev. E* **50**, 2293–2303 (1994).

58. Dauxois, T., Litvak-Hinenzon, A., MacKay, R. & Spanoudaki, A. (eds.). *Energy Localisation and Transfer in Crystals, Biomolecules and Josephson Arrays, Advanced Series in Nonlinear Dynamics*, Vol. 22. World Scientific, Singapore (2004).

59. Piazza, F. & Sanejouand, Y. H. Breather-mediated energy transfer in proteins. *Discrete and Continuous Dynamical Systems–Series S (DCDS-S)* **4**, 1247–1266 (2011).

60. Piazza, F. & Sanejouand, Y.-H. Discrete breathers in protein structures. *Phys. Biol.* **5**, 026001 (2008).

61. Piazza, F. & Sanejouand, Y. H. Long-range energy transfer in proteins. *Phys. Biol.* **6**, 046014 (2009).

Approaches to Intrinsically Disordered Proteins

Chapter 6

ABSINTH Implicit Solvation Model and Force Field Paradigm for Use in Simulations of Intrinsically Disordered Proteins

Anuradha Mittal, Rahul K. Das,
Andreas Vitalis, and Rohit V. Pappu

CONTENTS

INTRODUCTION

The conformational properties of intrinsically disordered proteins (IDPs) and intrinsically disordered regions (IDRs) within proteins are of considerable interest. Despite their inability to fold spontaneously as autonomous units (Dunker et al. 2001; Dyson 2011; Gsponer and Babu 2009; Tompa 2012; Uversky

2002, 2011) IDPs and IDRs play important functional roles in biology (Dunker 2007; Dunker et al. 2008; Dyson and Wright 2005; Forman-Kay and Mittag 2013; Liu, Faeder, and Camacho 2009; Romero, Obradovic, and Dunker 2004; Xie et al. 2007). IDPs and IDRs play prominent roles in cell signaling and transcription regulation and they serve as hubs in protein interaction networks (Dunker et al. 2005; Haynes et al. 2006; Singh, Ganapathi, and Dash 2007). Mutations within IDPs/IDRs are implicated in several diseases (Babu et al. 2011; Uversky, Oldfield, and Dunker 2008; Uversky et al. 2009).

As autonomous units, IDPs display considerable conformational heterogeneity under standard physiological conditions (aqueous solutions, pH 7.0–7.4, 150-mM monovalent salt, low millimolar concentrations of di- and multivalent ions, and temperatures in the 25–37°C range). This implies that distinct and disparate conformations of equivalent free energies are readily sampled via spontaneous fluctuations. In order to achieve a coherent understanding of how conformational heterogeneity leads to protein function, we need accurate descriptions of the conformations accessible to an IDP under given solution conditions. Computer simulations play an important role in describing the conformational ensembles of IDPs. They are used either *de novo* to offer quantitative insights regarding archetypal IDP sequences or in synergy with data from spectroscopic experiments, which serve as restraints, in order to provide interpretations of experimental data. In either mode, simulation results are reliable only if accurate descriptions of conformational ensembles are achievable. Additionally, it is necessary to be able to perform large numbers of independent simulations for a range of sequences if we are to obtain quantitative insights regarding the relationships between information encoded in amino acid sequences and the ensembles they sample under given solution conditions (Mao, Lyle, and Pappu 2013). Due to their conformational heterogeneity, IDPs create unique challenges for computer simulations: We desire an optimal combination of efficiency, accuracy, and throughput, and this cannot be achieved using standard molecular dynamics approaches or methods that are designed for protein structure prediction.

In this chapter, we discuss advances that leverage the power of enhanced sampling methods that are based on Monte Carlo methods and are made tractable through the development of next generation (NextGen) implicit solvation models and force field refinements. Previous reviews (Vitalis and Pappu 2009b) and articles have documented advances made on the sampling end and the insights generated from analysis of simulation results for archetypal IDP sequences (Mao, Lyle, and Pappu 2013). Here, we focus on NextGen implicit solvation models and paradigms for force field development that afford robustness in terms of predictive power and higher throughput in terms of the number of sequences one can simulate, thereby enabling us to unmask the rules that govern sequence-ensemble relationships.

In implicit as opposed to explicit solvent models, solvent-mediated interactions that affect protein conformations and interactions are treated using mean-field descriptions. These descriptions have a track record for describing functionally relevant conformational equilibria for proteins and domains that fold as autonomous units (Chen, Brooks, and Khandogin 2008; Cramer and Truhlar 2008; Ren et al. 2012). Here, we focus our narrative on the conceptual foundations, parameterization, and summary of results obtained using the ABSINTH implicit solvation model (in which ABSINTH stands for self-assembly of biomolecules studied by an implicit, novel, and tunable Hamiltonian) and force field paradigm (Vitalis and Pappu 2009a). The rest of the chapter is organized as follows: the underlying concepts of the implicit solvation models are introduced, the physics of different flavors of implicit solvation models are briefly discussed to lay out a platform for introducing ABSINTH, the physics of ABSINTH is discussed, the differences of ABSINTH from the previous implicit solvation models are discussed, salient features of ABSINTH are highlighted, the results from simulations of different archetypes of IDPs using ABSINTH are summarized, and the potential for improvements to ABSINTH are discussed.

IMPLICIT SOLVATION MODELS

The free energy of solvation ΔG_{solv} for a rigid solute is defined as the change in the free energy associated with the transfer of the solute from vacuum to the solvent. For the remainder of this chapter we will presume that the solvent of interest is an aqueous solution without dispersed ions. The free energy of solvation ΔG_{solv} is traditionally computed by decomposing the transfer process as shown in Equation 6.1:

$$\Delta G_{solv} = \left(\Delta G_{ch}^{sol} - \Delta G_{ch}^{vac}\right) + \left(\Delta G_{cav} + \Delta G_{disp}\right) \equiv \Delta G_{polar} + \Delta G_{nonpolar} \qquad (6.1)$$

Each of the terms in Equation 6.1 make up different legs of a thermodynamic cycle and the corresponding free energy changes are as follows:

1. ΔG_{ch}^{sol}: Free-energy change associated with charging the solute in the solvent
2. ΔG_{ch}^{vac}: Free-energy change associated with charging the solute in vacuum
3. ΔG_{cav}: Free-energy cost associated with forming a cavity corresponding to the shape and size of the solute within the solvent
4. ΔG_{disp}: Free-energy change associated with attractive dispersion interactions between the solute and solvent

Equation 6.1 shows that the overall solvation free energy can be written as the sum of two composite terms: ΔG_{polar}, the polar/electrostatic term is the difference between charging free energies for the rigid solute in solvent versus vacuum, and $\Delta G_{\text{nonpolar}}$, the nonpolar term that accounts for mean-field estimates of the dispersive attractive interactions between the solute and solvent as well as the free-energy cost associated with cavitation. The two most popular classes of implicit solvation models are the Poisson-Boltzmann model (Baker 2005) and the generalized Born model (Chen, Brooks, and Khandogin 2008; Chen, Im, and Brooks 2006; Feig and Brooks 2004; Sigalov, Fenley, and Onufriev 2006; Tanizaki and Feig 2005). In both approaches, computations of polar and nonpolar terms are decoupled from each other. The main differences between the two approaches center on the treatment of the polar/electrostatic term and both models can be combined with similar empirical approaches for the nonpolar term.

Poisson Boltzmann Model

The Poisson equation (PE) (Equation 6.2) of electrostatics provides a formalism to compute the charging free energy for a rigid solute in a solvent that is modeled as a continuum with a dielectric constant ε. The electrostatic potential of a rigid solute with an arbitrary charge distribution can be calculated by solving the Poisson equation shown in Equation 6.2.

$$\nabla \cdot [\varepsilon(\mathbf{r})\nabla\phi(\mathbf{r})] = -\rho(\mathbf{r}); \qquad (6.2)$$

Here, $\mathbf{r}$ denotes the position vector of solute atoms and $\rho(\mathbf{r})$ is the charge density of the solute. The solute defines a dielectric boundary with the solvent and this boundary can have a rather complex shape. For proteins in aqueous solvents, one typically sets the interior dielectric constant to be small, around 4, and the bulk dielectric constant to be that for water at 25°C (i.e., around 78). In Equation 6.2, $\varepsilon(\mathbf{r})$ refers to the position-dependent dielectric constant that is intended to capture inhomogeneities in the dielectric response that arise due to a nonuniform $\rho(\mathbf{r})$. Of course, an explicit position dependence has to be specified to capture these inhomogeneities, whereas in most cases one assumes two fixed values, one for the interior and the other for the bulk solvent. Finally, $\phi(\mathbf{r})$ is the electrostatic potential corresponding to $\rho(\mathbf{r})$ for the rigid solute in a continuum solvent and can be used directly to obtain the charging free energy for the solute in solvent. The Poisson-Boltzmann equation generalizes the Poisson equation and accounts for the presence of mobile ions in the solvent. The work done to place a test charge of the solution at a specific position around the solute is governed by the Boltzmann weight of the potential of mean force. The defining approximation made in arriving at the Poisson-Boltzmann equation is that the potential of mean force can be approximated by the final electrostatic potential due to the solute. The resulting form for the Poisson-Boltzmann equation is shown in Equation 6.3.

$$\nabla \cdot [\varepsilon(\mathbf{r})\nabla\phi(\mathbf{r})] = -\rho(\mathbf{r}) - \sum_s z_s c_s^\infty \exp\left[-\frac{z_s \phi(\mathbf{r})}{kT}\right] \tag{6.3}$$

In Equation 6.3, z_s is the charge of ion s, c_s^∞ is the bulk concentration of ion s, k is the Boltzmann constant, and T is the temperature. Equation 6.3 can be generalized to include the effects of ion correlations due to the finite sizes of mobile ions in solution. In the Poisson framework that underlies Equations 6.2 and 6.3, the polar component of the solvation free energy is modeled as the mean-field response of the dipolar continuum to the accumulation of point charges inside a low-dielectric cavity embedded in a high-dielectric medium. Analytical solutions of Equations 6.2 and 6.3 are nontrivial to obtain for complex geometries such as a protein of arbitrary shape and size. These equations are amenable to numerical solution through the use of finite difference methods implemented in a variety of software suites (Baker et al. 2001; Nicholls and Honig 1991; Rocchia, Alexov, and Honig 2001). A particular disadvantage derives from the computational cost of these calculations. For a given rigid solute, the electrostatic potential derived from solution to either Equation 6.2 or 6.3 yields an estimate for ΔG_{polar} using the following:

$$\Delta G_{\text{polar}} = \frac{1}{2}\sum_i q_i [\phi(\mathbf{r}) - \phi_{\text{vac}}(\mathbf{r})] \; ; \tag{6.4}$$

Here, $\phi(\mathbf{r})$ is the solution to either Equation 6.2 or 6.3, q_i is the charge on atom i of the solute, and $\phi_{\text{vac}}(\mathbf{r})$ is the electrostatic potential computed for the solute charge distribution in a vacuum.

Generalized Born Model

In light of the computational cost associated with numerical computation of $\phi(\mathbf{r})$ using Equations 6.2 or 6.3, further approximation is needed to utilize the Poisson framework for rapid estimation of ΔG_{polar}, which is necessary in molecular simulations. The generalized Born (GB) model provides an analytical approximation to the estimate for ΔG_{polar} by explicitly ignoring reaction field contributions and deploying the so-called Coulomb field approximation. In the GB model, the solvent is again treated as a continuum, and the exact solution of the Poisson equation for point charges situated at the center of a spherical cavity of radius α is used to estimate the solvation free energy in direct analogy with the formula proposed by Born. For a point charge q, the Born formula for ΔG_{polar} is

$$\Delta G_{\text{polar}} = -\frac{q^2}{2\alpha}\left(1-\frac{1}{\varepsilon}\right) \qquad (6.5)$$

Implicit in Equation 6.5 is the assumption that the dielectric constant within the interior of the spherical cavity is that of a vacuum (i.e., unity), and that the dielectric constant of the bulk solvent is ε.

The Born formula can be generalized to a polyatomic solute with multiple point charges by assuming each charge i to be embedded within its own spherical cavity of radius α_i. Accordingly, we have

$$\Delta G_{\text{polar}} = -\sum_{i=1}^{N}\frac{q_i^2}{2\alpha_i}\left(1-\frac{1}{\varepsilon}\right) - \frac{1}{2}\sum_{i=1}^{N}\sum_{j\neq i}\frac{q_iq_j}{r_{ij}}\left(1-\frac{1}{\varepsilon}\right) \equiv \Delta G_{\text{polar}}^{\text{DMFI}} + \Delta G_{\text{polar}}^{\text{screening}} \qquad (6.6)$$

where q_i and q_j are the charges atoms i and j, and r_{ij} is interatomic distance between atoms i and j. The self-energy of atom i is the electrostatic contribution to the free energy of solvation of the solute when only atom i bears a nonzero charge. The self-term $\Delta G_{\text{polar}}^{\text{DMFI}}$, which we refer to as a direct mean-field interaction (DMFI), approximates the net interactions between the high-dielectric solvent and an individual charge situated in the low-dielectric cavity defined by the entire solute. The alterations to the charging free energy due to the presence of charges on all other sites $j \neq i$ are accounted for by a standard Coulomb field approximation (ignoring contributions due to reaction field effects). The screening term denoted as $\Delta G_{\text{polar}}^{\text{screening}}$ in Equation 6.6 can be approximated empirically, as was done by Still et al. (1990), who merged the two terms in Equation 6.6 into a single compact equation of the form:

$$\Delta G_{\text{polar}} = -\frac{1}{2}\left(1-\frac{1}{\varepsilon}\right)\sum_{i=1}^{N}\sum_{j=1}^{N}\frac{q_iq_j}{f_{\text{GB}}(r_{ij})} \qquad (6.7)$$

$$f_{\text{GB}}(r_{ij}) = \left[r_{ij}^2 + \alpha_i\alpha_j\exp\left(-\frac{r_{ij}^2}{4\alpha_i\alpha_j}\right)\right]^{\frac{1}{2}} \qquad (6.8)$$

where α_i and α_j are the effective Born radii of atoms i and j, respectively. These are evaluated by integrating the electrostatic charging free-energy density over the cavity volume. In the GB formalism, the Born radius α_i is an estimate of the distance between the atomic site i and the dielectric boundary and is therefore a measure of the degree of burial of the atom in question.

The expression for $f_{\text{GB}}(r_{ij})$ essentially invokes a sigmoidal dependence (Hingerty et al. 1985; Warshel et al. 2006) for the dependence of the dielectric saturation on the pairwise separation r_{ij}, with the generalization from previous attempts being that each pairwise interaction is treated separately by the introduction of atom-specific Born radii that are modulated from their intrinsic values by the arrangement of all the other atoms within the polyatomic solute.

Almost all implementations of GB models rely on the Still et al. formula for f_{GB}; the main differences among different implementations arise from the approaches and assumptions used to estimate the Born radii. The accuracy of GB models relies on the accurate evaluation of Born radii as adjudged against numerical estimates obtainable from direct calculations of ΔG_{polar} using the Poisson framework (Onufriev, Case, and Bashford 2002). An intrinsic inaccuracy arises from the assumption of the Coulomb field approximation (i.e., from ignoring reaction field contributions to ΔG_{polar}) and these inaccuracies are fixable using methods such as the so-called GBr^6 method (Tjong and Zhou 2007) that implements the corrections proposed by Grycuk (Grycuk 2003).

Nonpolar Contribution

The $\Delta G_{nonpolar}$ term in Equation 6.1 includes a mean-field treatment of the contributions from dispersive interactions of the solvent with the solute and the free-energy cost associated with cavitation. To a first approximation, the first-solvation shell should dominate the contributions to $\Delta G_{nonpolar}$ and hence it can be modeled as being proportional to the average number of solvent molecules within the first-solvation shell. This leads to the use of the solvent accessible surface area (SASA) for evaluating $\Delta G_{nonpolar}$ according to the relationship

$$\Delta G_{nonpolar} = \sum_{i=1}^{N} \gamma_i A_i \qquad (6.9)$$

where the terms γ_i and A_i denote the microscopic surface tension and SASA, respectively, for atom i. The calculation $\Delta G_{nonpolar}$ can be made efficient using analytical approximations to SASA, such as the one proposed by Ferrara and Caflisch, which significantly reduces the computational cost without a major trade-off in accuracy (Ferrara, Apostolakis, and Caflisch 2002).

The central assumptions surrounding the use of SASA-based models for $\Delta G_{nonpolar}$ have been revisited using a systematic set of investigations that dissect the contributions of enthalpy and entropy to the solvation free energies of small and large as well as linear and cyclic hydrophobic solutes. The principal finding is that the relative solubility patterns cannot be explained by the variation of SASA with solute, but instead it requires consideration of solute-solvent dispersion interactions (Gallicchio, Kubo, and Levy 2000). A similar shortfall of SASA based terms was found when comparing solvation forces in simulations using explicit solvation and implicit solvation models (Wagoner and Baker 2006). Gallicchio and Levy developed an improved model that accounts for the differential contributions from SASA and solute-solvent dispersion interactions in their AGBNP model, which co-opts the GB framework for polar interactions (Gallicchio and Levy 2004).

EEF1 Model

The preceding discussion focused on models that explicitly decouple and independently model the charging and nonpolar contributions to the free energy of solvation. This decoupling leverages the rich formalism for estimating ΔG_{polar} that dates back to the seminal contributions of Born, Onsager, and Kirkwood. The approximations used to estimate $\Delta G_{nonpolar}$ are considerably less well-founded (Wagoner and Baker 2006). This leads to the possibility of persistent inaccuracies with strategies that are based on the decoupling approach laid out in Equation 6.1. Improvements to estimates of the ΔG_{polar} term can, in theory, be sought through certain types of experiments, whereas considerably fewer constraints are available from experiments with regard to improvements to the estimates of the $\Delta G_{nonpolar}$ term, especially with regard to the different contributions from solvation of atomic sites in polyatomic solutes. Lazaridis and Karplus (Lazaridis and Karplus 1999) introduced an effective energy function (EEF1) method wherein one estimates the contribution of distinct solvation groups to ΔG_{solv} by using group decompositions of experimentally measured free energies of solvation (Makhatadze and Privalov 1995) for small molecules as the reference free energies. In EEF1 the overall solvation process is modeled as the sum of a single DMFI term ΔG^{DMFI} and an empirical screening term $\Delta G_{polar}^{screening}$. The DMFI term is defined as

$$\Delta G^{DMFI} = \Delta G_{polar}^{DMFI} + \Delta G_{nonpolar} = \sum_i \Delta G_{solv}^i \tag{6.10}$$

where ΔG_{solv}^i is the free energy of solvation of group i within the solute. A polyatomic solute can be thought of as a concatenation of different solvation groups, where each solvation groups corresponds to a small chemical moiety. For group i the reference value for the free energy of solvation is denoted as ΔG_{ref}^i. Since solvent can be excluded from sites occupied by atoms of other solvation groups of the polyatomic solute, the true ΔG_{solv}^i for solvation group i can be calculated by estimating the degree to which group i is occluded from the solvent. Accordingly

$$\Delta G_{solv}^i = \Delta G_{ref}^i - \Delta G_{excl}^i \tag{6.11}$$

where ΔG_{excl}^i represents the expected diminution of ΔG_{ref}^i due to exclusion of solvent by neighboring groups. This term is computed as

$$\Delta G_{\text{excl}}^{i} = \sum_{j \neq i} f_i(r_{ij}) V_j \qquad (6.12)$$

where r_{ij} is the distance between solvation groups i and j and $f_i(r_{ij})$ is the free energy density that is a function of distance with its magnitude being the largest near the solute and decaying to zero away from the solute. Lazaridis and Karplus used a Gaussian function for $f_i(r_{ij})$; V_j, the volume occupied by solvation group j is evaluated as

$$V_j = \frac{4}{3} \pi R_j^3 - \frac{1}{2} \sum_k V_{jk}^{(2)} \qquad (6.13)$$

where R_j is the radius of atom j and $V_{jk}^{(2)}$ is the overlap volume between atoms j and k.

The screening of electrostatic interactions is modeled by a separate term, $\Delta G_{\text{polar}}^{\text{screening}}$, which in EEF1 is computed by a combination of a distance-dependent dielectric response and neutralization of side chains carrying a net charge. This assures that the short-range electrostatic attractions remain unaltered whereas long-range interactions are significantly diminished due to screening. The choices made in EEF1 for modeling the modulation of charge-charge interactions by the surrounding solvent become problematic for simulations of IDP sequences, which are often enriched in charged residues. This necessitated the pursuit of an improved, second-generation implicit solvation model that relies on the conceptual framework of EEF1 while simultaneously leveraging the conceptual strengths of the Poisson framework and improving on both approaches. The resultant ABSINTH implicit solvation model and force field paradigm are discussed in the following.

ABSINTH MODEL

In the ABSINTH approach (Vitalis and Pappu 2009a), each polyatomic solute is parsed into a set of solvation groups. Each of these groups corresponds to model compounds for which free energies of solvation have been measured. The choice of solvation groups is different from the EEF1 model, which pursues a finer parsing of the solvation groups based on empirical dissections of the free energies of solvation (Makhatadze, Lopez, and Privalov 1997; Makhatadze and Privalov 1994). The total energy associated with a rigid polyatomic solute that includes the biopolymer and solution ions is written as

$$E_{\text{total}} = W_{\text{solv}} + W_{\text{el}} + U_{\text{LJ}} + U_{\text{corr}} \qquad (6.14)$$

where W_{solv} represents the DMFI and captures the free-energy change associated with transferring the polyatomic solute into a mean-field solvent while accounting for the modulation of each solvation group's reference free energy of solvation due to occlusion from the solvent by atoms of the polyatomic solute—an intrinsically many-body effect. Further modulations to the free energy of solvation of the solute due to screened interactions with charged sites on the polyatomic solute are accounted for by the W_{el} term, wherein the effects of dielectric inhomogeneities are fully accounted for without making explicit assumptions regarding the distance or spatial dependencies of dielectric saturation. The term U_{LJ} is a standard 12-6 Lennard-Jones potential that models the joint contributions of steric exclusion and London dispersion whereas U_{corr} models specific torsion and bond angle-dependent stereoelectronic effects that are not captured by the U_{LJ} term.

W_{solv} and W_{el}

As noted in the preceding paragraph, the polyatomic solute is parsed into a set of nonoverlapping solvation groups such that each solvation group is a small molecule analog whose free energy of solvation is known from experimental measurements. W_{solv} is written as

$$W_{solv} = \Delta G^{DMFI} = \sum_{i=1}^{N_{SG}} \left[\sum_{k=1}^{n_i} \lambda_{ik} \upsilon_{ik}^{solv} \right] \Delta G_i^{solv} \tag{6.15}$$

where N_{SG} is the number of solvation groups in the system, ΔG_i^{solv} is the experimentally measured reference free energy of solvation for solvation group i, n_i is the number of atoms that belong to solvation group i, λ_{ik} is a multiplicative factor $(0 \leq \lambda_{ik} \leq 1)$ associated with the kth atom of solvation group i, and υ_{ik}^{solv} is solvation state $\left(0 \leq \upsilon_{ik}^{solv} \leq 1 \right)$ of the kth atom of solvation group i.

In the current implementation of ABSINTH, the solvation state υ_{ik}^{solv} is a non-linear stretched sigmoidal function of the solvent accessible volume fraction η_{ik} such that $0 \leq \upsilon_{ik}^{solv} \leq 1$. The value of υ_{ik}^{solv} approaches zero as the atom k in solvation group i becomes inaccessible to the solvent, and conversely it approaches unity as the site becomes fully accessible to the solvent. The degree of solvent accessibility is quantified using η_{ik}, which depends on the radius of the solvation shell as well as the van der Waals radius of atom k and all other overlapping atoms. The mapping between η_{ik} and υ_{ik}^{solv} is governed by values chosen for χ_d and τ_d, which are the parameters of the stretched sigmoidal function (see Figure 6.1). The values for χ_d and τ_d dictate the choice made for the stabilities of partially solvated states and free energy changes associated with transitioning between fully solvated and fully desolvated states.

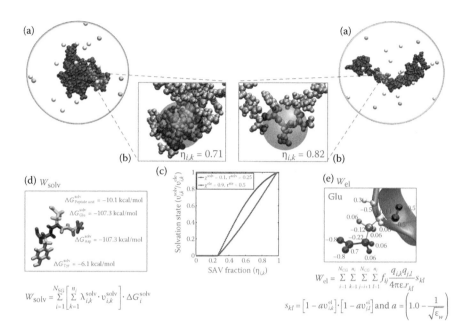

Figure 6.1 (a) Representative compact and expanded conformation drawn from the simulations of the sequence DP00503 (Das and Pappu 2013) at 298 K. Atoms are shown in space-filling representation and the models were drawn using the Visual Molecular Dynamics package (Humphrey, Dalke, and Schulten 1996). The polypeptide is shown in magenta and residues Y31, E32, and D33 are in orange. The sodium ions are shown in yellow and the chloride ions are shown in green. (b) Solvent-accessible volume fractions of the CA atom of the E32 peptide unit in the two representative conformations shown in (a). Atoms are shown in space-filling representation. The CA atom of E32 is shown in red and its mean-field solvation shell is shown in the gray sphere. The atoms within the solvation shell of CA atom of E32 are shown in cyan, atoms occupying a part of the solvation shell are shown in purple, and the atoms outside the solvation shell are shown in pink. Calculated solvent-accessible volume fractions of the CA atom of E32 in compact and expanded conformations shown in (a) are 0.71 and 0.82, respectively. (c) Mapping between SAV fractions and solvation states using stretched sigmoidal interpolation for the computation of W_{solv} and W_{el}. (d) Enlarged perspective of the residues Y31, E32, and D33 to show their parsing into solvation groups. The residues are depicted in the licorice representation and are color-coded depending on the model compounds. The three peptide units are modeled using N-methylacetamide and are shown in magenta, blue, and orange. The side chains of tyrosine, glutamate, and aspartate are modeled using p-cresol, propionic acid and acetic acid, respectively, and are colored cyan, green, and red. The reference free energies of solvation for the solvation groups are displayed. (e) Zoom of the residue E32 showing its charge groups. The polypeptide backbone is shown in cartoon representation and E32 is shown in ball-and-stick representation. The partial charges for all atoms comprising E32 are displayed and the atoms forming a particular charge group are indicated with same color.

While the W_{solv} term captures the sum of $\Delta G_{polar}^{DMFI} + \Delta G_{nonpolar}$ (see Equations 6.1 and 6.6), the term that remains (within the Coulomb field approximation) is $\Delta G_{polar}^{screening}$, and this is modeled using the W_{el} term, which is defined as

$$W_{el} = \sum_{i=1}^{N_{CG}} \sum_{k=1}^{n_i} \sum_{j=i+1}^{N_{CG}} \sum_{l=1}^{n_j} f_{ij} \frac{q_{ik}q_{jl}}{r_{kl}} s_{kl}$$

$$s_{kl} = \left[1 - a\upsilon_{ik}^{el}\right]\left[1 - a\upsilon_{jl}^{el}\right]$$

$$a = \left(1 - \frac{1}{\sqrt{\varepsilon}}\right) \tag{6.16}$$

where N_{CG} refers to the number of charge groups within the system. Unlike the identities of solvation groups, which are governed by experimental data for model compounds and hence independent of the molecular mechanics parameter set, the identities of charge groups, which are actually groups of point charges, are chosen based on the molecular mechanics force field from which the charges are extracted. Each charge group is net-neutral unless the functional group has a net charge at neutral pH. Examples of functional groups with a net charge are the amine, guanido, and carboxyl groups of charged side chains, the phosphate group on nucleic acid backbones, and mono- and polyvalent ions that are part of the solution milieu and include ions such as Na^+, K^+, Mg^{2+}, and Cl^-. The factor f_{ij} is set to zero if any pair of atoms k and l within groups i and j are connected either via a direct covalent bond or are part of a bond angle; otherwise, f_{ij} is set to unity. This choice has two consequences: First, no charge-charge interactions can be computed within a charge group, and second, interacting charge groups cannot have any pair of atoms separated by less than a single rotatable bond. This choice was made to ensure fidelity with the neutral groups paradigm that governs the parameterization of most charge sets—a paradigm that gets upended when the constraints on neutral groups are not explicitly specified (as done above). Inclusion of spurious charge-charge and charge-dipole interactions is inconvenient for parameterization and incompatible with the many-body coupling in Equation 6.16. By ensuring that W_{el} accounts only for nonlocal interactions, the design of ABSINTH departs from all other implicit solvation models and force field paradigms. The remaining elements in Equation 6.16 are n_i, n_j, ε, and υ_{ik}^{el}. These refer to the number of point charges within groups i and j, the bulk dielectric constant of the solvent, and the solvation state of the charge k from charge group i, respectively. In addition to the definition of charge groups and setting of f_{ij}, which are encoded by the parameter set for the charges, the parameters that govern the screening model are the solvation states of the charges; namely, υ_{ik}^{el}.

In direct analogy with the solvation states for atoms within solvation groups, $0 \leq \upsilon_{ik}^{el} \leq 1$, and the individual υ_{ik}^{el} are stretched sigmoidal functions of the solvent-accessible volume fractions η_{ik}. The dependence of υ_{ik}^{el} on η_{ik} is encoded by two separate parameters χ_s and τ_s and their values are distinct from those of χ_d and τ_d (see Figure 6.1). The choices for χ_s and τ_s as well as those for χ_d and τ_d are made empirically and they represent a particular choice of models for describing the free energies of partially solvated states.

It is worth noting that ABSINTH allows flexibility in the choice of a model used for dielectric saturation because this is governed by the choices made for χ_s and τ_s. The absence of an analytical model for describing the partially screened/descreened interactions necessitates empirical choices to be made for the values χ_s and τ_s, and of course these choices are also the choices made for χ_d and τ_d. Through systematic testing and calibration it was found that setting $\chi_d = 0.1$, $\tau_d = 0.25$, $\chi_s = 0.9$, and $\tau_s = 0.5$ helped preserve the folded states for a range of globular proteins at simulation temperatures spanning the range from 5°C to 40°C. In addition, these choices lead to reliable and reversible folding of several model peptide systems that were part of the initial and follow-up calibration studies (Vitalis and Pappu 2009a,b; Vitalis, Wang, and Pappu 2008).

The ABSINTH model is unique because it allows the inclusion of explicit solution ions of finite size for modeling the effects of ionic strength and specific effects of aqueous electrolyte solutions. This is particularly relevant when accounting for the effects of salt concentration on the conformational properties of polyelectrolyte and polyampholyte sequences, which tend to be the dominant archetypes among IDPs (Das and Pappu 2013; Mao et al. 2010). The choice of parameters χ_d, τ_d, χ_s, and τ_s that govern functional dependencies of υ_{ik}^{solv} and υ_{ik}^{el} on solvent-accessible volume fractions is purely empirical and reflects our lack of knowledge regarding the free energies of solvation of partially solvated species as well as a theoretical grounding for this process. The free energies of solvation of species with a net charge (including charged side chains, the phosphate groups along nucleic acid backbones, and mobile solution ions) are highly favorable. The free energies of solvation even of monovalent ions are roughly an order of magnitude larger than those of polar moieties. The screened/descreened electrostatic interactions between groups with a net charge can have energies that are either on par with or more favorable than the free energies of solvation. The equivalence of large energy scales can create salt bridging between desolvated and partially solvated ion pairs that represent spurious, albeit deep traps on the energy landscape. Consequently, in almost all deployments of the ABSINTH model in simulations, if a solvation group has a net charge and is part of a polymer, we lower the reference free energies of solvation by making the values of ΔG_i^{solv} more negative than the values inferred from experimental data for whole salts. This helps avoid the formation of spurious salt bridges and tilts the balance toward the solvation of charged groups. This choice appears to work well for describing the conformational properties of a large spectrum of IDP sequences that are enriched in charged residues. Finally, ABSINTH is also unique because no cutoffs or truncations are used when modeling electrostatic interactions between groups that carry a net charge—a feature that is enabled by using large spherical droplets for all simulations that also guard against finite size artifacts. Although the choice of modeling the full range of electrostatic interactions guards against deleterious consequences of mishandling these interactions, it poses a computational challenge, especially when we desire to model the effects of increased salt concentration on conformational properties.

Lennard-Jones Interactions and Stereospecific Effects

The effects of steric exclusion and solute-solute dispersive interactions are modeled using a 12-6 Lennard-Jones potential. In any simulation using ABSINTH that is based on the use of the CAMPARI simulation package (http://campari .sourceforge.net), the user is restricted to a fixed set of parameters for the Lennard-Jones potential that are designed for use with all molecular mechanics charge parameters that are supported in CAMPARI. This ensures a general robustness of the simulation results to the choice of parameters for charges. The parameters for the Lennard-Jones potential have been extensively calibrated for accuracy and are based on the original numbers proposed by Pauling and are designed to reproduce the heats of fusion and densities of small molecules in their crystalline form (Radhakrishnan et al. 2012; Tran, Wang, and Pappu 2005; Vitalis and Pappu 2009a). This particular choice ensures that the hard sphere radii are generally smaller than those that are part of standard molecular mechanics force fields. Bonded parameters are included for strong electronic effects that are impossible to describe approximately by steric interactions (Radhakrishnan et al. 2012; Vitalis and Pappu 2009a). Examples of the extensive efforts invested into the calibration of Lennard-Jones parameters include the early work (Tran, Wang, and Pappu 2005) that was geared toward modeling denatured state ensembles of a series of proteins, further optimization of these parameters to produce reliable and reversible folding of and experimental data for local conformational preferences of a series of model peptides (Vitalis and Pappu 2009a), demonstration that the choices of modified Lennard-Jones parameters preserve/improve the accuracy of free energies of solvation in explicit solvent simulations for a series of model compounds (Wyczalkowski, Vitalis, and Pappu 2010), the development of a transferable set of parameters for alkali and halide ions that are based on the use of crystal lattice parameters (Mao and Pappu 2012), and the reproduction of accurate descriptions of conformational properties of poly-L-proline polymers, which also required the inclusion of additional torsional and bond angle degrees of freedom and parameters to describe these degrees of freedom (Radhakrishnan et al. 2012). As with all models, the ABSINTH model encompasses a set of tunable parameters, the choices for which are dictated by a combination of constraints provided by experimental data, existing parameters from molecular mechanics force fields, and parameter exploration that is constrained by the outcomes of a series of test simulations. In ABSINTH, the charge parameters are treated as modular entities. As noted, we employ chemically accurate Lennard-Jones parameters that have undergone significant testing and optimization given the typical set of degrees of freedom sampled in the Monte Carlo simulations. They would most likely have to be adjusted for an adaptation of the model to Cartesian dynamics (Vitalis and Caflisch 2012; Vitalis and Pappu 2009b).

APPLICATIONS OF THE ABSINTH MODEL IN SIMULATIONS OF IDPs

The ABSINTH model was primarily developed for studying the self-assembly of biomolecules. Since its development, ABSINTH has evolved considerably and has been demonstrated to be successful at accurate modeling of the conformational ensembles of various polypeptide systems and of particular relevance is the deployment of ABSINTH in Monte Carlo simulations of various archetypal IDPs. Reproducibility of simulation results and easy access to the ABSINTH paradigm and sampling tools have been made possible by the distribution of CAMPARI, a freely downloadable software suite that is developed and maintained primarily by Andreas Vitalis (see http://campari.sourceforge .net). Prominent examples of the application of ABSINTH for simulations of IDPs include archetypes that are enriched in polar tracts, polyelectrolytic and polyampholytic IDPs, and modules of transcription factors that are enriched in basic residues. The model has afforded medium- to high-throughput atomistic simulations that are imperative for obtaining quantitative description of the coarse-grained conformational properties. Furthermore, quantitative corroboration of the predictions with the experimental data has increased confidence in the accuracy of the ABSINTH model. The following provides a brief and less than exhaustive survey of results obtained using the ABSINTH model for various IDP systems.

Polyglutamine: Aggregation of expanded polyglutamine tracts in proteins is associated with several neurodegenerative diseases. Insights regarding the conformational equilibria and the driving forces for aggregation of polyglutamine sequences are crucial for understanding the molecular mechanisms of polyglutamine pathogenesis. Vitalis et al. characterized the phase behavior of polyglutamine as a function of chain length and temperature (Vitalis,

Wang, and Pappu 2008). Several polyglutamine constructs, ranging between 5–45 residues, were simulated in this work. Monomeric polyglutamine tracts were shown to be intrinsically disordered for all chain lengths, in agreement with experimental data (Chen et al. 2001). Vitalis et al. further demonstrated that conformations with high β-content are thermodynamically disfavored for polyglutamine monomers and that the free-energy penalty associated with forming β-rich conformations increases with increasing chain length (Vitalis, Lyle, and Pappu 2009). Additionally, in accord with polymer physics theories (Pappu et al. 2008), the simulation results for polyglutamine constructs show evidence for continuous globule-to-coil transitions. The simulation results demonstrated that the stabilities of globular conformations, the sharpness of the globule-to-coil transitions, as well as the spontaneities of the intermolecular association increase with increasing chain length of polyglutamine.

The exon1 from huntingtin protein includes two flanking sequence modules: an amphipathic 17-residue N-terminal stretch or N17 and a 38-residue C-terminal proline-rich stretch or C38. Williamson et al. used atomistic simulations based on the ABSINTH model to interrogate the effects of N17 on polyglutamine conformations and intermolecular associations (Williamson et al. 2010). The N17 module shows preference for helical conformations in accord with results from circular dichroism (Williamson et al. 2010), and it undergoes a polyglutamine length dependent helix-to-coil transition to increase the intramolecular interface between its hydrophobic groups and the polyglutamine tract, also in agreement with the experimental measurements of intra-N17 distances (Thakur et al. 2009). Simulations showed that the N17 unfolds and adsorbs on and polyglutamine. This generates a coarse-grained structure for N17-Q_n that resembles a patchy colloid, which in turn decreases the frequency of nonspecific intermolecular associations—a prediction that was recently confirmed by the experiments (Crick et al. 2013).

Polyelectrolytic and polyampholytic IDPs: A majority of IDPs are enriched in charged residues. Two coarse-grained parameters, namely, the net charge per residue (NCPR) and the fraction of charged residues (FCRs) are useful for quantitative classification of IDP properties. The NCPR and FCR are defined as $|f_+ - f_-|$ and $(f_+ + f_-)$ where f_+ and f_- refer to the fractions of positive- and negatively charged residues, respectively. In calculating f_+ and f_- we assume that the charge states of amino acid side chains are governed by their pK_a values at neutral pH—an assumption that bears revisiting. IDPs can be either polyelectrolytic or polyampholytic based on their charge composition. Polyelectrolytic IDPs are characterized by presence or abundance of only one type of charge and hence have NCPR ≈ FCR whereas polyampholytic IDPs consist of both kinds of charges, and NCPR ≈ 0 for these sequences.

Atomistic simulations based on the ABSINTH model (Mao et al. 2010) were used to characterize the conformational ensembles of arginine-rich polyelectrolytes. For these polyelectrolytic IDPs, NCPR acts as the discriminating order parameter for transitions between collapsed globules and swollen coils. The predicted translational diffusion constants from the simulations were in quantitative agreement with those measured from the fluorescence correlation spectroscopy experiments. A total of 21 protamine sequences, ranging from 24–48 residues, were simulated in this work.

Approximately 95% of sequences in intrinsically disordered proteomes are polyampholytes. Atomistic simulations of synthetic sequence permutants of polyampholytic (Glu-Lys)$_{25}$ and several naturally occurring polyampholytic IDP/IDR sequences were used to demonstrate that combination of FCR and the linear distribution of oppositely charged residues define the conformational properties of polyampholytic IDPs (Das and Pappu 2013). A parameter κ was introduced to quantify the patterning of opposite charges; κ approaches zero for sequences where opposite charges are all mixed and approaches one where opposite charges are segregated. It was shown that among strong polyampholytes the sequences with low κ form swollen coils while those with high κ form hairpinlike conformations. Furthermore, it was demonstrated that alteration of κ in strong polyampholytes yields the *de novo* design of sequences with altered conformational properties.

MoREs/MoRFs: The dogma that IDPs possess uniformly high disorder has come under scrutiny by the discovery of molecular recognition elements/features (MoREs/MoRFs) (Mohan et al. 2006; Vacic et al. 2007). These are short linear motifs (SLiMs) (Davey et al. 2012) within IDPs that have propensities to sample secondary structures in their unbound states and IDPs often use these to recognize their partners. MoREs or MoRFs undergo disorder-to-order transitions upon binding to their partners and hence quantitative predictions of their propensities to preorganize in unbound states are useful for understanding both the thermodynamics and kinetics of coupled folding and binding and the mechanisms of specificity in molecular recognition through IDPs (Das, Mittal, and Pappu 2013).

Basic region leucine zippers (bZIPs) (Amoutzias et al. 2007) are a family of bipartite transcription factors that, upon binding to the deoxyribonucleic acid (DNA), dimerize to form a coiled coil helix. The sequences of bZIPs are modular. They have a basic region that is responsible for binding to the cognate site and a leucine zipper region that promotes the dimerization. Experiments and bioinformatics predictions suggest that monomeric basic regions of the bZIPs should be uniformly disordered, undergoing disorder-to-order transitions upon binding to the DNA. Atomistic simulations of basic regions from 15 different bZIPs using the ABSINTH model (Das, Crick, and Pappu 2012) showed that the sequences of basic regions of bZIPs encode sequence-specific helicities. Despite significant similarities in the DNA-binding motifs, the intrinsic helicities are modulated by the N-terminal sequence contexts of the basic regions of bZIPs. The DNA binding motifs (DBMs) within the basic regions of bZIPs are therefore α-MoREs and their sequence context, specifically, the eight-residue segments directly N-terminal to DBMs modulate their helicities. The accuracy of helicity predictions from simulations was tested using circular dichroism experiments. The predicted α-helicities from the simulations agree reasonably well with the measured α-helicities from the experiments (Das, Crick, and Pappu 2012), demonstrating the utility of the ABSINTH model for quantitative prediction of MoREs. A total of 64 sequences, ranging from 8 to 28 residues, were simulated in this work.

PROSPECTS FOR IMPROVEMENTS AND ONGOING DEVELOPMENTS

Constant pH engine: The protonation states of the charged side chains depend on their environments. As the IDPs are abundant in charged residues, pK_a shifts of ionizable groups are likely to have significant effects on their electrostatic interactions, which in turn will have significant impact on conformational properties and functions. For example, alterations to protonation states can either engender a coil-to-globule transition or a disorder-to-order transition (Mao, Lyle, and Pappu 2013). In its current implementation, the ABSINTH model uses a fixed set of charges. This needs to be generalized to perform simulations at constant pH thus enabling the alteration of protonation states for ionizable groups. We envision an adaptation of constant pH methods that are well suited for use with the ABSINTH paradigm and the Monte Carlo sampling strategy (Baptista, Teixeira, and Soares 2002; Machuqueiro and Baptista 2006; Mongan and Case 2005).

Improved modeling of partial desolvation: The ABSINTH model utilizes two distinct stretched sigmoidal functions that are used to map the solvent-accessible volume fractions to values for solvation states of atoms within solvation and charge groups, respectively. While the fully solvated and fully desolvated states are well-defined limits, the free energies of partially solvated/desolvated states are likely to depend on the solvent-solute interactions between first solvation shell molecules and the solute atoms. It is unclear if the current interpolation scheme is well suited for describing the physics of partially solvated/desolvated states. Highly charged systems pose a particular challenge, and this is especially relevant when modeling protein-nucleic acid interactions, a recurrent theme with IDPs. One of many stringent tests of the model used in ABSINTH for interpolating between fully solvated and fully desolvated limits is the quantitative reproduction of the concentration dependence of mean activity coefficients as a function of salt concentration. Unpublished results suggest that recasting the interpolation model by increasing the metastability and degeneracy of partially solvated states using a rather simple stair-step model for interpolation might afford improved accuracy in terms of describing the thermodynamic properties of aqueous electrolyte solutions while maintaining or improving the accuracy of ABSINTH for simulations of proteins and nucleic acids. A detailed investigation of an improved, NextGen ABSINTH model is currently underway, and this is being pursued in conjunction with the development of a constant pH engine.

CONCLUSIONS

In this chapter, we have provided an overview of implicit solvation models and focused our narrative on the paradigm that underlies the ABSINTH model while surveying some of its uses in simulations of IDPs. Simulations of conformational properties, self-assembly, and functional interactions of IDPs are topics of growing interest. Since its development and introduction in 2009, the ABSINTH model has proven to be the only model that is suitable for accurate medium- to high-throughput simulations of a range of IDPs. The simulation results have received quantitative certification through systematic assessments of the predictions using experiments that directly query conformational properties of IDPs. This has helped in identification of rules that govern sequence-ensemble relationships for archetypal IDPs. In addition to providing open access to all of the published simulation results to ensure reproducibility, full access to the ABSINTH model and the code base is available through the CAMAPRI simulation software suite. Of course, challenges loom on the horizon and these reflect the many complexities associated with systems such as IDPs that are characterized by high degrees of conformational heterogeneity (Lyle, Das, and Pappu 2013). In addition to the increased flexibility we desire for ABSINTH (see discussion regarding the handling of ionizable groups and partially solvated states), two challenges stand out and are motivated by the biophysical aspects of IDPs. The first pertains to accurate and efficient simulation of protein-protein and protein-nucleic acid interactions with the goal of answering important questions pertaining to the mechanisms and driving forces for coupled folding and binding. The second challenge pertains to the simulation of phase transitions of IDPs, which will require the development of multiscale methods. Systematic coarse-graining is essential to understand the synergy between and the renormalization of collective coordinates describing phase transitions mediated by homo- and heterotypic interactions. The ABSINTH model will be a key ingredient for this approach both by providing information from atomistic simulations and by natural extensions of the framework to a coarser scale.

REFERENCES

Amoutzias, G. D., A. S. Veron, J. Weiner, 3rd, M. Robinson-Rechavi, E. Bornberg-Bauer, S. G. Oliver, and D. L. Robertson. 2007. One billion years of bZIP transcription factor evolution: Conservation and change in dimerization and DNA-binding site specificity. *Mol Biol Evol* 24 (3):827–835.

Babu, M. M., R. van der Lee, N. S. de Groot, and J. Gsponer. 2011. Intrinsically disordered proteins: Regulation and disease. *Curr Opin Struct Biol* 21 (3):432–440.

Baker, N. A. D. Sept, S. Joseph, M. J. Holst, and J. A. McCammon. 2001. Electrostatics of nanosystems: Applications to the microtubulue and the ribosome. *Pro Natl Acad Sci U S A* 98 (18):10037–10041.

Baker, N. A. 2005. Improving implicit solvent simulations: A Poisson-centric view. *Curr Opin Struct Biol* 15 (2):137–143.

Baptista, A. M., V. H. Teixeira, and C. M. Soares. 2002. Constant-pH molecular dynamics using stochastic titration. *J Chem Phys* 117 (9):4184–4200.

Chen, J., C. L. Brooks, 3rd, and J. Khandogin. 2008. Recent advances in implicit solvent-based methods for biomolecular simulations. *Curr Opin Struct Biol* 18 (2):140–148.

Chen, J., W. Im, and C. L. Brooks, 3rd. 2006. Balancing solvation and intramolecular interactions: Toward a consistent generalized Born force field. *J Am Chem Soc* 128 (11):3728–3736.

Chen, S., V. Berthelier, W. Yang, and R. Wetzel. 2001. Polyglutamine aggregation behavior in vitro supports a recruitment mechanism of cytotoxicity. *J Mol Biol* 311 (1):173–182.

Cramer, C. J., and D. G. Truhlar. 2008. A universal approach to solvation modeling. *Acc Chem Res* 41 (6):760–768.

Crick, S. L., K. M. Ruff, K. Garai, C. Frieden, and R. V. Pappu. 2013. Unmasking the roles of N- and C-terminal flanking sequences from exon 1 of huntingtin as modulators of polyglutamine aggregation. *Proc Natl Acad Sci U S A* 110 (50):20075–20080.

Das, R. K., S. L. Crick, and R. V. Pappu. 2012. N-terminal segments modulate the alpha-helical propensities of the intrinsically disordered basic regions of bZIP proteins. *J Mol Biol* 416 (2):287–299.

Das, R. K., A. Mittal, and R. V. Pappu. 2013. How is functional specificity achieved through disordered regions of proteins? *Bioessays* 35 (1):17–22.

Das, R. K., and R. V. Pappu. 2013. Conformations of intrinsically disordered proteins are influenced by linear sequence distributions of oppositely charged residues. *Proc Natl Acad Sci U S A* 110 (33):13392–13397.

Davey, N. E., K. Van Roey, R. J. Weatheritt, G. Toedt, B. Uyar, B. Altenberg, A. Budd, F. Diella, H. Dinkel, and T. J. Gibson. 2012. Attributes of short linear motifs. *Mol Biosyst* 8 (1):268–281.

Dunker, A. K. 2007. Another window into disordered protein function. *Structure* 15 (9):1026–1028.

Dunker, A. K., M. S. Cortese, P. Romero, L. M. Iakoucheva, and V. N. Uversky. 2005. Flexible nets. The roles of intrinsic disorder in protein interaction networks. *FEBS J* 272 (20):5129–5148.

Dunker, A. K., J. D. Lawson, C. J. Brown, R. M. Williams, P. Romero, J. S. Oh, C. J. Oldfield, A. M. Campen, C. M. Ratliff, K. W. Hipps, J. Ausio, M. S. Nissen, R. Reeves, C. Kang, C. R. Kissinger, R. W. Bailey, M. D. Griswold, W. Chiu, E. C. Garner, and Z. Obradovic. 2001. Intrinsically disordered protein. *J Mol Graph Model* 19 (1):26–59.

Dunker, A. K., I. Silman, V. N. Uversky, and J. L. Sussman. 2008. Function and structure of inherently disordered proteins. *Curr Opin Struct Biol* 18 (6):756–764.

Dyson, H. J. 2011. Expanding the proteome: Disordered and alternatively folded proteins. *Q Rev Biophys* 44 (4):467–518.

Dyson, H. J., and P. E. Wright. 2005. Intrinsically unstructured proteins and their functions. *Nat Rev Mol Cell Biol* 6 (3):197–208.

Feig, M., and C. L. Brooks, 3rd. 2004. Recent advances in the development and application of implicit solvent models in biomolecule simulations. *Curr Opin Struct Biol* 14 (2):217–224.

Ferrara, P., J. Apostolakis, and A. Caflisch. 2002. Evaluation of a fast implicit solvent model for molecular dynamics simulations. *Proteins* 46 (1):24–33.

Forman-Kay, J. D., and T. Mittag. 2013. From sequence and forces to structure, function, and evolution of intrinsically disordered proteins. *Structure* 21 (9):1492–1499.

Gallicchio, E., M. M. Kubo, and R. M. Levy. 2000. Enthalpy-entropy and cavity decomposition of alkane hydration free energies: Numerical results and implications for theories of hydrophobic solvation. *J Phys Chem B* 104 (26):6271–6285.

Gallicchio, E., and R. M. Levy. 2004. AGBNP: An analytic implicit solvent model suitable for molecular dynamics simulations and high-resolution modeling. *J Comput Chem* 25 (4):479–499.

Grycuk, T. 2003. Deficiency of the Coulomb-field approximation in the generalized Born model: An improved formula for Born radii evaluation. *J Chem Phys* 119 (9):4817–4826.

Gsponer, J., and M. M. Babu. 2009. The rules of disorder or why disorder rules. *Prog Biophys Mol Biol* 99 (2–3):94–103.

Haynes, C., C. J. Oldfield, F. Ji, N. Klitgord, M. E. Cusick, P. Radivojac, V. N. Uversky, M. Vidal, and L. M. Iakoucheva. 2006. Intrinsic disorder is a common feature of hub proteins from four eukaryotic interactomes. *PLoS Comput Biol* 2 (8):e100.

Hingerty, B. E., R. H. Ritchie, T. L. Ferrell, and J. E. Turner. 1985. Dielectric effects in bio-polymers—The theory of ionic saturation revisited. *Biopolymers* 24 (3):427–439.

Humphrey, W., A. Dalke, and K. Schulten. 1996. VMD: Visual molecular dynamics. *J Mol Graph Model* 14 (1):33–38.

Lazaridis, T., and M. Karplus. 1999. Effective energy function for proteins in solution. *Proteins* 35 (2):133–152.

Liu, J., J. R. Faeder, and C. J. Camacho. 2009. Toward a quantitative theory of intrinsically disordered proteins and their function. *Proc Natl Acad Sci U S A* 106 (47):19819–19823.

Lyle, N., R. K. Das, and R. V. Pappu. 2013. A quantitative measure for protein conformational heterogeneity. *J Chem Phys* 139 (12):121907.

Machuqueiro, M., and A. M. Baptista. 2006. Constant-pH molecular dynamics with ionic strength effects: Protonation-conformation coupling in decalysine. *J Phys Chem B* 110 (6):2927–2933.

Makhatadze, G. I., M. M. Lopez, and P. L. Privalov. 1997. Heat capacities of protein functional groups. *Biophys Chem* 64 (1–3):93–101.

Makhatadze, G. I., and P. L. Privalov. 1994. Hydration effects in protein unfolding. *Biophys Chem* 51 (2–3):291–309.

Makhatadze, G. I., and P. L. Privalov. 1995. Energetics of protein structure. *Adv Prot Chem* 47:307–425.

Mao, A. H., S. L. Crick, A. Vitalis, C. L. Chicoine, and R. V. Pappu. 2010. Net charge per residue modulates conformational ensembles of intrinsically disordered proteins. *Proc Natl Acad Sci U S A* 107 (18):8183–8188.

Mao, A. H., N. Lyle, and R. V. Pappu. 2013. Describing sequence-ensemble relationships for intrinsically disordered proteins. *Biochem J* 449:307–318.

Mao, A. H., and R. V. Pappu. 2012. Crystal lattice properties fully determine short-range interaction parameters for alkali and halide ions. *J Chem Phys* 137 (6):064104.

Mohan, A., C. J. Oldfield, P. Radivojac, V. Vacic, M. S. Cortese, A. K. Dunker, and V. N. Uversky. 2006. Analysis of molecular recognition features (MoRFs). *J Mol Biol* 362 (5):1043–1059.

Mongan, J., and D. A. Case. 2005. Biomolecular simulations at constant pH. *Curr Opin Struct Biol* 15 (2):157–163.

Nicholls, A., and B. Honig. 1991. A rapid finite difference algorithm, utilizing successive over-relaxation to solve the Poisson–Boltzmann equation. *J Comput Chem* 12 (4):435–445.

Onufriev, A., D. A. Case, and D. Bashford. 2002. Effective Born radii in the generalized Born approximation: The importance of being perfect. *J Comput Chem* 23 (14):1297–1304.

Pappu, R. V., X. Wang, A. Vitalis, and S. L. Crick. 2008. A polymer physics perspective on driving forces and mechanisms for protein aggregation. *Arch Biochem Biophys* 469 (1):132–141.

Radhakrishnan, A., A. Vitalis, A. H. Mao, A. T. Steffen, and R. V. Pappu. 2012. Improved atomistic Monte Carlo simulations demonstrate that poly-L-proline adopts heterogeneous ensembles of conformations of semi-rigid segments interrupted by kinks. *J Phys Chem B* 116 (23):6862–6871.

Ren, P., J. Chun, D. G. Thomas, M. J. Schnieders, M. Marucho, J. Zhang, and N. A. Baker. 2012. Biomolecular electrostatics and solvation: A computational perspective. *Q Rev Biophys* 45 (4):427–491.

Rocchia, W., E. Alexov, and B. Honig. 2001. Extending the applicability of the nonlinear Poisson–Boltzmann equation: Multiple dielectric constants and multivalent ions. *J Phys Chem B* 105 (28):6507–6514.

Romero, P., Z. Obradovic, and A. K. Dunker. 2004. Natively disordered proteins: Functions and predictions. *Appl Bioinformatics* 3 (2–3):105–113.

Sigalov, G., A. Fenley, and A. Onufriev. 2006. Analytical electrostatics for biomolecules: Beyond the generalized Born approximation. *J Chem Phys* 124 (12):124902.

Singh, G. P., M. Ganapathi, and D. Dash. 2007. Role of intrinsic disorder in transient interactions of hub proteins. *Proteins* 66 (4):761–765.

Tanizaki, S., and M. Feig. 2005. A generalized Born formalism for heterogeneous dielectric environments: Application to the implicit modeling of biological membranes. *J Chem Phys* 122 (12):124706.

Thakur, A. K., M. Jayaraman, R. Mishra, M. Thakur, V. M. Chellgren, I. J. L. Byeon, D. H. Anjum, R. Kodali, T. P. Creamer, J. F. Conway, A. M. Gronenborn, and R. Wetzel. 2009. Polyglutamine disruption of the huntingtin exon 1 N terminus triggers a complex aggregation mechanism. *Nat Struct Molec Biol* 16 (4):380–389.

Tjong, H., and H. X. Zhou. 2007. GBr(6): A parameterization-free, accurate, analytical generalized Born method. *J Phys Chem B* 111 (11):3055–3061.

Tompa, P. 2012. Intrinsically disordered proteins: A 10-year recap. *TIBS* 37 (12):509–516.

Tran, H. T., X. L. Wang, and R. V. Pappu. 2005. Reconciling observations of sequence-specific conformational propensities with the generic polymeric behavior of denatured proteins. *Biochemistry* 44 (34):11369–11380.

Uversky, V. N. 2002. Natively unfolded proteins: A point where biology waits for physics. *Protein Sci* 11 (4):739–756.

Uversky, V. N. 2011. Intrinsically disordered proteins from A to Z. *Int J Biochem Cell Biol* 43 (8):1090–1103.

Uversky, V. N., C. J. Oldfield, and A. K. Dunker. 2008. Intrinsically disordered proteins in human diseases: Introducing the D2 concept. *Annu Rev Biophys* 37:215–246.

Uversky, V. N., C. J. Oldfield, U. Midic, H. Xie, B. Xue, S. Vucetic, L. M. Iakoucheva, Z. Obradovic, and A. K. Dunker. 2009. Unfoldomics of human diseases: Linking protein intrinsic disorder with diseases. *BMC Genomics* 10 (Suppl 1):S7.

Vacic, V., C. J. Oldfield, A. Mohan, P. Radivojac, M. S. Cortese, V. N. Uversky, and A. K. Dunker. 2007. Characterization of molecular recognition features, MoRFs, and their binding partners. *J Proteom Res* 6 (6):2351–2366.

Vitalis, A., and A. Caflisch. 2012. 50 Years of Lifson-Roig models: Application to molecular simulation data. *J Chem Theor Comput* 8 (1):363–373.

Vitalis, A., N. Lyle, and R. V. Pappu. 2009. Thermodynamics of β-sheet formation in polyglutamine. *Biophys J* 97 (1):303–311.

Vitalis, A., and R. V. Pappu. 2009a. ABSINTH: A new continuum solvation model for simulations of polypeptides in aqueous solutions. *J Comput Chem* 30 (5):673–699.

Vitalis, A., and R. V. Pappu. 2009b. Methods for Monte Carlo simulations of biomacromolecules. *Annu Rep Comput Chem* 5:49–76.

Vitalis, A., X. L. Wang, and R. V. Pappu. 2008. Atomistic simulations of the effects of polyglutamine chain length and solvent quality on conformational equilibria and spontaneous homodimerization. *J Mol Biol* 384 (1):279–297.

Wagoner, J. A., and N. A. Baker. 2006. Assessing implicit models for nonpolar mean solvation forces: The importance of dispersion and volume terms. *Proc Natl Acad Sci U S A* 103 (22):8331–8336.

Warshel, A., P. K. Sharma, M. Kato, and W. W. Parson. 2006. Modeling electrostatic effects in proteins. *Biochim Biophys Acta-Proteins Proteom* 1764 (11):1647–1676.

Williamson, T. E., A. Vitalis, S. L. Crick, and R. V. Pappu. 2010. Modulation of polyglutamine conformations and dimer formation by the N-terminus of huntingtin. *J Mol Biol* 396 (5):1295–1309.

Wyczalkowski, M. A., A. Vitalis, and R. V. Pappu. 2010. New estimators for calculating solvation entropy and enthalpy and comparative assessments of their accuracy and precision. *J Phys Chem B* 114 (24):8166–8180.

Xie, H., S. Vucetic, L. M. Iakoucheva, C. J. Oldfield, A. K. Dunker, Z. Obradovic, and V. N. Uversky. 2007. Functional anthology of intrinsic disorder. 3. Ligands, post-translational modifications, and diseases associated with intrinsically disordered proteins. *J Proteome Res* 6 (5):1917–1932.

Chapter 7

Intrinsically Disordered Protein
A Thermodynamic Perspective

Jing Li, James O. Wrabl, and Vincent J. Hilser

CONTENTS

INTRODUCTION: THERMODYNAMIC STABILITY, STRUCTURE, AND DISORDER

Historical View of Intrinsic Disorder as an All-or-None Phenomenon

The first crystal structures of protein molecules represented breakthroughs of immense importance that transformed our collective understanding of the physical processes sustaining life (Kendrew et al. 1958; Muirhead and Perutz 1963). An unintended side effect of those discoveries, however, was the cementing of the biological paradigm of "one gene, one protein, one structure, one function" (Dickerson 1972; Lehninger 1975). Further investigation over the course of decades demonstrated that this notion was insufficient to describe all aspects of the molecular function of proteins. Examples where this reigning paradigm presently fails include (but are not limited to) dynamic allostery (Cooper and Dryden 1984; Tzeng and Kalodimos 2011), protein as ensemble (Boehr et al. 2009; Hilser et al. 2006), functional moonlighting (Copley 2012; Jeffery 2009), metamorphic proteins (Bryan and Orban 2010; Murzin 2008), frameshift alternative translation (Michel et al. 2012), and intrinsically disordered proteins (Dunker and Obradovic 2001; Uversky 2013).

Of these, perhaps the most revisionary concept to emerge is that of intrinsic disorder because it conflicts most directly with the intuitively appealing notion that structure is function (Fuxreiter and Tompa 2012). Indeed, the defining characteristic of an intrinsically disordered protein is that it is experimentally structureless (Bracken et al. 2004), yet some examples of such proteins seem to have more functionality than many structured proteins (Babu et al. 2012)! A second challenge to the reigning paradigm is the generally large fraction of proteins thought to be intrinsically disordered (Tompa et al. 2006; Ward et al. 2004), especially in eukaryotic proteomes. Since indispensable cellular functions such as transcription (Liu et al. 2006) and signaling (Hilser 2013) are nonetheless known to be mediated by intrinsically disordered proteins, the notion of structureless protein cannot be ignored.

Confronted with the sharp contrasts in the observable characteristics of globular and disordered proteins, it is natural to assume a working classification: either a protein is structured or it is not. A seminal resource to this developing field was (and continues to be) the DisProt database (Sickmeier et al. 2007) that collects and annotates experimentally verified structured and disordered regions. The dichotomy is simultaneously bolstered by the key observations that disorder can be recognized from primary sequence, either from high charge/hydrophobicity ratio (Uversky et al. 2000) or the presence of low amino acid type complexity (Weathers et al. 2007). The resulting plethora of useful disorder prediction methods (Monatstyrskyy et al. 2011) are mostly based on these two sources of information. Although no method is 100% accurate, the best meta predictors, such as PONDR-FIT (Xue et al. 2010), combine many of these all-or-none algorithms at the residue level and then employ a neural network, machine learning, or jury vote for the final classification (Ishida and Kinoshita 2008; Mizanty et al. 2010; Schlessinger et al. 2009; Walsh et al. 2011). Thus, the all-or-none view of intrinsic disorder permeates both current thinking and the most effective disorder prediction.

An Alternative View of Disorder as Part of an Energetic Continuum

However, even within a protein classified as disordered, certain nuances appear to exist. These so-called "flavors" of disorder are also thought to be encoded by amino acid sequence (Huang et al. 2012; Mao et al. 2013; Vucetic et al. 2003). Functional attributes of such diversity include molecular recognition, molecular assembly, protein modification, and entropic chain effects (Dunker et al. 2008). At the same time, a recent review places these apparent subtypes of disorder along a continuum of structural space (Uversky 2013). Thus, within the single category of disorder, a huge range of measureable characteristics are observed.

How can we make sense of all this diversity? A modest proposal is suggested, borrowed from concepts long employed in the analysis of structured proteins: that of transitions and energy barriers between two thermodynamic states (Barrick 2009). A continuum in structure suggests a continuum in energy, since energy determines structure. This proposal pinpoints intrinsic disorder as the unfolded state in the energy landscape of a potentially foldable protein: perhaps intrinsically disordered protein is an extremely destabilized globular protein.

What aspects of experiment can be rationalized by this view? First, if the energy barrier is small enough, apparent disorder will be experimentally observed instead of the folded structure. In fact, a difference of only 3 kcal/mol in energy, approximately 5 kT under physiological conditions, is enough to promote an order/disorder transition that becomes visible (or invisible) to experiment (Figure 7.1, red box). In fact, such phenomena are likely already occurring but are confounding interpretation (Mohan et al. 2009). Second, the well-entrenched concept of structure as function is again applicable; the functional state is just not the physiologically dominant state. Put another way, the specificity of amino acid sequence for the functional structure is emphasized instead of the stability of the functional structure (Lattman and Rose 1993). Third, this view guides valuable experimental resources away from cataloging myriad types of disorder and toward stabilization of the underlying structure. A recent example of such experimental refocusing is the folding of the intrinsically disordered N-terminal domain of glucocorticoid receptor using osmolytes (Li et al. 2012). Finally, consideration of energetics can unify the dichotomy of "structured" or "disordered" with the realization that the favorable and unfavorable

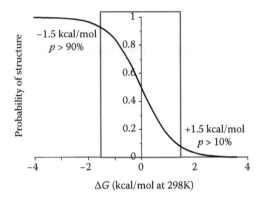

Figure 7.1   Observable disorder depends on thermodynamic stability, ΔG. The y-axis, probability of stable structure, is computed from the x-axis according to the standard relation: $\exp[-\Delta G/RT]/(1 + \exp[-\Delta G/RT])$, where R is the gas constant and T is 298K. The red box indicates a region where a small change in ΔG of approximately 3 kcal/mol results in a large reduction of more than 80% population of the structured state. It is possible that many biologically intrinsically disordered proteins fall within this region and therefore experimentally appear unstructured, but are nonetheless close in energy to a stable folded structure.

interactions between atoms dictates where a protein lies on the structural continuum as well as its binding and catalytic properties. In short, the physics of structure and function must be unchanging, whether folded or unfolded.

MODELING THERMODYNAMIC STABILITY OF STRUCTURED AND DISORDERED REGIONS IN PROTEINS

Prediction of Thermodynamics from Sequence Using the eScape Algorithm

How can thermodynamic information be gleaned from large numbers of intrinsically disordered proteins, especially given their remarkable intractability to experiment? Clearly, a computational approach is needed. One such approach is to employ the eScape (energy landscape) algorithm, introduced by Gu & Hilser (Gu and Hilser 2008). This software estimates native state thermodynamic stability (ΔG) and its component solvation enthalpy (ΔH) and conformational entropy (ΔS_{conf}) from the amino acid sequence alone. Crucially, the calculation uses a trained linear regression model that is fast (typically ~1 s per protein) and accurate (~80% cross-validation accuracy with an error of ± 2 kcal/mol) (Gu and Hilser 2008) so proteome-scale estimates could be performed (Gu and Hilser 2009).

In more detail, eScape relies on the postulates that protein thermodynamics are composed of separable local and global contributions, that the local contributions are dominant, and that they can be estimated from sequence alone. Logical support for these postulates comes, for example, from experimentally measured side chain hydrophobic solvation energies (Nozaki and Tanford 1971) that imply thermodynamic boundaries for stable globular protein structure. Also important is the finding that primary-sequence-dependent thermodynamic quantities are greatly influenced by nearest-neighbor effects rather than by single independent residue contributions (Creamer et al. 1995, 1997; Gu and Hilser 2008). Thus, in the training of the eScape algorithm, the observed position-specific thermodynamics of a reference database of globular proteins were assigned to the corresponding tripeptides centered at the particular position.

The information in this list, or library, of tripeptides and their associated energetic values is used in two distinct ways during the eScape prediction. First, the library defines minimum and maximum observed values for the energetics of every tripeptide in the input protein sequence (Figure 7.2). These preliminary

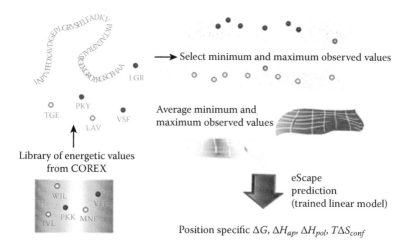

Figure 7.2 Thermodynamic stability ΔG and its enthalpic and entropic components can be computationally estimated from amino acid sequence by the eScape algorithm. Energetic values ΔG, ΔH, ΔS were extracted from experimentally validated COREX ensemble-based analyses of many diverse protein crystal structures. These values were assigned to the corresponding tripeptide fragments of the original amino acid sequence (lower left to upper left panels). For a given tripeptide, the energetic values observed in different protein contexts were recorded and the extreme values were tabulated (upper right panel). A linear prediction model was trained from the average values, resulting in approximately 80% cross-validated accuracy relative to the original data (bottom right panel).

boundaries are then averaged across the amino acid sequence with a sliding window of five residues. Second, the averaged boundaries become input features to the linear regression model, which as mentioned above, was also trained using the library.

The thermodynamic data used for the tripeptide library were themselves obtained from a second calculation that was ultimately validated by multiple experiments (Liu et al. 2012). This second calculation was a structural thermodynamic enumeration of the protein conformational ensemble, the well-known COREX algorithm (Hilser and Freire 1996; Hilser et al. 2006). Importantly, the nature of the COREX calculation and its relationship to experimental hydrogen exchange data provide insight into the uniqueness of the considered approach and its applicability to characterizing disorder. As noted previously (Hilser and Freire 1997; Milne et al. 1999), hydrogen exchange (*HX*) of an individual residue *j* in a protein under *EX2* conditions provides an equilibrium constant between the states in which a residue *j* is structured (ordered) and the states in which it is unfolded (or disordered):

$$PF_{HX,j} = \frac{k_{\text{int},j}}{k_{ex,j}} = \frac{\sum P_{ordered}}{\sum P_{disordered}} \tag{7.1}$$

In Equation 7.1, the protection factor (*PF*) is an experimental quantity that can be directly converted to a free energy:

$$\Delta G_{HX,j} = -RT \ln(PF_{HX,j}) \tag{7.2}$$

The free energy in Equation 7.2 describes the difference in energy between the states in which that residue *j* is ordered and the states in which it is disordered. In other words, the protection factor measured by *HX* that is calculated and validated (Liu et al. 2012) by COREX provides a means of describing the relative stability of a particular position that corresponds to how much time that residue spends either in an ordered or disordered state. The experimental population of disorder directly contributes to the measurement; therefore the propensity for a particular residue to be intrinsically disordered is directly part of the COREX calculation. As such, the calculation provides a graded (i.e., continuous) description of population that corresponds to the actual stability of that site.

While the COREX algorithm requires primary sequence and tertiary structural information as input, the eScape algorithm requires only primary sequence information. The COREX algorithm explicitly includes any global or context-dependent effects of structure on thermodynamics. The agreement between eScape and COREX does imply that some global thermodynamic information can be retained in the sequence-based estimate.

The output from both programs, however, is similar: residue-specific estimates of thermodynamic stability. Importantly, although trained on the thermodynamics of folded globular proteins, eScape permits estimation of these difficult to measure quantities for the case of intrinsically disordered or denatured proteins (Gu and Hilser 2008).

Analyzing the Energetics of Structured and Disordered Residues

Is the experimental instability of intrinsically disordered protein actually captured by the eScape energetic estimates? The answer is yes, and this conclusion was arrived at through analysis of large collections of structured and disordered proteins, as described in the following.

The amino acid sequences of these proteins were obtained from PDB-Select-25 (November 2009 release, [Griep and Hobohm 2010]) and DisProt (v4.9, Sickmeier et al. 2007). Full-length sequences were individually run with eScape and the thermodynamic parameters were parsed by amino acid type. Specifically, the energetics of disordered regions, only if annotated as such in the DisProt comment line, were retained after the eScape analysis on the complete sequence, as structured regions influence the calculation for neighboring disordered regions. In summary, 1222 disordered regions extracted from 523 annotated DisProt sequences, and 4823 structured proteins from the Protein Data Bank (Berman et al. 2000) were analyzed. As the total number of structured residues in this initial collection was greater than the total number of disordered residues, the number of structured residues was randomly reduced such that the amino acid counts of both structured and disordered sets were of comparable size: approximately 40,000 total residues each.

The amino acid frequencies in these final sets were not significantly different from commonly used background frequencies (Robinson and Robinson 1991), exhibiting Spearman rank order correlation coefficients (Press et al. 2007) of approximately $r_s = 0.9$, $p = 2 \times 10^{-4}$ (data not shown). Importantly, although the PDB-Select-25 sequences were, by definition, filtered to a maximum pairwise sequence identity of 25%, the DisProt sequences, by design, appeared to contain a small level of redundancy that was not explicitly filtered.

Modeling the Energetics of Residue Stability with Normal Distributions

Distributions of the eScape native state stabilities as a function of amino acid type were collated for the structured and intrinsically disordered sets. In this context, "native state" refers to the parameterization mode of the eScape algorithm that simulates physiological conditions under which globular proteins are expected to be structured. (This is in contrast to the other optional param-eterization mode that simulates unfolding conditions.)

It was observed that all these stability distributions, whether structured or disordered, were unimodal (Figure 7.3). Several common two-parameter analytical functions were attempted as fits to this data, including normal, gamma, extreme value, and inverse Gaussian (Johnson et al. 1994). A variable bin width procedure, filling each bin with at least 50 data points, was employed and the chi-squared value between expected and observed points per bin was minimized (Mathematica 8.0, Wolfram Research). The statistical likelihood, or p-value, of the observed fit as due to chance could then be computed from standard chi-square statistics. Statistically, most of the 40 structured and dis-ordered sets were best modeled by normal distributions. An example of the binned data for structured and disordered alanine residue stabilities and the optimal fits to the normal distributions generated by this procedure are dis-played in Figure 7.3.

Only one of the 20 structured distributions could be rejected as normal at the $p < 0.05$ level, while 11 of the 20 disordered distributions could be rejected at that same level (Table 7.1). At first glance, this latter result might imply that the disordered stabilities were not well modeled by normal distributions. However,

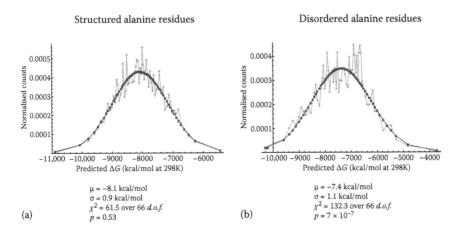

(a) Structured alanine residues

$\mu = -8.1$ kcal/mol
$\sigma = 0.9$ kcal/mol
$\chi^2 = 61.5$ over 66 d.o.f.
$p = 0.53$

(b) Disordered alanine residues

$\mu = -7.4$ kcal/mol
$\sigma = 1.1$ kcal/mol
$\chi^2 = 132.3$ over 66 d.o.f.
$p = 7 \times 10^{-7}$

Figure 7.3 For most structured and disordered amino acid types, distributions of eScape predicted native state stability are statistically normal. The examples of ala-nine residues, extracted from either PDB-Select structures (a) or from the annotated disordered regions of DisProt entries (b) are displayed. Note that the statistical failure of this disordered distribution is likely due to redundant sampling of nearly identical entries in the DisProt database (e.g., spikes at specific data values) rather than due to the inappropriateness of the normal distribution. The structured distribution contains 3346 alanine residues and the disordered distribution contains 3344.

TABLE 7.1   ESTIMATES OF MEAN AND STANDARD DEVIATION FROM NORMAL FITS TO PREDICTED NATIVE STATE STABILITY DISTRIBUTIONS OF STRUCTURED AND DISORDERED RESIDUES FOR 20 AMINO ACID TYPES

Structured Residue	μ	σ	P-Value	Disordered Residue	μ	σ	P-Value
A	−8.08	0.92	0.53	A	−7.37	1.14	7.75e−7
C	−8.04	1.00	0.90	C	−7.62	1.11	0.87
D	−8.10	0.98	0.58	D	−7.67	1.00	0.23
E	−8.19	0.97	0.98	E	−7.55	1.04	1.39e−3
F	−8.32	0.98	0.94	F	−7.91	0.98	0.14
G	−7.78	0.96	0.06	G	−7.07	1.08	4.24e−14
H	−8.22	1.06	0.72	H	−7.60	1.17	0.13
I	−8.00	0.95	0.36	I	−7.70	0.90	0.03
K	−8.07	0.96	0.17	K	−7.48	1.06	0.04
L	−8.16	0.95	0.61	L	−7.74	0.95	3.22e−3
M	−8.25	0.98	0.22	M	−7.84	1.01	0.39
N	−8.07	0.99	0.69	N	−7.61	1.03	2.95e−4
P	−7.80	1.02	0.35	P	−6.99	1.11	8.30e−13
Q	−8.26	1.04	0.72	Q	−7.72	1.14	1.67e−5
R	−8.34	0.98	0.78	R	−8.01	1.16	0.06
S	−8.15	0.94	0.81	S	−7.54	0.98	3.96e−3
T	−7.95	0.97	0.48	T	−7.33	1.02	2.79e−3
V	−7.98	0.93	0.02	V	−7.53	1.02	0.07
W	−8.00	0.97	0.97	W	−7.78	1.15	0.11
Y	−8.31	0.99	0.99	Y	−7.91	1.12	0.71

a closer look at the data (Figure 7.3b) suggested that these apparently high chi-squared values were rather due to the localized effects of redundancy in the disordered data: spikes in counts of a particular energetic value due to nearly duplicate amino acid sequences in the DisProt database. If these spikes were provisionally ignored, the underlying distribution for the sets that did not statistically pass nonetheless appeared reasonably Gaussian (Figure 7.3b). Due to the simplicity and familiarity of the normal distribution, it was adopted to consistently model these energetic data for both structured and disordered residue types. Optimally fit parameters, means μ, and standard deviations σ describing the eScape stabilities are listed in Table 7.1.

Intrinsically Disordered Residues Exhibit Lower Predicted Stability

Regardless of whether the residue stability data belonged to the structured or intrinsically disordered sets, the spread of any distribution was approximately identical ($\sigma \sim 1.0$). The predicted values for the stabilities of both disordered and structured residues were usually negative, indicating an inability of the eScape algorithm to predict the absolute value of energetic stability accurately. However, for every amino acid type, a striking difference was noted between the means of the structured and disordered distributions: every disordered distribution exhibited a mean that was less stable than the structured counterpart (Table 7.2). These differences, $\Delta\mu$, strongly supported the notion that, in

TABLE 7.2   DIFFERENCE IN MEANS OF NATIVE STATE STABILITY $\Delta\mu$ (kcal/mol) BETWEEN DISORDERED AND STRUCTURED DISTRIBUTIONS FOR 20 AMINO ACID TYPES

Residue	$\Delta\mu$
A	0.71
C	0.42
D	0.43
E	0.64
F	0.41
G	0.71
H	0.61
I	0.30
K	0.59
L	0.42
M	0.41
N	0.46
P	0.81
Q	0.54
R	0.33
S	0.61
T	0.62
V	0.45
W	0.22
Y	0.40

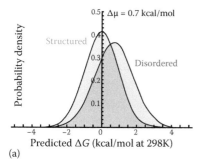

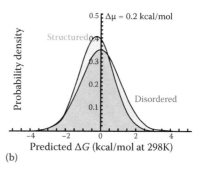

(a)

(b)

Figure 7.4   Disordered distributions are predicted to be less stable on average than structured distributions. Examples of the amino acids glycine (a) and tryptophan (b) exhibiting the extremes in stability difference are displayed. In each panel, the probability densities defined by the best fit normal distributions (Table 7.1) are shown. Structured distribution is colored blue, and disordered distribution is colored red. The difference in means of the two distributions, $\Delta\mu$ in units of kcal/mol, is labeled. For all 20 amino acid types, the disordered distribution is less stable on average than the structured one.

a relative manner, eScape energetic predictions accurately reflected the experimentally observed disorder.

The largest difference, $\Delta\mu = 0.71$ kcal/mol, was observed for glycine and the smallest difference, $\Delta\mu = 0.22$ kcal/mol, was observed for tryptophan (Figure 7.4). In Figure 7.4, relative adjustment of the probability distributions so that the mean stability over all structured and disordered residues was set to zero makes it clear that substantial populations of structured and disordered residues *did not* overlap in stability. This finding suggested that eScape might be able to detect regions of a polypeptide chain with thermodynamic propensity for intrinsic disorder. Although the individual energetic contributions per amino acid were only on the order of a few tenths of a kcal/mol, summation of many such contributions could easily provide several kcal/mol, and therefore the energetic basis of a population switch from macroscopic order to disorder, as shown in Figure 7.1.

BASIC NATIVE STATE STABILITY PREDICTOR OF DISORDER FROM AMINO ACID SEQUENCE

Differences in Mean Stability Correlate with Disorder Propensity and Physical Property Scales

The two residues glycine and tryptophan have extreme physicochemical properties. The former residue, with the smallest side chain, can nominally occupy the greatest amount of Ramachandran conformational space while

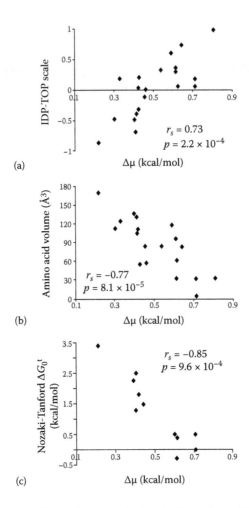

Figure 7.5   Robust correlation between $\Delta\mu$ for the 20 amino acid types and several amino acid property scales. In each panel, the Spearman rank-order correlation coefficient r_s and its statistical significance p are displayed. (a) Correlation with TOP-IDP propensity scale for intrinsic disorder demonstrating that differences in predicted stability are proportional to an accepted disorder propensity scale. (b) Correlation with amino acid volume demonstrating that differences in predicted stability, and thus, disorder, are related to amino acid size. (c) Correlation with Nozaki-Tanford hydrophobicity scale demonstrating that experimental hydrophobicity is related to predicted stability, and thus, disorder.

the latter residue is the largest by volume and among the most hydrophobic. Did the predicted mean energetic differences correlate with other amino acid property scales? Although few known scales matched, there were at least two that were significantly correlated among the hundreds contained in the Japanese Amino Acid Database Index (Tomii and Kanehisa 1996). These two were fundamental, and previously exploited, physical properties for structure prediction: amino acid side chain volume (Figure 7.5b; Grantham 1974) and experimental side chain hydrophobicity (Figure 7.5c; Nozaki and Tanford 1971).

These two scales could be rationalized as part of the sequence-dependent characteristics, such as the charge/hydrophobicity ratio mentioned above, that have long been known to be the basis of effective disorder prediction from primary sequence. However, the fact that these physical properties could now be linked to predicted thermodynamic stability hinted at a larger truth: that the fundamental determinant of intrinsic disorder is actually energy, of which primary sequence properties are merely a surrogate. Indeed, pioneering work (Dosztanyi et al. 2005) has attempted to exploit this for disorder prediction, albeit in an aggregate, instead of a residue position specific, manner.

In this vein, the $\Delta\mu$ values were also significantly correlated (Figure 7.5a) to a newer, machine-learning derived scale, the TOP-IDP scale (Campen et al. 2008), specifically trained to mathematically maximize discrimination between disordered and structured residue positions. Although this scale outperformed all other known amino acid scales for the task of disorder discrimination (Campen et al. 2008), any physical basis for its effectiveness was obscure. Figure 7.5a, again suggested that this optimized disorder propensity scale may actually be a surrogate for the underlying thermodynamic origins of structure and disorder.

Development of a Stability-Based Probability Model for Disorder Prediction

Although an amino acid sequence predictor of intrinsic disorder could be envisaged from the scale of mean stability differences, the unique importance of having access to the predicted thermodynamics of the polypeptide would be to estimate populations of the conformational states of the system. Thus, it was desired to also leverage the information in the spread and overlap of the distributions displayed in Figure 7.4. A first approach to energetic disorder prediction was attempted by integration over these observed probability distributions. The simple scheme was composed of two parts. First, for a given predicted stability value, ΔG, the disorder distribution area to the left of this value was defined to be proportional to the equilibrium population of disorder while the structured distribution area to the right of this value was defined to be proportional to the equilibrium population of structure. Second, the ratio of the two populations gave the disorder propensity at the residue position. This scheme is formalized by the following.

In general, an assumed unfolding reaction between intrinsically disordered (U) and structured (F) measured at every residue position can be written as

$$[F] \Leftrightarrow [U] \tag{7.3}$$

The position of the equilibrium, K, which specifies whether the residue position is structured or disordered, is given by the standard relation:

$$e^{-\frac{\Delta G}{RT}} = K_U = \frac{[U]}{[F]} = \frac{P_U}{P_F} \tag{7.4}$$

where P_F and P_U are the populations of structured and intrinsically disordered, respectively. (It is important to note that the ΔG value in Equation 7.4 refers to the true instability of the system and not necessarily the eScape native state stability value.)

The population of intrinsic disorder is then given by

$$P_U = \frac{K_U}{1+K_U} \tag{7.5}$$

Equation 7.5 can be expanded to obtain:

$$P_U = \frac{\dfrac{P_U}{P_F}}{1+\dfrac{P_U}{P_F}} = \frac{P_U}{P_F+P_U} \approx \frac{P_U'}{P_F'+P_U'} \tag{7.6}$$

The new variables P_U' and P_F' have been introduced on the extreme right-hand side of Equation 7.6 to indicate that they are approximations to the true populations. These new variables are proposed to be computed from the cumulative distribution functions (CDF) of the normal distributions defined in Table 7.1, as follows:

$$P_U' = CDF(x \mid \mu_D, \sigma_D) \tag{7.7}$$

$$P_F' = 1 - CDF(x \mid \mu_S, \sigma_S) \tag{7.8}$$

In Equations 7.7 and 7.8, x refers to the native state ΔG value output by eScape (in kcal/mol), μ_D and σ_D refer to the disordered means and standard deviations in Table 7.1, and μ_S and σ_S refer to the structured means and standard deviations in Table 7.1. Thus, Equations 7.7 and 7.8 are dependent on amino acid type.

The logic for this scheme is based on the reasonable conjecture that the area under the eScape native state ΔG distribution is proportional to the equilibrium disordered or structured population of a particular residue. The degree of nonoverlap of the empirical disordered and structured distributions would therefore be related to the propensity for intrinsic disorder.

For example, a residue exhibiting an eScape ΔG that is relatively quite negative, say, –10 kcal/mol, the area (measured from negative infinity) under its structured stability distribution will be small and the complement (Equation 7.8) will be quite large, favoring a structured conformation. At the same time, the area (measured from negative infinity) under its disordered distribution will be small, disfavoring the disordered conformation (Equation 7.7). The ratio of Equation 7.6 for this residue will be quite small, and thus its propensity for intrinsic disorder under this scheme will be quite low. In contrast, residues with a high propensity for disorder will have more positive eScape ΔG values, and their Equation 7.6 ratios will be larger in magnitude. In all cases, a wider separation between the structured and disordered normal distributions results in greater sensitivity to intrinsic disorder propensity prediction. An appealing aspect is that the physical reality of a Boltzmann distribution, where distinct conformations could simultaneously be populated according to their energy, could be partially captured as a predicted equilibrium: Equation 7.6 is naturally bounded between zero and one.

To test this predictor, the amino acid sequences of 25 proteins (Table 7.3) were chosen at random from the human proteome (NCBI CCDS Release 6 [Pruitt et al. 2009]). The native state eScape values ΔG were obtained for each protein from a publicly available portal (http://best.bio.jhu.edu/eScape). For every residue, the *CDF* values in Equations 7.7 and 7.8 were evaluated at $x = \Delta G$ for the appropriate normal distributions parameterized by the means and standard deviations listed in Table 7.1. These calculations were facilitated by *gnumeric* spreadsheet software (version 1.8.2, https://projects.gnome .org/gnumeric). The disorder propensity was evaluated for each residue from Equation 7.6. To facilitate interpretation, a smoothed curve was obtained over the entire protein by averaging over a window size of 21 positions. Results were quantitatively compared with the predictions returned by the state-of-the-art PONDR-FIT server (Xue et al. 2010).

Table 7.3 lists the Spearman rank-order correlations r_s between the residue-specific disorder prediction results of the two methods. Clearly, both robustly agree: 24 of 25 proteins were positively correlated and 22 of the positive correlations were significant to at least the 1% level (Table 7.3). The median value of

TABLE 7.3   ROBUST CORRELATION r_s BETWEEN PONDR-FIT AND STABILITY DISTRIBUTION-BASED DISORDER PROBABILITY PROFILES FOR 25 RANDOMLY CHOSEN HUMAN PROTEIN SEQUENCES

Protein	Length	r_s	P-Value
CFHR1	330	0.32	$<10^{-6}$
NELFA	528	0.38	$<10^{-6}$
SLC35G5	339	0.24	1.1×10^{-5}
ST18	1047	0.09	5.7×10^{-3}
EEF1D	647	0.34	$<10^{-6}$
ATXN2L	1044	0.26	$<10^{-6}$
HS3ST3A1	406	0.30	$<10^{-6}$
CETN1	172	0.54	$<10^{-6}$
1DH3B	383	0.27	$<10^{-6}$
LAMP5	280	0.07	2.4×10^{-1}
CYP27A1	531	0.48	$<10^{-6}$
BST1	318	0.44	$<10^{-6}$
SORBS2	1100	0.01	6.5×10^{-1}
WHSC1L1	645	0.23	$<10^{-6}$
DKK3	350	0.13	1.4×10^{-2}
ZNF384	518	0.20	4.0×10^{-6}
GCHFR	84	0.68	$<10^{-6}$
KIF22	665	0.31	$<10^{-6}$
MRO	248	−0.07	2.9×10^{-1}
MUM1	711	0.47	$<10^{-6}$
SIGLEC6	442	0.27	$<10^{-6}$
MMP9	707	0.44	$<10^{-6}$
METTL3	580	0.13	1.3×10^{-3}
GRM7	915	0.11	1.4×10^{-3}
CRAT	605	0.10	1.1×10^{-2}

the correlations was approximately 0.5 (Figure 7.6a), a substantial positive correlation out of a possible range of −1 to +1 (Press et al. 2007). This level of non-random agreement justifies the use of thermodynamic stability information as an important contributor to the prediction of intrinsic disorder (although there is obviously room for improvement).

A further example of the methods' position-specific agreement comes from a protein of medical relevance as well as of research interest to this laboratory, human androgen receptor (Figure 7.6b). Like other members of the steroid hormone receptor family, the N-terminus of the protein is thought to be functionally important as well as intrinsically disordered while the C-terminus contains structured ligand-binding and deoxyribonucleic acid (DNA)-binding domains (McEwan et al. 2007). The agreement is excellent and highly significant, with a correlation coefficient of $r_s = 0.59$. Tantalizingly, several regions of the empirically disordered N-terminus are predicted by the proposed energy-based scheme to nonetheless exhibit characteristics of structured protein. These regions, for example the one between residues 50 and 100, may suggest new hypotheses for experimental test: Is such an intrinsically disordered region energetically poised to respond to a transient binding event (Hilser and Thompson 2007)?

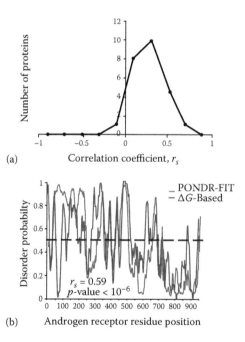

(a)

(b)

Figure 7.6   Robust correlation between PONDR-FIT and stability distribution-based disorder probability profiles. (a) Histogram of Spearman correlation coefficients r_s for 25 randomly chosen human protein sequences, listed in Table 7.3. Nearly all correlations are significantly and substantially positive. (b) Disorder probability profiles for the additional medically important case of human androgen receptor. The correlation between the stability distribution predictor and PONDR-FIT is strong and significant.

MORE SOPHISTICATED SUPPORT VECTOR MACHINE PREDICTORS OF DISORDER BASED ON MULTIPLE THERMODYNAMIC PARAMETERS

Support Vector Machine Predictor Based on Native State ΔG Alone

In an effort to increase the accuracy of the disorder predictor, a support vector machine (SVM) was employed. SVM is a well-known supervised machine-learning method that is ideally suited to binary classification of data (Press et al. 2007) and has in fact been effectively used for prediction of intrinsic disorder (Vullo et al. 2006; Ward et al. 2004). Training of the SVM on known examples results in, depending on the dimensionality of the data, one or more support vectors. These mathematically define one or more linear planes, or hyperplanes, that optimally recapitulate the known classes subject to a maximum margin specified by a user-adjustable parameter c. The support vectors thus represent an objective model that can be used to classify (predict) data not in the training set. The particular implementation used in this work was SVM^{perf} (Joachims 2005, 2006; Joachims and Chun-Nam 2009).

The two matched-size data sets of known structured and disordered residues, described above in the section titled "Analyzing the Energetics of Structured and Disordered Residues," were used to perform a total of five rounds of cross-validation training and testing. The matched sets minimized sampling bias in the training (Disfani et al. 2012). Every amino acid was trained and tested independently so that every residue type received a separate SVM classification model. As all energetic values output by eScape are in the same thermodynamic units of kcal/mol, no additional standardization of the enthalpic or entropic values relative to stability was necessary. In each round, approximately one-fifth of each structured and disordered residue type was held out of the training. After the training was complete, the classification model defined by the found support vectors was used to test the set that was held out.

During each round of training and testing, the accuracy and area under the true-positive/false positive curve were automatically recorded by the SVM^{perf} software. The results were averaged over the five rounds and the 20 amino acids as a measure of the effectiveness of the classifier. Finally, the margin parameter c was systematically varied by powers of 10 to crudely but rapidly optimize the effectiveness if possible.

Using only the eScape native state ΔG values as information, the best SVM classifiers achieved an average area under the curve of approximately 65% and an average accuracy of just over 60% (Figure 7.7, filled circles). This means that, on average, about 60% of the residues in the testing sets were predicted correctly. (These results were obtained by smoothing the raw value of the

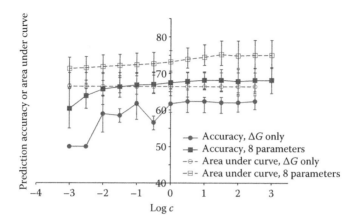

Figure 7.7   SVM training results for energy-based disorder predictor. The *x*-axis represents *c*, the value determining the width of the prediction margin. The *y*-axis represents prediction accuracy or area under curve, quantities related to the effectiveness of the disorder prediction. Values are averaged over five cross-validated runs. The results demonstrate that using eight eScape thermodynamic parameters achieves higher disorder prediction effectiveness than native state stability only, and that the highest effectiveness is obtained at larger values of *c*.

SVM output over a window size of 21 residues, which improved the numbers by approximately 5% as compared to the absence of smoothing.) Parameters defining these models for every residue type are listed in Table 7.4.

Although this performance indicated the presence of nonrandom information training the SVM, the accuracy fell far short of state-of-the-art methods. Application of the SVM model to the 25 randomly chosen human proteins tested in Table 7.3 resulted in a similar practical accuracy of 62 ± 8% on average (Table 7.5, middle column). The relatively poor performance of the ΔG-only SVM was likely due to the substantial overlap apparent in the probability distributions (e.g., the overlap between red and blue for Tryptophan in Figure 7.4b), which hindered a crisp one-dimensional linear separation between structured and disordered.

SVM Predictor Based on Eight Native and Denatured State Thermodynamic Quantities

Because stability is a combination of enthalpy and entropy, the eScape algorithm was intentionally developed to predict these component quantities as well (Gu and Hilser 2008). Moreover, the algorithm was parameterized to simulate a nonphysiological denatured state as well as the physiological native state, so several additional thermodynamic values not considered in the previous

TABLE 7.4   OPTIMAL SVM TRAINING PARAMETERS (LOG c = 1.0) FOR eScape NATIVE STATE STABILITY (ΔG) BASED DISORDER PREDICTOR

Amino Acid	b	w1 (Native ΔG)
A	−4.45	0.59
C	−4.32	0.55
D	−4.97	0.63
E	−5.49	0.70
F	−5.13	0.63
G	−5.49	0.75
H	−4.98	0.64
I	−4.67	0.59
K	−5.42	0.70
L	−5.22	0.66
M	−5.08	0.63
N	−5.08	0.65
P	−5.48	0.75
Q	−4.90	0.63
R	−4.41	0.55
S	−5.52	0.71
T	−5.42	0.72
V	−5.10	0.66
W	−4.04	0.53
Y	−4.62	0.57

SVM could potentially be included in the training. In detail, conformational entropy (ΔS_{conf}), apolar enthalpy (ΔH_{ap}), polar enthalpy (ΔH_{pol}), and stability (ΔG), a total of eight thermodynamic quantities in both native and denatured states, could contribute to a multidimensional SVM model. Would these additional variables possibly improve the effectiveness of a more sophisticated classifier of intrinsic disorder?

The cross-validated training and testing were performed as described in the Support Vector Machine Predictor Based on Native State ΔG Alone section, and a substantial increase in both area under the curve and accuracy were indeed observed. These increases were between 5% and 10% as compared to the stability only model (after 10% net improvement by smoothing

TABLE 7.5 POSITION-SPECIFIC AGREEMENT BETWEEN PONDR-FIT AND OPTIMAL THERMODYNAMIC SVM DISORDER PREDICTION RESULTS FOR 25 RANDOMLY CHOSEN HUMAN PROTEINS

Protein	Prediction Agreement (Native ΔG Only) (%)	Prediction Agreement (Eight Energetic Parameters) (%)
CFHR1	51	68
NELFA	57	65
SLC35G5	57	92
ST18	61	61
EEF1D	65	75
ATXN2L	73	77
HS3ST3A1	53	76
CETN1	66	77
1DH3B	58	76
LAMP5	70	62
CYP27A1	66	73
BST1	70	60
SORBS2	46	67
WHSC1L1	66	85
DKK3	62	59
ZNF384	58	62
GCHFR	69	59
KIF22	58	65
MRO	80	81
MUM1	71	85
SIGLEC6	49	68
MMP9	66	73
METTL3	53	70
GRM7	63	74
CRAT	64	74

TABLE 7.6   OPTIMAL SVM TRAINING PARAMETERS (LOG c = 1.5) FOR INTRINSIC DISORDER PREDICTOR BASED ON EIGHT eScape THERMODYNAMIC PARAMETERS

Amino Acid	b	w1 (Native ΔG)	w2 (Native ΔH_{ap})	w3 (Native ΔH_{pol})	w4 (Native $T\Delta S_{conf}$)	w5 (Denatured ΔG)	w6 (Denatured ΔH_{ap})	w7 (Denatured ΔH_{pol})	w8 (Denatured $T\Delta S_{conf}$)
A	-7.25	1.13	-0.14	-0.88	1.51	-1.02	-0.61	-0.05	-0.86
C	-3.96	1.51	0.13	-1.19	1.16	-2.11	-1.04	0.74	-1.75
D	-1.96	0.47	-0.43	-0.46	0.28	-0.83	-0.38	-0.23	-0.91
E	-3.33	0.71	-0.40	-0.67	0.69	-0.29	-0.34	-0.37	-0.36
F	-5.30	0.90	-0.12	-0.40	-0.58	-1.72	-0.71	-0.09	-1.15
G	-3.23	0.86	-0.27	-0.56	0.47	-1.29	-0.78	0.04	-1.30
H	-3.54	0.38	-0.49	-0.09	-0.93	-2.06	-0.87	0.56	-1.77
I	-3.91	0.58	-0.29	-0.47	0.28	-1.35	-0.91	0.04	-1.14
K	-6.66	0.11	-0.61	-0.04	-0.34	-1.23	-0.56	0.09	-0.89
L	-5.00	0.63	-0.36	-0.39	-0.04	-1.02	-0.65	-0.23	-0.77
M	-4.62	0.82	-0.21	-0.43	0.15	-1.92	-0.67	0.92	-1.77
N	-4.64	0.15	-0.55	-0.17	0.03	-1.00	-0.76	-0.14	-0.84
P	-0.06	0.15	-0.35	-0.05	-0.02	0.08	-0.06	-0.26	-0.29
Q	-5.84	0.01	-0.63	-0.18	0.43	-0.75	-0.36	-0.27	-0.63
R	-6.41	0.31	-0.56	-0.29	-0.55	-1.57	-0.55	0.27	-0.96
S	-6.60	0.77	-0.35	-0.47	0.37	-1.80	-0.53	0.42	-1.54
T	-6.85	0.60	-0.35	-0.26	-0.03	-1.50	-0.60	0.10	-1.16
V	-4.77	0.61	-0.31	-0.50	0.66	-1.49	-1.17	0.19	-1.30
W	-4.05	0.92	0.01	-0.78	1.18	-0.92	-1.39	-0.38	-0.67
Y	-6.50	0.82	-0.33	-0.67	0.63	-1.00	0.46	0.10	-0.80

over a 21-residue window) depending on the exact values of c (Figure 7.7, filled squares). Obtained parameters for the optimal support vector model are listed in Table 7.6.

The practical accuracy on the 25 randomly chosen human proteins also demonstrated substantial improvement over the stability-only model (Table 7.5, right column). These achieved 71% ± 9% on average, with several examples reaching accuracies greater than 80%. The particular case of human androgen receptor, introduced in Figure 7.6b, exhibited 78% agreement with PONDR-FIT. Although currently anecdotal, sustained performance at this higher level would be on par with state-of-the-art methods for intrinsic disorder prediction (Deng et al. 2012; Monatstyrskyy et al. 2011).

CONCLUSIONS AND OUTLOOK: THE IMPORTANCE OF PROTEIN THERMODYNAMICS FOR INTRINSIC DISORDER

Two broad conclusions can be drawn, with interesting implications. First, thermodynamic stability information, estimated from amino acid sequence alone, provided residue-specific information about a protein's propensity for intrinsic disorder. Although this would be expected from any accurately parameterized atomic energy function, the ability of the crude but continuous eScape algorithm to recapitulate the results of binary disorder prediction implies that the sequence signature of intrinsic disorder is encoded not at a side-chain level, but rather at a more fundamental, energetic level. Second, the improvement of the SVM predictors, by including additional thermodynamic parameters, suggests that knowledge of the enthalpy and entropy, together comprising stability, provides additional information about disorder.

It may be surprising that eScape is effective for this task in spite of being parameterized on globular proteins. However, other disorder prediction approaches also based on energy are similarly parameterized (Dosztanyi et al. 2005). This nonintuitive observation hints at a deeper unifying principle: that structure, or lack of it, is governed by the energetics of the system. Thus, viewing intrinsic disorder from the standpoint of thermodynamics unifies efforts to model and predict disordered as well as structured proteins. We therefore believe that continued progress in disorder prediction will come most naturally from improvement of conformational sampling and potential functions rather than from refinement and combination of existing sequence or library based methods. Indeed, exciting new advances in this area of all-atom modeling are being reported (Lindorff-Larsen et al. 2012; Mao et al. 2013).

The biggest gain from cutting edge computational methods, however, will be not from knowledge of accurate energy, but rather from the ability to predict conformational population from the stability for both disordered and structured states. Such knowledge may permit rational manipulation of the functional protein ensemble (Liu et al. 2007), which will be crucial for next-generation drug design (Kar et al. 2010; Rezaei-Ghaleh et al. 2012).

ACKNOWLEDGMENTS

Figure 7.2 was prepared by kind courtesy of Jenny Gu. Grant support from NSF (MCB-0446050) and NIH (R01-GM063747) is gratefully acknowledged.

REFERENCES

Babu, M.M. et al. (2012) Versatility from protein disorder, *Science*, **337**, 1460–1461.

Barrick, D. (2009) What have we learned from the studies of two-state folders, and what are the unanswered questions about two-state protein folding? *Phys. Biol.*, **6**, 15001.

Berman, H.M. et al. (2000) The protein data bank, *Nucleic Acids Res.*, **28**, 235–242.

Boehr, D.D. et al. (2009) The role of dynamic conformational ensembles in biomolecular recognition, *Nat. Chem. Biol.*, **11**, 789–796.

Bracken, C. et al. (2004) Combining prediction, computation, and experiment for the characterization of protein disorder, *Curr. Opin. Struct. Biol.*, **14**, 570–576.

Bryan, P.N. and Orban, J. (2010) Proteins that switch folds, *Curr. Opin. Struct. Biol.*, **20**, 482–488.

Campen, A. et al. (2008) TOP-IDP scale: A new amino acid scale measuring propensity for intrinsic disorder, *Protein Pept. Lett.*, **15**, 956–963.

Cooper, A. and Dryden, D.T.F. (1984) Allostery without conformational change, *Eur. Biophys. J.*, **11**, 105–109.

Copley, S.D. (2012) Moonlighting is mainstream: Paradigm adjustment required, *Bioessays*, **34**, 578–588.

Creamer, T.P. et al. (1995) Modeling unfolded states of peptides and proteins, *Biochemistry*, **34**, 16245–16250.

Creamer, T.P. et al. (1997) Modeling unfolded states of proteins and peptides. II. Backbone solvent accessibility, *Biochemistry*, **36**, 2832–2835.

Deng, X. et al. (2012) A comprehensive overview of computational protein disorder prediction methods, *Mol. Biosyst.*, **8**, 114–121.

Dickerson, R.E. (1972) X-ray studies of protein mechanisms, *Annu. Rev. Biochem.*, **41**, 815–842.

Disfani, F.M. et al. (2012) MoRFpred, a computational tool for sequence-based prediction and characterization of short disorder-to-order transitioning binding regions in proteins, *Bioinformatics*, **28**, i75–i83.

Dosztanyi, Z. et al. (2005) The pairwise energy content estimated from amino acid composition discriminates between folded and intrinsically unstructured proteins, *J. Mol. Biol.*, **347**, 827–839.

Dunker, A.K. and Obradovic, Z. (2001) The protein trinity—Linking function and disorder, *Nat. Biotechnol.*, **19**, 805–806.

Dunker, A.K. et al. (2008) Function and structure of inherently disordered proteins, *Curr. Opin. Struct. Biol.*, **18**, 756–764.

Fuxreiter, M. and Tompa, P. (2012) Fuzzy complexes: A more stochastic view of protein function, *Adv. Exp. Med. Biol.*, **725**, 1–14.

Grantham, R. (1974) Amino acid difference formula to help explain protein evolution, *Science*, **185**, 862–864.

Griep, S. and Hobohm, U. (2010) PDBselect 1992–2009 and PDBfilter-select, *Nucleic Acids Res.*, **38**, D318–D319.

Gu, J. and Hilser, V.J. (2008) Predicting the energetics of conformational fluctuations in proteins from sequence: A strategy for profiling the proteome, *Structure*, **16**, 1627–1637.

Gu, J. and Hilser, V.J. (2009) Sequence-based analysis of protein energy landscapes reveals nonuniform thermal adaptation within the proteome, *Mol. Biol. Evol.*, **26**, 2217–2227.

Hilser, V.J. (2013) Signaling from disordered proteins, *Nature*, **498**, 308–310.

Hilser, V.J. and Freire, E. (1996) Structure-based calculation of the equilibrium folding pathway of proteins. Correlation with hydrogen exchange protection factors, *J. Mol. Biol.*, **262**, 756–772.

Hilser, V.J. and Freire, E. (1997) Predicting the equilibrium protein folding pathway: Structure-based analysis of staphylococcal nuclease, *Proteins*, **27**, 171–183.

Hilser, V.J. et al. (2006) A statistical thermodynamic model of the protein ensemble, *Chem. Rev.*, **106**, 1545–1558.

Hilser, V.J. and Thompson, E.B. (2007) Intrinsic disorder as a mechanism to optimize allosteric coupling in proteins, *Proc. Natl. Acad. Sci. U. S. A.*, **104**, 8311–8315.

Huang, F. et al. (2012) Subclassifying disordered proteins by the CH-CDF plot method, *Pac. Symp. Biocomput.*, **17**, 128–139.

Ishida, T. and Kinoshita, K. (2008) Prediction of disordered regions in proteins based on the meta approach, *Bioinformatics*, **24**, 1344–1348.

Jeffery, C.J. (2009) Moonlighting proteins—An update, *Mol. Biosyst.*, **5**, 345–360.

Joachims, T. (2005) A support vector method for multivariate performance measures, in *Proceedings of the International Conference on Machine Learning*, Bonn, Germany, August 7–11, 2005. Association of Computing Machinery, New York, pp. 377–384.

Joachims, T. (2006) Training linear SVMs in linear time, in *Proceedings of the ACM Conference on Knowledge Discovery and Data Mining*, Philadelphia, PA, August 20–23, 2006. Association of Computing Machinery, New York, pp. 217–226.

Joachims, T. and Chun-Nam, J.Y. (2009) Sparse kernel SVMs via cutting-plane training, in *Proceedings of the European Conference on Machine Learning*, Bled, Slovenia, September 7–11, 2009. Springer, Berlin, p. 8.

Johnson, N.L. et al. (1994) *Continuous Univariate Distributions*, New York: John Wiley & Sons.

Kar, G. et al. (2010) Allostery and population shift in drug discovery, *Curr. Opin. Pharmacol.*, **10**, 715–722.

Kendrew, J.C. et al. (1958) A three-dimensional model of the myoglobin molecule obtained by x-ray analysis, *Nature*, **181**, 662–666.

Lattman, E.E. and Rose, G.D. (1993) Protein folding—What's the question? *Proc. Natl. Acad. Sci. U. S. A.*, **90**, 439–441.

Lehninger, A.L. (1975) *Biochemistry*, New York: Worth Publishers.

Li, J. et al. (2012) Thermodynamic dissection of the intrinsically disordered N-terminal domain of human glucocorticoid receptor, *J. Biol. Chem.*, **287**, 26777–26787.

Lindorff-Larsen, K. et al. (2012) Structure and dynamics of an unfolded protein examined by molecular dynamics simulation, *J. Am. Chem. Soc.*, **134**, 3787–3791.

Liu, J. et al. (2006) Intrinsic disorder in transcription factors, *Biochemistry*, **45**, 6873–6888.

Liu, T. et al. (2007) Functional residues serve a dominant role in mediating the cooperativity of the protein ensemble, *Proc. Natl. Acad. Sci. U. S. A.*, **104**, 4347–4352.

Liu, T. et al. (2012) Quantitative assessment of protein structural models by comparison of H/D exchange MS data with exchange behavior accurately predicted by DXCOREX, *J. Am. Soc. Mass Spectrom.*, **23**, 43–56.

Mao, A.H. et al. (2013) Describing sequence-ensemble relationships for intrinsically disordered proteins, *Biochem. J.*, **449**, 307–318.

McEwan, I.J. et al. (2007) Natural disordered sequences in the amino terminal domain of nuclear receptors: Lessons from the androgen and glucocorticoid receptors, *Nucl. Recept. Signal.*, **5**, e001.

Michel, A.M. et al. (2012) Observation of dually decoded regions of the human genome using ribosome profiling data, *Genome Res.*, **22**, 2219–2229.

Milne, J.S. et al. (1999) Experimental study of the protein folding landscape: Unfolding reactions in cytochrome c, *J. Mol. Biol.*, **290**, 811–822.

Mizanty, M.J. et al. (2010) Improved sequence-based prediction of disordered regions with multilayer fusion of multiple information sources, *Bioinformatics*, **26**, i489–i496.

Mohan, A. et al. (2009) Influence of sequence changes and environment on intrinsically disordered proteins, *PLoS Comput. Biol.*, **5**, e1000497.

Monatstyrskyy, B. et al. (2011) Evaluation of disorder predictions in CASP9, *Proteins*, **79**, 107–118.

Muirhead, H. and Perutz, M.F. (1963) Structure of haemoglobin: A three-dimensional Fourier synthesis of reduced human haemoglobin at 5.5 A resolution, *Nature*, **199**, 633–638.

Murzin, A.G. (2008) Metamorphic proteins, *Science*, **320**, 1725–1726.

Nozaki, Y. and Tanford, C. (1971) The solubility of amino acids and two glycine peptides in aqueous ethanol and dioxane solutions. Establishment of a hydrophobicity scale, *J. Biol. Chem.* **246**, 2211–2217.

Press, W.H. et al. (2007) *Numerical Recipes: The Art of Scientific Computing*, New York: Cambridge University Press.

Pruitt, K.D. et al. (2009) The consensus coding sequence (CCDS) project: Identifying a common protein-coding gene set for the human and mouse genomes, *Genome Res.*, **19**, 1316–1323.

Rezaei-Ghaleh, N. et al. (2012) Intrinsically disordered proteins: From sequence and conformational properties toward drug discovery, *Chembiochem*, **13**, 930–950.

Robinson, A.B. and Robinson, L.R. (1991) Distribution of glutamine and asparagine residues and their near neighbors in peptides and proteins, *Proc. Natl. Acad. Sci. U. S. A.*, **88**, 8880–8884.

Schlessinger, A. et al. (2009) Improved disorder prediction by combination of orthogonal approaches, *PLoS One*, **4**, e4433.

Sickmeier, M. et al. (2007) DisProt: The database of disordered proteins, *Nucleic Acids Res.*, **35**, D786–D793.

Tomii, K. and Kanehisa, M. (1996) Analysis of amino acid indices for prediction of protein structure and function, *Protein Eng.*, **9**, 27–36.

Tompa, P. et al. (2006) Prevalent structural disorder in *E. coli* and *S. cerevisiae* proteomes, *J. Proteome Res.*, **5**, 1996–2000.

Tzeng, S.R. and Kalodimos, C.G. (2011) Protein dynamics and allostery: An NMR view, *Curr. Opin. Struct. Biol.*, **21**, 62–67.

Uversky, V.N. (2013) A decade and a half of protein intrinsic disorder: Biology still waits for physics, *Protein Sci.*, **22**, 693–724.

Uversky, V.N. et al. (2000) Why are "natively unfolded" proteins unstructured under physiologic conditions? *Proteins*, **41**, 415–427.

Vucetic, S. et al. (2003) Flavors of protein disorder, *Proteins*, **52**, 573–584.

Vullo, A. et al. (2006) Spritz: A server for the prediction of intrinsically disordered regions in protein sequences using kernel machines, *Nucleic Acids Res.*, **34**, W164–W168.

Walsh, I. et al. (2011) CSpritz: Accurate prediction of protein disorder segments with annotation for homology, secondary structure, and linear motifs, *Nucleic Acids Res.*, **39**, W190–W196.

Ward, J.J. et al. (2004) Prediction and functional analysis of native disorder in proteins from the three kingdoms of life, *J. Mol. Biol.*, **337**, 635–645.

Weathers, E.A. et al. (2007) Insights into protein structure and function from disorder-complexity space, *Proteins*, **66**, 16–28.

Xue, B. et al. (2010) PONDR-FIT: A meta-predictor of intrinsically disordered amino acids, *Biochim. Biophys. Acta*, **1804**, 996–1010.

Chapter 8

Long Molecular Dynamics Simulations of Intrinsically Disordered Proteins Reveal Preformed Structural Elements for Target Binding

Elio Cino, Mikko Karttunen, and Wing-Yiu Choy

CONTENTS

INTRODUCTION

The intensive studies of intrinsically disordered proteins (IDPs) in the past two decades have revolutionized our view of the protein-structure relationship [1–8]. Unlike folded proteins, which adopt well-defined structures in their native states, IDPs exist as a dynamic ensemble of distinct conformations that are

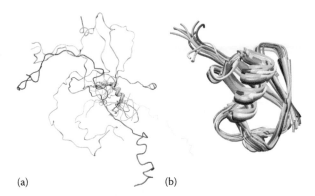

(a)   (b)

Figure 8.1 Structures of an IDP (a) and a well-folded protein (b). The NMR ensemble structures of the IDP (Thylakoid soluble phosphoprotein TSP9, PDB id: 2FFT [9]) do not overlay well because its intrinsic dynamic properties allow exchange between different conformations over time. On the other hand, the NMR ensemble structures of the well-folded protein (Ubiquitin, PDB id: 1D3Z [10]) illustrate that a similar structure is maintained over time.

differentially populated [11–15]. The unique structural and dynamic characters of these proteins are owing to the unusual amino acid composition of their primary sequences. Compared to folded proteins, IDPs have relatively high percentages of charged and polar amino acids as well as structure-breaking residues such as glycine and proline [3,16,17]. Hydrophobic and aromatic content is also lower in IDPs, rendering them unable to form a stable hydrophobic core. As a result, IDPs are more dynamic than their well-folded counterparts (Figure 8.1).

IDPs comprise >30% of the eukaryotic proteome [3,18,19]. The abundance of IDPs in organisms suggests that they are essential for numerous functions. Indeed, IDPs are found to be involved in crucial signaling and regulatory functions in cells [20–24]. The structural and dynamic properties of IDPs are well suited to their roles in signaling pathways, where reversible binding and the ability to interact with multiple partners are essential [1,4,6]. Clearly, IDPs are a biologically functional class of proteins that perform important functions while not conforming to the traditional structure-function paradigm.

TARGET RECOGNITION MECHANISMS OF IDPs

Protein-protein interaction (PPI) networks are the foundation on which cells carry out their functions. IDPs perform crucial functions, such as signal transduction and transcription [6,25–28], through extensive interactions with other proteins in these networks. Compared to well-folded proteins, however, there is relatively limited data describing the molecular mechanisms by which IDPs interact with binding partners [8,29–34].

Extensive structural characterizations of IDPs illustrate that even though these proteins lack stable tertiary structures, they frequently contain elements of secondary structure along their sequences that are crucial for their target recognition [30,35,36]. In contrast to PPIs between globular proteins, which typically involve larger interaction surfaces that are discontinuous in sequence, IDPs are frequently found to interact with targets via short, ~6 amino acid, stretches that are continuous in sequence, called linear motifs (LMs) [37–39]. These short segments of IDPs that contain residual structure may act as molecular recognition features (MoRFs) for binding to their targets [35,40].

Besides the structural features, the dynamic properties of IDPs that are intimately related to the timescale of conformational exchange within the ensemble also govern partner binding and how these proteins function. Since different structures in the ensemble can participate in the interactions with distinct targets, the rate of exchange between conformers can have significant impact on their molecular recognitions [29,34,41]. Therefore, to have better understanding of how IDPs function, it is important to characterize the structural and dynamic features of these interaction hot spots and identify their target recognition mechanisms.

There are at least two major mechanisms by which MoRFs mediate PPIs (Figure 8.2). For some IDPs, the interaction hot spots contain preformed structural elements (PSEs) that resemble the bound state conformations [30,34], while others may couple conformational changes/folding with target binding [8,32,42,43]. For IDPs that bind using PSEs, the bound state structure is already formed (or highly populated) in the free state. While in the coupled folding and binding model, the IDP undergoes a disorder-to-order transition upon binding to a target. It is important to realize that these two interaction methods represent opposite ends of the binding mode continuum. In most cases, binding of IDPs is probably modulated by a combination of these two mechanisms [44,45].

Tremendous efforts have been devoted to develop methodologies, both computationally and experimentally, to identify and characterize PSEs or regions with high secondary structure propensities in IDPs [11,15,46–54]. Bioinformatics approaches have been established for predicting disorder tendency as well as identifying potential binding regions of IDPs [25,55–60]. Interaction hot spots in IDPs often have distinct sequence characteristics compared to their surroundings, with the primary difference being an increased hydrophobic content, which may promote local structure formations [30,39]. In addition to bioinformatics approaches, several biophysical techniques, in particular nuclear magnetic resonance (NMR), have also proven to be useful for the structural characterization of IDPs [11,15,47,50]. Although these experimental methodologies can yield a wealth of data, they have their limitations.

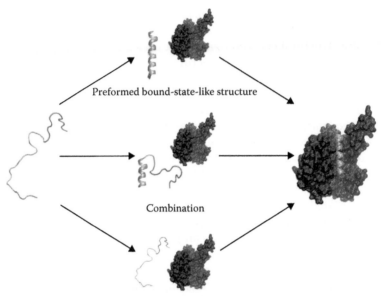

Figure 8.2   Binding mechanisms of IDPs. An IDP can interact with binding partners by either folding into a bound-state-like conformation prior to binding (top), encountering the binding partner and then folding (bottom), or a combination of these two mechanisms (middle). (Adapted from Sugase, K. et al., *Nature* 447, 1021–1025, 2007.)

For instance, NMR data collected on an IDP undergoing fast conformational exchange on the NMR timescale are averaged over the entire ensemble of conformations sampled by the protein. Therefore, unlike for folded proteins, it is inappropriate to determine a single conformation to represent the disordered state. Another commonly used experimental technique for protein structure determination is x-ray crystallography. There are many structures of IDP binding regions in complex with well-folded partners [61–64]; however, their dynamic nature makes acquiring diffracting crystals of IDPs in their free state extremely challenging, if not impossible. Therefore, while bound state structures are useful for identifying possible PSEs, they do not provide detailed information of the free state forms. Molecular dynamics (MD) simulations have been used to overcome some of the issues related to experimental characterization of IDPs and to complement the experimental data [65]. In particular, MD simulations have been employed to study the conformational transition of IDPs in solution and the mechanisms by which they interact with targets [44,66–68]. The results have shed light on the roles of PSEs in regulating the binding of IDPs.

MD SIMULATIONS OF IDPs

Through advancements in computing power and methodologies, MD simulations have become an invaluable tool for investigating the structure, dynamics, and target-binding of IDPs [44,65–77]. From a historical perspective, the first protein MD simulation was performed in 1975 on the bovine pancreatic trypsin inhibitor (BPTI) and lasted just a few picoseconds [78]: modest in current terms, but a major landmark and breakthrough at that time. Since then, advancements in computing power, algorithms, and methods have allowed for MD simulations long enough to study complex processes such as folding of small proteins. Here, the focus is on MD simulations; however, there are also other simulation and modeling methods for IDPs and proteins [12,14,79–81]. A couple of excellent reviews discussing some of the current challenges and successes in simulations of protein folding are provided by Freddolino et al. [82] and Lane et al. [83].

Methodologies Used in MD Simulations of IDPs

The development of high-performance simulation codes such as NAMD [84], GROMACS [85], AMBER [86], CHARMM [87] and others have made large-scale simulations accessible to a large number of research groups worldwide. The longest timescale MD simulations to date have been performed using the specialized Anton supercomputer and have been up to milliseconds in length [74,88–90]. Such trajectories have provided important insights into the fluctuations of protein structure and dynamics over time. For instance, millisecond timescale MD simulations have been used to fold ubiquitin, providing atomic-level details of its folding pathway [89]. In another study, 200 μs long simulations performed on Anton showed that under acid-denatured conditions, bovine acyl-coenzyme A binding protein (ACBP) still contains substantial residual helical structure in regions that form helical structure in the native state [74], which are in reasonable agreement with the experimental observations by NMR [54].

Access to specialized resources such as Anton or MDGRAPE [91] systems is limited, but there are other means for achieving long timescale MD simulations. By using (cost-effective) graphics processing units (GPUs), simulations can be accelerated 10- to 100-fold relative to purely central processing unit (CPU) based implementations [92]. As a result, desktop systems equipped with modern GPUs can perform calculations that would otherwise require a much larger computing cluster: In our tests using the GPU-accelerated GROMACS code, we were able to reach 72 ns per day (72,000 atoms) using a $240 GPU card installed on a regular personal computer (PC). Using GPUs comes at a cost, however: Codes need to be rewritten for GPU simulations and the GPU architecture poses its own limitations for codes and parallelization.

In addition to hardware, method and algorithm advancements are also playing a crucial role in achieving long timescale simulations on standard architectures. Folding@home (http://folding.stanford.edu) [93] and Rosetta@home (http://http://boinc.bakerlab.org/rosetta/) [94] use a different and quite revolutionary strategy: They use distributed computing that aims to computationally fold proteins [83,94]. These methods utilize idle CPU/GPU time on computers and even gaming consoles to fold proteins. Because of the large number of users worldwide, aggregate simulation times on the millisecond timescale have been attained [83]. Rosetta@home is also the base platform for Foldit [95], an easy-to-learn game that uses the puzzle-solving ability of humans to fold proteins.

Another common way to address computational limitations is through simulation techniques that enhance conformational sampling. MD- and Monte Carlo based methods such as replica exchange [96], parallel tempering [97], activation-relaxation technique (ART) [98], metadynamics [99], Go modeling [100,101], transition path sampling [102–104], and string methods [105] are some of the approaches being used to enhance sampling and extract free energies. For example, in temperature replica exchange simulations, copies of the system are simulated at different temperatures and exchanged according to the Metropolis criterion. As a result, replica exchange allows the system to sample high- and lower-energy conformations more rapidly than in fixed temperature simulations [67,75,76,106,107]. In comparison, metadynamics simulations can improve conformational sampling by discouraging the system to revisit previous steps [99,108,109]. This is accomplished though careful selection of collective variables (reaction coordinates) that describe particular state(s). These techniques have been applied recently in the MD simulation of conformation ensemble of IDPs. As an example, Mittal et al. used replica exchange MD (REMD) simulations to probe structural ensemble of a peptide derived from the N-terminal transactivation domain (TAD) of p53 (residues 15–29), which harbors the MDM2 binding site [106]. The results illustrate that residues 19–23 of the TAD peptide sample conformations that are highly similar to the bound-state structure observed in the TAD/MDM2 complex. Moreover, the free-energy profile of the TAD peptide is relatively flat, suggesting that the peptide can undergo rapid interconversion without having to overcome high energy barriers [106]. In addition to using single-resolution traditional MD, coarse-grained modeling techniques have been applied to the protein problem [110]. Coarse-graining, although an active field or research of its own, offers the possibility to reach even more realistic systems sizes and simulation times [111]. Another new technique is the application of Bayesian modeling to IDPs [112]. This technique is different in the sense that it combines both experimental and computational data and tries to calculate the probability density for the states.

Force Fields Used in MD Simulations

All simulations need a force field. A force field is the parameterization that contains all interactions, bonded and nonbonded, for all the atoms and molecules in a simulation. It is impossible to determine which force field is the absolute best one: a certain force field may give a good agreement with experimental data for helical proteins but not necessarily for beta or mixed alpha/beta proteins (or vice versa): The choice of force field is of crucial importance [113–121]; it is well documented that different force fields have preferences toward certain types of secondary structures. In the other words, the secondary structure propensities of protein/peptide systems derived from the MD simulations may depend on the force fields used. The bias can be particularly problematic in IDPs because the sampled conformations are usually with small free-energy differences. This potential problem has been recognized and a substantial amount of work by several groups has been devoted to resolving these issues [122–124].

It has become increasingly common to use peptides encoding small structural elements with folding times on the microsecond timescale for force field optimization. Such trajectories provide reasonable sampling of conformations and sufficient length to examine the stability of the force field [113,117,125,126]. Some of the extensively studied test systems for protein-folding simulations include a peptide encoding a hairpin motif from the globular protein GB1 [125,127] and the designed 20-residue Trp-Cage protein [128–130]. Even though using self-folding motifs of globular proteins like the GB1 hairpin is a viable approach to decrease system sizes and observe folding events, care must be taken to ensure the motif does indeed fold properly in the absence of the rest of the protein. An alternative approach for force field validation of protein folding is to use naturally occurring PSEs extracted from IDP sequences [113]. Because PSEs typically consist of a short segment of amino acids, it is often possible to observe their folding in reasonable length (microsecond timescale) of MD simulations. Moreover, comparisons can be made to experimental data, such as NMR chemical shifts, coupling constants, and spin relaxation measurements [114].

In a recent work, 10 biomolecular force fields were compared with respect to the folding of a β-hairpin-forming peptide derived from nuclear factor erythroid 2-related factor 2 (Nrf2) [113]. In its free state, this motif has been shown to have structural resemblance to its bound state conformation with Keap1, indicating that it functions as an important PSE for target binding [64]. Starting from an extended conformation, the amino acid sequence encoding this hairpin has been shown to fold into a structure consistent with experimental data in <1 μs [44,113]. However, when comparing the folding of this structural element with commonly used force fields, differences were observed. Although many of the force fields reproduced experimentally determined free-state contacts and secondary structure, some did not (Figure 8.3). Furthermore, clear

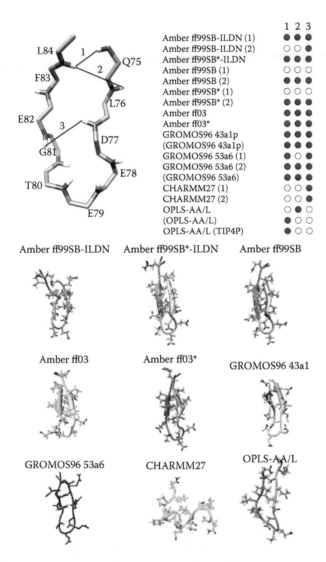

Figure 8.3 Folding of the Nrf2 hairpin with different force fields. Top: Formation of experimentally determined free state contacts. Bottom: Structures from the MD simulations with different force fields after about 1 μs. (Reprinted with permission from Cino, E. A. et al., 2012, 2725–2740. Copyright 2012, American Chemical Society.)

differences were evident between duplicate simulations with the same force field and starting structure, but with different initial velocities. This highlights the importance of performing multiple replicate simulations and considering enhanced sampling techniques. Overall, PSEs may be ideal candidates for force field testing; the results obtained from folding simulations of such elements should be useful for improving bimolecular force fields [113].

In addition to the force field, attention should be paid to protocol issues. Some tricks of the trade, while perfectly fine under some circumstances and often set as the default options in software packages, can lead to significant errors under some other conditions. Such issues are typically related to the treatment of charges since computation of the long-range Coulomb interactions is time-consuming. Such concerns apply to all molecular simulations. We will not discuss these issues further but refer the reader of the some of the recent articles discussing these matters [131–135]. Artifacts aside, we would also like to point out that at best, simulations are a predictive tool and they can be even used to elaborate the precise molecular level mechanisms in experiments [136].

MD SIMULATIONS REVEAL PREFORMED STRUCTURAL ELEMENTS IN IDPs

MD simulations have yielded important details about the structure and dynamics of IDPs, which have helped to understand their mechanisms of target interactions. In an earlier study of the binding of intrinsically disordered p27 to the cyclin A/Cdk2 complex, Sivakolundu et al. used a combination of MD and NMR to investigate the roles of preformed structural elements in molecular recognition [68]. MD results show that p27 exhibits intrinsically folded structural units (IFSUs) within the regions that are known to bind to its targets. In particular, the IFSUs in the LH domain and the domains 2.1 and 2.3, which mediate specific interactions with cyclin A/Cdk2, adopt bound-state-like conformation in the absence of targets. However, the simulation time was relatively short (~100 ns).

MD simulations and NMR experiments have also been employed to identify preformed bound-state-like structures in the free-state ensemble of a fragment of the disordered p21 [77]. Interestingly, despite being a small protein, p21 interact with a large (~25) number of targets. The peptide examined in this study contains PSEs that resemble bound state conformations with two partners, proliferating cell nuclear antigen (PCNA) and calmodulin. The authors proposed that the ability of p21 to have such a diverse, yet specific, set of interactions relies upon its intrinsic disorder coupled with residual structures.

In a recent investigation, microsecond MD simulations were performed to compare the free-state structures and dynamics of two IDPs, prothymosin α

(PTMA) and NRF2 [31,44]. These two proteins interact with a common partner, Keap1, in order to control the cellular response to oxidative stress. By
conducting microsecond timescale MD simulations, structural features of the
PSEs located in the Keap1 binding regions of PTMA and NRF2 were identified.
In the absence of Keap1, the PSEs had clear resemblance to their bound state
structures. NRF2, which binds to Keap1 with a higher affinity than PTMA,
formed more stable PSEs with lower root-mean-square deviations (RMSDs) to
its bound state structure compared to PTMA [44] (Figure 8.4). It appears that
the extents of bound-state-like structures that are formed in the absence of
binding partner have important implications in dictating the thermodynamics
of binding of these proteins. Detailed analysis of the MD trajectories suggested
that NRF2 is able to form more intrachain hydrogen bonds compared to PTMA,
which contribute to the stabilization of bound-state-like β-hairpin structure.
Additionally, NRF2 contains more hydrophobic amino acids surrounding its
PSE, which may also promote structure formation.

The conformational propensities of IDPs in their uncomplexed (free) states
can provide insights into their target-binding affinity. As mentioned in the section titled "Methodologies Used in MD Simulations of IDPs," Mittal et al. [106]
used REMD to simulate the structural ensemble of a peptide derived from the
N-terminal TAD of p53 (residues 15–29) and the P27L mutational variant. The
simulations showed that the TAD peptide sampled conformations that are highly
similar to the bound-state helical structure observed in the p53-TAD/MDM2 complex. Interestingly, the P27L mutant, which exhibits a stronger binding affinity to
MDM2, had a higher helical propensity compared to the wild-type TAD peptide.
This is in good agreement with the experimental observations by NMR and circular dichroism (CD) spectroscopy [106]. The finding suggests that the more favorable binding affinity of P27L might arise in part from it having a more defined
free-state structure, which would decrease the entropic loss upon binding.

In another study, Knott and Best performed REMD simulations using optimized water models (TIP4P/2005) with Amber ff03w force field on the free state
of the intrinsically disordered NCBD in order to gain insight into its binding
mechanism to another IDP, activation domain from the p160 transcriptional co-
activator for thyroid hormone and retinoid receptors (ACTR) [67]. In the complex
form, the ACTR binding region of nuclear-receptor co-activator binding domain
of the transcriptional co-activator CBP (NCBD) adopts three helical structures,
αI, αII, and αIII. REMD simulations of NCBD in the free form indicate that residues in αI and αIII, the two regions that directly mediate the interaction with
ACTR, all display high helical propensity. On the other hand, the helical structure of residues in αII, which is involved in forming tertiary structure with αI
and αIII, are less pronounced. The authors suggest that the two proteins interact
by the so-called binding interface preference mechanism: While the existence of
bound-state-like structures in αI and αIII facilitate the interaction with ACTR,

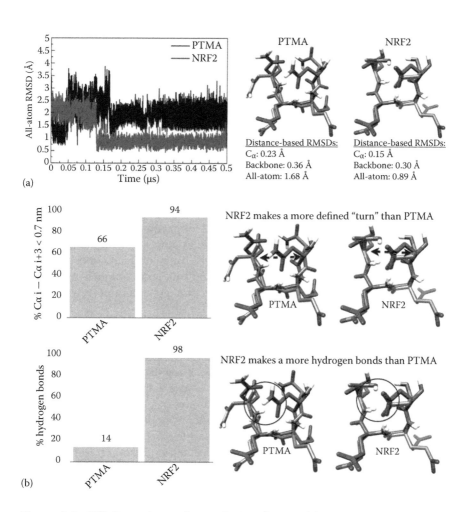

Figure 8.4 PSE formation and contributing factors. (a) PTMA and NRF2 form PSEs with different extents of bound state resemblance. Left: RMSDs to the bound state structures during the MD trajectories. Right: Snapshots from the MD simulations (gray) overlaid with their bound state structures (pink). (b) Contributing factors to explain the different extents of preformed structure in PTMA and NRF2. Top: Percentage of structures from the MD simulations with an end-to-end distance less than 0.7 nm. Bottom: Percentage of MD structures with one or more intraturn hydrogen bonds. (Adapted from Cino, E. A. et al., *PLoS One* 6, e27371, 2011.)

the lack of native-like structure in αII allows the protein to remain disorder in the unbound state.

EXAMPLES OF ONE-TO-MANY AND MANY-TO-ONE SIGNALING STUDIED BY MD SIMULATIONS

IDPs Act as Hub Proteins: One-to-Many Signaling

The structural plasticity of IDPs confers many functional advantages [28,137]. Due to the lack of stable tertiary fold, IDPs can participate in interactions with multiple targets by using variable LMs along the protein sequences. Moreover, since IDPs exist as an ensemble of fast interconverting conformers [11,15], the same region of a protein can bind to distinct partners by adopting different conformations. Physically, this implies the existence of a large number of degenerate or near-degenerate states separated by small free-energy barriers. These characteristics allow many IDPs to bind to a large number of targets, either simultaneously or sequentially, so function as hubs in protein-protein interaction networks [6,26–28,137].

The transcription factor p53 is an IDP that participates in one-to-many binding and has been extensively studied using MD simulations. Because of its role as the "guardian" of the genome, p53 is possibly one of the most intensively studied proteins in the past two decades [138]. It regulates the expression of a battery of genes involved in deoxyribonucleic acid (DNA) repair, cell cycle control, and apoptosis in response to the cellular stresses [139]. Structural studies of this tumor suppressor revealed that it adopts modular domain structure and functions as a tetramer [140]. In particular, the N-terminal transactivation domain (residues 1–61) and the C-terminal regulatory domain (residues 356–393) of p53 have been identified as intrinsically disordered and act as hubs in protein-protein interaction network by binding to multiple partners [137,140].

To gain insight into the molecular mechanisms by which p53 interacts with different partners, Allen et al. performed MD simulations on the C-terminal domain (CTD) of p53 and five known binding partners, including S100B, Sir2, CBP, Set9, and cyclinA/Cdk2 [69]. Experimental studies have shown that CTD adopts distinct structures upon binding to different targets, ranging from α-helix, β-strand to unstructured conformations [141–145]. For instance, NMR structure of the CTD peptide (residues 367–388) in complex with S100B revealed that a large part of the peptide (residues 376–387) adopts a helical structure upon binding [145]. Meanwhile, residues 380–385 of p53 were found to form a β-strand structure when in complex with Sir2 [141].

MD simulations of p53 CTD and the five binding partners in their free states and complex forms revealed that the disordered CTD recognize different targets via distinct mechanisms [69]. In the free-state simulations, p53 CTD was highly flexible. It sampled different secondary structures transiently without adopting conformations that resemble any particular bound-state structures. Meanwhile, the binding partners exhibited subtle changes in structure and dynamics upon binding p53 CTD. In particular, the dynamics of hydrophobic residues located in the p53-binding pockets of the partners in general increased upon binding p53, while the fluctuations of the charge or polar residues surrounding the p53 binding pockets decreased in the complex form. The authors suggested that the conformation of p53 CTD induced by the binding partners depends on the size of the hydrophobic binding site of the partners and the number of hydrogen bonds that can be formed in the interfaces [69].

Binding of IDPs to Protein Hubs: Many-to-One Signaling

In addition to acting as hubs (one-to-many signaling), IDPs often preferentially bind to hub proteins (many-to-one signaling) [28,137]. One of the best-known many-to-one hubs for IDP interactions is the 14-3-3 family. These ~30-kDa dimeric proteins adopt rigid structures capable of binding hundreds of ligands that together may comprise 0.6% of the human proteome [146,147]. Interestingly, it is estimated that over 90% of the 14-3-3 binding partners are completely or partially disordered [146,147]. Binding is governed by phosphoserine- and phosphothreonine-containing motifs on the partners. Several distinct consensus motifs (modes) with varying binding affinities have been discovered across the broad spectrum of 14-3-3 ligands. Moreover, several crystal structures are available of disordered proteins bound to 14-3-3s [148–151]. Another protein with a preference for binding disordered partners is the N-terminal β-propeller domain (TD) domain of clathrin. Interactions between this hub and several disordered partners, including amphiphysins, AP180 and SNX9, have been demonstrated [152,153].

In a recent work, Cino et al. have identified the Kelch domain of Keap1 as a protein hub that preferentially binds to the disordered regions of its partners. These binding partners share similar Keap1-binding motifs but have a wide range of affinities [154]. Through protein sequence analysis, disorder predictions, biophysical experiments, and MD simulations, the factors that determine the affinities and specificities of individual IDPs to the Kelch domain of Keap1 were investigated.

The Kelch domain is a ~32-kDa β-propeller located near the C-terminus of Keap1 [155]. At least 10 different proteins have been shown to interact with Kelch and most of these partners interact via disordered regions harboring similar binding motifs [154]. Of all of the Kelch partners, the most studied is NRF2 (see also the section titled "MD Simulations Reveal Preformed Structural Elements in IDPs"), a key transcription factor for coordinating cellular responses to oxidative stress [156]. It contains high- and low-affinity Kelch binding motifs and can bind two Kelch molecules [64]. The two-site binding promotes the ubiquitination of NRF2 and subsequently its degradation [157]. Besides NRF2, WTX [158], p62 [159], PGAM5 [160], PALB2 [161], FAC1 [162], PTMA [62], IKKβ [163], and many others [164] have also been found to interact with Kelch. Aside from the NRF2–Kelch interaction, the functions of the other protein-protein interactions are less understood. However, there is evidence that these interactions are also involved in regulating the oxidative stress response. For example, several of the Kelch partners are able to disrupt the low-affinity NRF2–Kelch interaction to promote cytoprotective gene expression. Additionally, PTMA have been shown to function as a vehicle for shuttling Keap1 into the nucleus [165]. The structures of the binding regions of NRF2, p62, and PTMA in complex with Kelch have been determined and show that they all adopt similar β-turn structure in bound state [62,64,159].

While the various Kelch partners share similar binding motifs, they have a wide range of binding affinities, spanning more than 2 orders of magnitude (Figure 8.5). These proteins share high sequence similarity in a six-residue

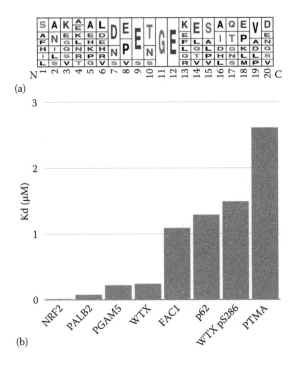

(a)

(b)

Figure 8.5 Binding motifs and affinities of the Keap1 partners [154]. (a) Frequency plot showing the relative amino acid preferences of the various partners around the Kelch binding site (positions 7–12). (b) Dissociation constants of the interactions, determined by isothermal titration calorimetry.

stretch, corresponding to the DEETGE of NRF2 that comprise the Kelch domain-binding interface. Outside of this six-residue stretch, however, there are no clear sequence similarities between the different proteins. NMR experiments on the 20-mer peptides indicated regions of compactness around the binding sites of many of the partners, consistent with beta turns and hairpins being formed. In addition, there also appeared to be a connection between greater amounts of free-state structure formation and increased binding affinity [154].

MD simulations were used to investigate factors stabilizing these PSEs and the relations between structure/dynamics to binding affinity (details of the MD simulations are described in Box 8.1 and in [44,113,154]). Proton pairs in close contacts identified in the MD trajectories of the free-state peptides were in good agreement with experimentally observed nuclear overhauser effects (NOEs) by NMR (Figure 8.6). Overall, the MD simulations illustrate that several, if not all, of the Kelch domain-binding peptides analyzed contain β-turn-like structure at their binding sites. These structures likely display resemblance to their bound state conformations to certain extents.

BOX 8.1   DETAILS OF MD SIMULATIONS [44,113,154]

The starting structures for the simulations were generated solely based on the amino sequences of the various partners. A simulated annealing procedure was used to generate random structures that were used in the simulations. Although the binding motifs comprise only ~6 residues, surrounding amino acids have been shown to be important for modulating Kelch binding affinity, possibly by enhancing bound-state-like structure formation prior to binding [44]. As a result, the motifs were extended on either end up to ~20 residues. MD simulations were performed in explicit solvent using a force field and parameters that showed good agreement with experimental data for the NRF2 hairpin [113]. In summary, all species in the simulations were represented with the GROMOS96 53a6 or 43a1 [166] force fields under periodic boundary conditions in a cubic box with a side length of 6 nm. Protonation states of all ionizable residues were chosen based on their most probable state at pH 7. Each system was neutralized and brought to an ionic strength of 0.1 M with sodium (Na^+) and chloride (Cl^-) ions. Protein and nonprotein atoms were coupled to their own temperature baths that were kept constant at 310 K using the Parrinello-Donadio-Bussi v-rescale algorithm [167]. Pressure was maintained isotropically at 1 bar using the Parrinello-Rahman barostat [168]. A 2-fs timestep was used. Prior to the production runs, the energy of each system was minimized using the steepest descents algorithm. Initial atom velocities were taken from a Maxwellian distribution at 310 K. All bond lengths were constrained using the linear constraint solver (LINCS) algorithm [169]. A 1.0-nm cutoff was used for Lennard-Jones interactions. Dispersion corrections for energy and pressure were applied. Electrostatic interactions were calculated using the particle mesh Ewald (PME) method [170] with 0.12-nm grid-spacing and a 1.0-nm real-space cutoff. Charge groups were not used (single-atom charge groups). Each simulation was run for at least 1 μs [131,132].

MD analysis also provided insights into the varied binding affinities of different targets to Kelch. For instance, the WTX peptide adopted a turn/bend conformation that was consistent with the NMR data but when phosphorylated the position was shifted from the expected location. The distortion was caused by interactions between the pS286 residue and a lysine on the opposite side of the turn. From the MD data, it appeared that S286 phosphorylation enhances the structure formation in the free state. This was supported by experimental binding data [154], which revealed that the binding of WTX pS286 peptide to Kelch had the smallest entropy change. However, this peptide interacted with the Kelch domain with the least favorable enthalpy, which suggests that the peptide conformation induced by phosphorylation is not optimal for binding.

The MD data further revealed the link between free-state structure and binding affinity. Interestingly, circular variance values of the φ and ψ angles, a measure of backbone dynamics extracted from the MD simulations, correlated

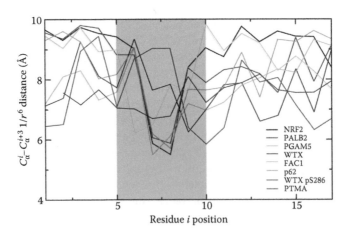

Figure 8.6 Regions of compactness in the MD simulations of the Kelch partners. The $C_\alpha^i - C_\alpha^{i+3}$ $1/r^6$ averaged distances were calculated over the last 0.5 μs of the MD trajectories. Gray shading indicates the major areas of compactness identified from NMR experiments. (Adapted from Cino, E. A. et al., *Sci Rep* 3, 2305, 2013.)

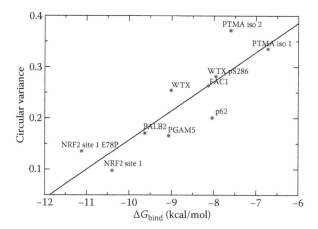

Figure 8.7 Correlation of peptide dynamics and binding free energy. (Adapted from Cino, E. A. et al., *Sci Rep* 3, 2305, 2013.)

well (r^2 = 0.75) with binding free energy (Figure 8.7). The trend illustrates that the binding affinity is heavily determined by the free-state dynamics. Based on this correlation between free-state dynamics and binding free energy, a peptide with a higher affinity for the Kelch domain than any of the natural peptides (identified to date) was designed [154]. A single-point mutation, E78P, was made to the natural, 20-mer NRF2 peptide to further enhance its β-turn propensity. Isothermal titration calorimetry (ITC) measurements showed that the E78P mutation indeed increases the binding affinity of the peptide (7 ± 1 nM) compared to the natural sequence by 3–4-fold [154].

Overall, the findings suggest that different disordered proteins interact with the Kelch domain via preformed β-turn structures resembling the bound state conformations. However, the binding regions are differentially stabilized by intraturn contacts and interactions between residues in close proximity to the binding region. The extent of turn stabilization is likely an important factor in modulating binding affinity [154].

FUTURE DIRECTIONS

MD simulations are becoming established tools for detecting and characterizing PSEs of IDPs and it is expected that they will become one of the go-to tools for studying PPIs. As with any methods, there are some challenges. For example, in establishing MD simulations as more accurate, predictive, and efficient tools, sampling methods, algorithms, and the development of transferable force fields are areas of great importance and active development. IDPs are a challenge for both experiments and simulations due to small free-energy barriers and degenerate states: Designability principles that have been used with globular protein models [171] cannot be easily applied for IDPs because there is no unique folded state.

While it is generally good practice to perform MD simulations of biological systems under physiological temperatures, pressure, and ionic strengths, the effects of other molecules are usually neglected. In living organisms, macromolecules such as proteins are not simply solvated in vast amounts of water and ions. The cellular environment is highly crowded with various macromolecules. In such crowded environments, the amount of water available to solvate a protein may actually be limited, causing it to behave differently than *in vitro*; the importance and unique behavior of water on processes in confined regions, such as ligand binding to a receptor, has been recently demonstrated [172]. It has also been shown that the crowded cellular environment may act to selectively stabilize target-binding regions of IDPs [173,174]. Although it may be difficult at this stage to perform simulations in the presence of the entire complement of cellular macromolecules, efforts are underway to include crowding agents in modern force fields. These agents should be able to mimic the excluded volume effect that occurs in the crowded cellular environment. Several groups have performed simulations with crowding agents [175–177]. Although the agents are often crudely parameterized as hard spheres with repulsive potentials to all species within the system, such models provide important insights into the effects of hydrodynamics and confinement on diffusive processes and conformations of macromolecules and processes. To improve the molecular fidelity, and thus making them more quantitative in terms of molecular level interactions, it is important to develop more detailed descriptions of the crowding agents, such as crowding agents with attractive and repulsive nonbonded potentials to different system components or spheres with specific functional groups that can sequester water. Such work, along with continued force field development and progression of advanced sampling techniques, will be crucial for establishing MD simulations as trusted routine tools for studying IDPs and other biomolecules.

ACKNOWLEDGMENTS

Computational resources were provided by Compute Canada and SharcNet (www.sharcnet.ca).

Financial support has been provided by the Natural Sciences and Engineering Research Council (NSERC) of Canada [MK], Canadian Institutes of Health Research (CIHR) [WYC], Ontario Early Researcher Award Program [WYC & MK], and University of Waterloo [MK].

REFERENCES

1. Wright, P. E., and Dyson, H. J. (1999) *J Mol Biol* **293**, 321–331.
2. Uversky, V. N., Gillespie, J. R., and Fink, A. L. (2000) *Proteins* **41**, 415–427.
3. Dunker, A. K., Lawson, J. D., Brown, C. J., Williams, R. M., Romero, P., Oh, J. S., Oldfield, C. J., Campen, A. M., Ratliff, C. M., Hipps, K. W. et al. (2001) *J Mol Graph Model* **19**, 26–59.
4. Dyson, H. J., and Wright, P. E. (2002) *Curr Opin Struct Biol* **12**, 54–60.
5. Tompa, P. (2002) *Trends Biochem Sci* **27**, 527–533.
6. Dunker, A. K., Cortese, M. S., Romero, P., Iakoucheva, L. M., and Uversky, V. N. (2005) *FEBS J* **272**, 5129–5148.
7. Dyson, H. J., and Wright, P. E. (2005) *Nat Rev* **6**, 197–208.
8. Wright, P. E., and Dyson, H. J. (2009) *Curr Opin Struct Biol* **19**, 31–38.
9. Song, J., Lee, M. S., Carlberg, I., Vener, A. V., and Markley, J. L. (2006) *Biochemistry* **45**, 15633–15643.
10. Cornilescu, G., Marquardt, J. L., Ottiger, M., and Bax, A. (1998) *J Am Chem Soc* **120**, 6836–6837.
11. Eliezer, D. (2009) *Curr Opin Struct Biol* **19**, 23–30.
12. Fisher, C. K., and Stultz, C. M. (2011) *Curr Opin Struct Biol* **21**, 426–431.
13. Jensen, M. R., Ruigrok, R. W., and Blackledge, M. (2013) *Curr Opin Struct Biol* **23**, 426–435.
14. Krzeminski, M., Marsh, J. A., Neale, C., Choy, W. Y., and Forman-Kay, J. D. (2013) *Bioinformatics* **29**, 398–399.
15. Mittag, T., and Forman-Kay, J. D. (2007) *Curr Opin Struct Biol* **17**, 3–14.
16. Radivojac, P., Iakoucheva, L. M., Oldfield, C. J., Obradovic, Z., Uversky, V. N., and Dunker, A. K. (2007) *Biophys J* **92**, 1439–1456.
17. Vucetic, S., Brown, C. J., Dunker, A. K., and Obradovic, Z. (2003) *Proteins* **52**, 573–584.
18. Pentony, M. M., Ward, J., and Jones, D. T. (2010) *Methods Mol Biol* **604**, 369–393.
19. Xue, B., Dunker, A. K., and Uversky, V. N. (2012) *J Biomol Struct Dyn* **30**, 137–149.
20. Dunker, A. K., Silman, I., Uversky, V. N., and Sussman, J. L. (2008) *Curr Opin Struct Biol* **18**, 756–764.
21. Dunker, A. K., and Uversky, V. N. (2008) *Nat Chem Biol* **4**, 229–230.
22. Follis, A. V., Galea, C. A., and Kriwacki, R. W. (2012) *Adv Exp Med Biol* **725**, 27–49.
23. Iakoucheva, L. M., Brown, C. J., Lawson, J. D., Obradovic, Z., and Dunker, A. K. (2002) *J Mol Biol* **323**, 573–584.
24. Minezaki, Y., Homma, K., Kinjo, A. R., and Nishikawa, K. (2006) *J Mol Biol* **359**, 1137–1149.
25. Dosztanyi, Z., Csizmok, V., Tompa, P., and Simon, I. (2005) *Bioinformatics* **21**, 3433–3434.
26. Haynes, C., Oldfield, C. J., Ji, F., Klitgord, N., Cusick, M. E., Radivojac, P., Uversky, V. N., Vidal, M., and Iakoucheva, L. M. (2006) *PLoS Comput Biol* **2**, e100.
27. Shimizu, K., and Toh, H. (2009) *J Mol Biol* **392**, 1253–1265.
28. Uversky, V. N., Oldfield, C. J., and Dunker, A. K. (2008) *Annu Rev Biophys* **37**, 215–246.
29. Boehr, D. D., Nussinov, R., and Wright, P. E. (2009) *Nat Chem Biol* **5**, 789–796.
30. Fuxreiter, M., Simon, I., Friedrich, P., and Tompa, P. (2004) *J Mol Biol* **338**, 1015–1026.
31. Khan, H., Cino, E. A., Brickenden, A., Fan, J., Yang, D., and Choy, W. Y. (2013) *J Mol Biol* **425**, 1011–1027.

32. Sugase, K., Dyson, H. J., and Wright, P. E. (2007) *Nature* **447**, 1021–1025.
33. Tompa, P., and Fuxreiter, M. (2008) *Trends Biochem Sci* **33**, 2–8.
34. Tsai, C. J., Ma, B., Sham, Y. Y., Kumar, S., and Nussinov, R. (2001) *Proteins* **44**, 418–427.
35. Mohan, A., Oldfield, C. J., Radivojac, P., Vacic, V., Cortese, M. S., Dunker, A. K., and Uversky, V. N. (2006) *J Mol Biol* **362**, 1043–1059.
36. Lee, S. H., Kim, D. H., Han, J. J., Cha, E. J., Lim, J. E., Cho, Y. J., Lee, C., and Han, K. H. (2012) *Curr Protein Pept Sci* **13**, 34–54.
37. Das, R. K., Mao, A. H., and Pappu, R. V. (2012) *Sci Signal* **5**, pe17.
38. Davey, N. E., Van Roey, K., Weatheritt, R. J., Toedt, G., Uyar, B., Altenberg, B., Budd, A., Diella, F., Dinkel, H., and Gibson, T. J. (2012) *Mol Biosyst* **8**, 268–281.
39. Mészáros, B., Tompa, P., Simon, I., and Dosztányi, Z. (2007) *J Mol Biol* **372**, 549–561.
40. Vacic, V., Oldfield, C. J., Mohan, A., Radivojac, P., Cortese, M. S., Uversky, V. N., and Dunker, A. K. (2007) *J Proteome Res* **6**, 2351–2366.
41. Mittag, T., Kay, L. E., and Forman-Kay, J. D. (2010) *J Mol Recognit* **23**, 105–116.
42. Kiefhaber, T., Bachmann, A., and Jensen, K. S. (2012) *Curr Opin Struct Biol* **22**, 21–29.
43. Receveur-Brechot, V., Bourhis, J. M., Uversky, V. N., Canard, B., and Longhi, S. (2006) *Proteins* **62**, 24–45.
44. Cino, E. A., Wong-ekkabut, J., Karttunen, M., and Choy, W. Y. (2011) *PLoS One* **6**, e27371.
45. Espinoza-Fonseca, L. M. (2009) *Biochem Biophys Res Commun* **382**, 479–482.
46. Baker, J. M., Hudson, R. P., Kanelis, V., Choy, W. Y., Thibodeau, P. H., Thomas, P. J., and Forman-Kay, J. D. (2007) *Nat Struct Mol Biol* **14**, 738–745.
47. Dyson, H. J., and Wright, P. E. (2005) *Methods Enzymol* **394**, 299–321.
48. Lacy, E. R., Filippov, I., Lewis, W. S., Otieno, S., Xiao, L., Weiss, S., Hengst, L., and Kriwacki, R. W. (2004) *Nat Struct Mol Biol* **11**, 358–364.
49. Mokhtarzada, S., Yu, C., Brickenden, A., and Choy, W. Y. (2011) *Biochemistry* **50**, 715–726.
50. Schneider, R., Huang, J. R., Yao, M., Communie, G., Ozenne, V., Mollica, L., Salmon, L., Jensen, M. R., and Blackledge, M. (2012) *Mol Biosyst* **8**, 58–68.
51. Uversky, V. N., and Dunker, A. K. (2012) *Intrinsically Disordered Protein Analysis*, Totowa, NJ: Humana Press.
52. Yi, S., Boys, B. L., Brickenden, A., Konermann, L., and Choy, Y. W. (2007) *Biochemistry* **46**, 13120–13130.
53. Gall, C., Xu, H., Brickenden, A., Ai, X., and Choy, W. Y. (2007) *Protein Sci* **16**, 2510–2518.
54. Moradi, M., Babin, V., Roland, C., and Sagui, C. (2012) *PLoS Comput Biol* **8**, e1002501.
55. Deng, X., Eickholt, J., and Cheng, J. (2012) *Mol Biosyst* **8**, 114–121.
56. Ishida, T., and Kinoshita, K. (2008) *Bioinformatics* **24**, 1344–1348.
57. Prilusky, J., Felder, C. E., Zeev-Ben-Mordehai, T., Rydberg, E. H., Man, O., Beckmann, J. S., Silman, I., and Sussman, J. L. (2005) *Bioinformatics* **21**, 3435–3438.
58. Romero, P., Obradovic, Z., Li, X., Garner, E. C., Brown, C. J., and Dunker, A. K. (2001) *Proteins* **42**, 38–48.
59. Ward, J. J., McGuffin, L. J., Bryson, K., Buxton, B. F., and Jones, D. T. (2004) *Bioinformatics* **20**, 2138–2139.
60. Xue, B., Dunbrack, R. L., Williams, R. W., Dunker, A. K., and Uversky, V. N. (2010) *Biochim Biophys Acta* **1804**, 996–1010.

61. Kussie, P. H., Gorina, S., Marechal, V., Elenbaas, B., Moreau, J., Levine, A. J., and Pavletich, N. P. (1996) *Science* **274**, 948–953.
62. Padmanabhan, B., Nakamura, Y., and Yokoyama, S. (2008) *Acta Crystallogr Sect F Struct Biol Cryst Commun* **64**, 233–238.
63. Russo, A. A., Jeffrey, P. D., Patten, A. K., Massague, J., and Pavletich, N. P. (1996) *Nature* **382**, 325–331.
64. Tong, K. I., Katoh, Y., Kusunoki, H., Itoh, K., Tanaka, T., and Yamamoto, M. (2006) *Mol Cell Biol* **26**, 2887–2900.
65. Rauscher, S., and Pomes, R. (2010) *Biochem Cell Biol* **88**, 269–290.
66. Chen, H.-F. (2009) *PLoS One* **4**, e6516.
67. Knott, M., and Best, R. B. (2012) *PLoS Comput Biol* **8**, e1002605.
68. Sivakolundu, S. G., Bashford, D., and Kriwacki, R. W. (2005) *J Mol Biol* **353**, 1118–1128.
69. Allen, W. J., Capelluto, D. G. S., Finkielstein, C. V., and Bevan, D. R. (2010) *J Phys Chem B* **114**, 13201–13213.
70. Espinoza-Fonseca, L. M. (2009) *FEBS Lett* **583**, 556–560.
71. Espinoza-Fonseca, L. M., Ilizaliturri-Flores, I., and Correa-Basurto, J. (2012) *Mol Biosyst* **8**, 1798–1805.
72. Ganguly, D., Zhang, W., and Chen, J. (2012) *Mol Biosyst* **8**, 198–209.
73. Higo, J., Nishimura, Y., and Nakamura, H. (2011) *J Am Chem Soc* **133**, 10448–10458.
74. Lindorff-Larsen, K., Trbovic, N., Maragakis, P., Piana, S., and Shaw, D. E. (2012) *J Am Chem Soc* **134**, 3787–3791.
75. Sethi, A., Tian, J., Vu, D. M., and Gnanakaran, S. (2012) *Biophys J* **103**, 748–757.
76. Wu, K.-P., Weinstock, D. S., Narayanan, C., Levy, R. M., and Baum, J. (2009) *J Mol Biol* **391**, 784–796.
77. Yoon, M. K., Venkatachalam, V., Huang, A., Choi, B. S., Stultz, C. M., and Chou, J. J. (2009) *Protein Sci* **18**, 337–347.
78. Levitt, M., and Warshel, A. (1975) *Nature* **253**, 694–698.
79. Choy, W. Y., Mulder, F. A., Crowhurst, K. A., Muhandiram, D. R., Millett, I. S., Doniach, S., Forman-Kay, J. D., and Kay, L. E. (2002) *J Mol Biol* **316**, 101–112.
80. Kashtanov, S., Borcherds, W., Wu, H., Daughdrill, G. W., and Ytreberg, F. M. (2012) *Methods Mol Biol* **895**, 139–152.
81. Marsh, J. A., and Forman-Kay, J. D. (2011) *Proteins* **80**, 556–572.
82. Freddolino, P. L., Harrison, C. B., Liu, Y. X., and Schulten, K. (2010) *Nat Phys* **6**, 751–758.
83. Lane, T. J., Shukla, D., Beauchamp, K. A., and Pande, V. S. (2013) *Curr Opin Struct Biol* **23**, 58–65.
84. Phillips, J. C., Braun, R., Wang, W., Gumbart, J., Tajkhorshid, E., Villa, E., Chipot, C., Skeel, R. D., Kale, L., and Schulten, K. (2005) *J Comput Chem* **26**, 1781–1802.
85. Hess, B., Kutzner, C., van der Spoel, D., and Lindahl, E. (2008) *J Chem Theory Comput* **4**, 435–447.
86. Case, D. A., Cheatham, T. E., Darden, T., Gohlke, H., Luo, R., Merz, K. M., Onufriev, A., Simmerling, C., Wang, B., and Woods, R. J. (2005) *J Comput Chem* **26**, 1668–1688.
87. Brooks, B. R., Brooks, C. L., Mackerell, A. D., Nilsson, L., Petrella, R. J., Roux, B., Won, Y., Archontis, G., Bartels, C., Boresch, S. et al. (2009) *J Comput Chem* **30**, 1545–1614.
88. Piana, S., Lindorff-Larsen, K., Dirks, R. M., Salmon, J. K., Dror, R. O., and Shaw, D. E. (2012) *PLoS One* **7**, e39918.

89. Piana, S., Lindorff-Larsen, K., and Shaw, D. E. (2013) *Proc Natl Acad Sci U S A* **110**, 5915–5920.

90. Shaw, D. E., Maragakis, P., Lindorff-Larsen, K., Piana, S., Dror, R. O., Eastwood, M. P., Bank, J. A., Jumper, J. M., Salmon, J. K., Shan, Y. et al. (2010) *Science* **330**, 341–346.

91. Nerukh, D., Okimoto, N., Suenaga, A., and Taiji, M. (2012) *J Phys Chem Lett* **3**, 3476–3479.

92. Stone, J. E., Phillips, J. C., Freddolino, P. L., Hardy, D. J., Trabuco, L. G., and Schulten, K. (2007) *J Comput Chem* **28**, 2618–2640.

93. Pande, V. S., Baker, I., Chapman, J., Elmer, S. P., Khaliq, S., Larson, S. M., Rhee, Y. M., Shirts, M. R., Snow, C. D., Sorin, E. J. et al. (2003) *Biopolymers* **68**, 91–109.

94. Das, R., Qian, B., Raman, S., Vernon, R., Thompson, J., Bradley, P., Khare, S., Tyka, M. D., Bhat, D., Chivian, D. et al. (2007) *Proteins* **69 Suppl 8**, 118–128.

95. Cooper, S., Khatib, F., Treuille, A., Barbero, J., Lee, J., Beenen, M., Leaver-Fay, A., Baker, D., Popovic, Z., and Players, F. (2010) *Nature* **466**, 756–760.

96. Sugita, Y., and Okamoto, Y. (1999) *Chem Phys Lett* **314**, 141–151.

97. Earl, D. J., and Deem, M. W. (2005) *Phys Chem Chem Phys* **7**, 3910–3916.

98. St-Pierre, J. F., and Mousseau, N. (2012) *Proteins* **80**, 1883–1894.

99. Laio, A., and Parrinello, M. (2002) *Proc Natl Acad Sci U S A* **99**, 12562–12566.

100. Ueda, Y., Taketomi, H., and Go, N. (1978) *Biopolymers* **17**, 1531–1548.

101. Clementi, C., Nymeyer, H., and Onuchic, J. N. (2000) *J Mol Biol* **298**, 937–953.

102. Best, R. B., and Hummer, G. (2006) *Phys Rev Lett* **96**, 228104.

103. Bolhuis, P. G., Chandler, D., Dellago, C., and Geissler, P. L. (2002) *Annu Rev Phys Chem* **53**, 291–318.

104. Juraszek, J., and Bolhuis, P. G. (2006) *Proc Natl Acad Sci U S A* **103**, 15859–15864.

105. Weinan, E., Ren, W. Q., and Vanden-Eijnden, E. (2005) *J Phys Chem B* **109**, 6688–6693.

106. Mittal, J., Yoo, T. H., Georgiou, G., and Truskett, T. M. (2013) *J Phys Chem B* **117**, 118–124.

107. Ostermeir, K., and Zacharias, M. (2013) *Biochim Biophys Acta* **1834**, 847–853.

108. Limongelli, V., Bonomi, M., and Parrinello, M. (2013) *Proc Natl Acad Sci U S A* **110**, 6358–6363.

109. Michel, J., and Cuchillo, R. (2012) *PLoS One* **7**, e41070.

110. Smith, W. W., Schreck, C. F., Hashem, N., Soltani, S., Nath, A., Rhoades, E., and O'Hern, C. S. (2012) *Phys Rev E* **86**, 041910.

111. Murtola, T., Bunker, A., Vattulainen, I., Deserno, M., and Karttunen, M. (2009) *Phys Chem Chem Phys* **11**, 1869–1892.

112. Fisher, C. K., Huang, A., and Stultz, C. M. (2010) *J Am Chem Soc* **132**, 14919–14927.

113. Cino, E. A., Choy, W. Y., and Karttunen, M. (2012) *J Chem Theory Comput* **8**, 2725–2740.

114. Beauchamp, K. A., Lin, Y.-S., Das, R., and Pande, V. S. (2012) *J Chem Theory Comput* **8**, 1409–1414.

115. Best, R. B., and Hummer, G. (2009) *J Phys Chem B* **113**, 9004–9015.

116. Best, R. B., Zhu, X., Shim, J., Lopes, P. E., Mittal, J., Feig, M., and Mackerell, A. D., Jr. (2012) *J Chem Theory Comput* **8**, 3257–3273.

117. Lindorff-Larsen, K., Maragakis, P., Piana, S., Eastwood, M. P., Dror, R. O., and Shaw, D. E. (2012) *PLoS One* **7**, e32131.

118. Matthes, D., and de Groot, B. L. (2009) *Biophys J* **97**, 599–608.

119. Monticelli, L., and Tieleman, D. P. (2013) *Methods Mol Biol* **924**, 197–213.

120. Vymetal, J., and Vondrasek, J. (2013) *J Chem Theory Comput* **9**, 441–451.
121. Schlesier, T., and Diezemann, G. (2013) *J Phys Chem B* **117**, 1862–1871.
122. Mackerell, A. D. (2004) *J Comput Chem* **25**, 1584–1604.
123. Best, R. B. (2012) *Curr Opin Struct Biol* **22**, 52–61.
124. Sorin, E. J., and Pande, V. S. (2005) *Biophys J* **88**, 2472–2493.
125. Best, R. B., and Mittal, J. (2011) *Proteins* **79**, 1318–1328.
126. Georgoulia, P. S., and Glykos, N. M. (2013) *J Phys Chem B* **117**, 5522–5532.
127. Best, R. B., and Mittal, J. (2011) *Proc Natl Acad Sci U S A* **108**, 11087–11092.
128. Kannan, S., and Zacharias, M. (2009) *Proteins* **76**, 448–460.
129. Pitera, J. W., and Swope, W. (2003) *Proc Natl Acad Sci U S A* **100**, 7587–7592.
130. Zhou, R. (2003) *Proc Natl Acad Sci U S A* **100**, 13280–13285.
131. Wong-Ekkabut, J., and Karttunen, M. (2012) *J Chem Theory Comput* **8**, 2905–2911.
132. Wong-Ekkabut, J., Miettinen, M. S., Dias, C., and Karttunen, M. (2010) *Nat Nanotechnol* **5**, 555–557.
133. Karttunen, M., Rottler, J., Vattulainen, I., and Sagui, C. (2008) *Curr Top Membr* **60**, 49–89.
134. Yonetani, Y. (2005) *Chem Phys Lett* **406**, 49–53.
135. Bonthuis, D. J., Rinne, K. F., Falk, K., Kaplan, C. N., Horinek, D., Berker, A. N., Bocquet, L., and Netz, R. R. (2011) *J Phys Condens Matter* **23**, 184110.
136. Repakova, J., Holopainen, J. M., Karttunen, M., and Vattulainen, I. (2006) *J Phys Chem B* **110**, 15403–15410.
137. Oldfield, C. J., Meng, J., Yang, J. Y., Yang, M. Q., Uversky, V. N., and Dunker, A. K. (2008) *BMC Genomics* **9 Suppl 1**, S1.
138. Efeyan, A., and Serrano, M. (2007) *Cell Cycle* **6**, 1006–1010.
139. Bieging, K. T., and Attardi, L. D. (2012) *Trends Cell Biol* **22**, 97–106.
140. Joerger, A. C., and Fersht, A. R. (2008) *Annu Rev Biochem* **77**, 557–582.
141. Avalos, J. L., Celic, I., Muhammad, S., Cosgrove, M. S., Boeke, J. D., and Wolberger, C. (2002) *Mol Cell* **10**, 523–535.
142. Chuikov, S., Kurash, J. K., Wilson, J. R., Xiao, B., Justin, N., Ivanov, G. S., McKinney, K., Tempst, P., Prives, C., Gamblin, S. J. et al. (2004) *Nature* **432**, 353–360.
143. Lowe, E. D., Tews, I., Cheng, K. Y., Brown, N. R., Gul, S., Noble, M. E., Gamblin, S. J., and Johnson, L. N. (2002) *Biochemistry* **41**, 15625–15634.
144. Mujtaba, S., He, Y., Zeng, L., Yan, S., Plotnikova, O., Sachchidanand, Sanchez, R., Zeleznik-Le, N. J., Ronai, Z., and Zhou, M. M. (2004) *Mol Cell* **13**, 251–263.
145. Rustandi, R. R., Baldisseri, D. M., and Weber, D. J. (2000) *Nat Struct Biol* **7**, 570–574.
146. Bustos, D. M. (2012) *Mol Biosyst* **8**, 178–184.
147. Jin, J., Smith, F. D., Stark, C., Wells, C. D., Fawcett, J. P., Kulkarni, S., Metalnikov, P., O'Donnell, P., Taylor, P., Taylor, L. et al. (2004) *Curr Biol* **14**, 1436–1450.
148. Aitken, A. (2006) *Semin Cancer Biol* **16**, 162–172.
149. Gardino, A. K., Smerdon, S. J., and Yaffe, M. B. (2006) *Semin Cancer Biol* **16**, 173–182.
150. Obsil, T., and Obsilova, V. (2011) *Semin Cell Dev Biol* **22**, 663–672.
151. Yaffe, M. B., Rittinger, K., Volinia, S., Caron, P. R., Aitken, A., Leffers, H., Gamblin, S. J., Smerdon, S. J., and Cantley, L. C. (1997) *Cell* **91**, 961–971.
152. Miele, A. E., Watson, P. J., Evans, P. R., Traub, L. M., and Owen, D. J. (2004) *Nat Struct Mol Biol* **11**, 242–248.
153. Zhuo, Y., Ilangovan, U., Schirf, V., Demeler, B., Sousa, R., Hinck, A. P., and Lafer, E. M. (2010) *J Mol Biol* **404**, 274–290.

154. Cino, E. A., Killoran, R. C., Karttunen, M., and Choy, W. Y. (2013) *Sci Rep* **3**, 2305.
155. Itoh, K., Wakabayashi, N., Katoh, Y., Ishii, T., Igarashi, K., Engel, J. D., and Yamamoto, M. (1999) *Genes Dev* **13**, 76–86.
156. Itoh, K., Chiba, T., Takahashi, S., Ishii, T., Igarashi, K., Katoh, Y., Oyake, T., Hayashi, N., Satoh, K., Hatayama, I. et al. (1997) *Biochem Biophys Res Commun* **236**, 313–322.
157. Tong, K. I., Kobayashi, A., Katsuoka, F., and Yamamoto, M. (2006) *Biol Chem* **387**, 1311–1320.
158. Camp, N. D., James, R. G., Dawson, D. W., Yan, F., Davison, J. M., Houck, S. A., Tang, X., Zheng, N., Major, M. B., and Moon, R. T. (2012) *J Biol Chem* **287**, 6539–6550.
159. Komatsu, M., Kurokawa, H., Waguri, S., Taguchi, K., Kobayashi, A., Ichimura, Y., Sou, Y.-S., Ueno, I., Sakamoto, A., Tong, K. I. et al. (2010) *Nat Cell Biol* **12**, 213–223.
160. Lo, S.-C., and Hannink, M. (2006) *J Biol Chem* **281**, 37893–37903.
161. Ma, J., Cai, H., Wu, T., Sobhian, B., Huo, Y., Alcivar, A., Mehta, M., Cheung, K. L., Ganesan, S., Kong, A.-N. T. et al. (2012) *Mol Cell Biol* **32**, 1506–1517.
162. Strachan, G. D., Morgan, K. L., Otis, L. L., Caltagarone, J., Gittis, A., Bowser, R., and Jordan-Sciutto, K. L. (2004) *Biochemistry* **43**, 12113–12122.
163. Kim, J.-E., You, D.-J., Lee, C., Ahn, C., Seong, J. Y., and Hwang, J.-I. (2010) *Cell Signal* **22**, 1645–1654.
164. Hast, B. E., Goldfarb, D., Muluaney, K. M., Hast, M. A., Siesser, P. F., Yan, F., Hayes, D. N., and Major, M. B. (2013) *Cancer Res* **73**, 2199–2210.
165. Niture, S. K., and Jaiswal, A. K. (2009) *J Biol Chem* **284**, 13856–13868.
166. Oostenbrink, C., Villa, A., Mark, A. E., and Van Gunsteren, W. F. (2004) *J Comput Chem* **25**, 1656–1676.
167. Bussi, G., Donadio, D., and Parrinello, M. (2007) *J Chem Phys* **126**, 014101.
168. Parrinello, M., and Rahman, A. (1981) *J Appl Phys* **52**, 7182–7190.
169. Hess, B., Bekker, H., Berendsen, H. J. C., and Johannes, J. G. E. M. (1997) *J Comput Chem* **18**, 1463–1472.
170. Darden, T., York, D., and Pedersen, L. (1993) *J Chem Phys* **98**, 10089–10093.
171. Wong, P., and Frishman, D. (2006) *PLoS Comput Biol* **2**, e40.
172. Kaszuba, K., Rog, T., Bryl, K., Vattulainen, I., and Karttunen, M. (2010) *J Phys Chem B* **114**, 8374–8386.
173. Cino, E. A., Karttunen, M., and Choy, W. Y. (2012) *PLoS One* **7**, e49876.
174. Szasz, C. S., Alexa, A., Toth, K., Rakacs, M., Langowski, J., and Tompa, P. (2011) *Biochemistry* **50**, 5834–5844.
175. Cho, E. J., and Kim, J. S. (2012) *Biophys J* **103**, 424–433.
176. Echeverria, C., and Kapral, R. (2010) *J Chem Phys* **132**, 104902.
177. Minh, D. D., Chang, C. E., Trylska, J., Tozzini, V., and McCammon, J. A. (2006) *J Am Chem Soc* **128**, 6006–6007.

Chapter 9

Multiscale Simulations of Large Conformational Changes of Disordered and Ordered Proteins Induced by Their Partners

Yong Wang, Xiakun Chu, and Jin Wang

CONTENTS

INTRODUCTION

Proteins realize their functions via interacting with other biomolecules. A classical example is the binding of a protein and its substrates during enzyme catalysis. To describe such a process, a lock and key notion was originally proposed by Emil Fischer more than a century ago (Fischer 1894). Subsequently, the lock and key mechanism was extended to explain the rigid biomolecular docking. However, the rigid body assumption was challenged by increasing evidence that biomolecular binding is often accompanied by local or global conformational adjustment of the proteins as well as their partners. This highlights the importance of conformational flexibility in protein function and has led to a new relationship between function, structure, and dynamics. In particular, a class of proteins have been found to be able to carry out their functions without the need of well-defined structures under physiological conditions. These proteins are known as intrinsically disordered proteins (IDPs). IDPs often change their conformations significantly when binding to their partners, giving a hint that the flexibility or dynamics should also determine the realization of function (Dunker et al. 2001; Dyson and Wright 2002, 2005; Papoian 2008; Wright and Dyson 1999).

To assess the interplay between conformational changes of the naturally ordered as well as disordered proteins and their binding to their partners, two extreme binding mechanisms were proposed. One is the induced-fit model (Koshland 1958), whereby the conformational changes are induced by binding and occur after the recognition. The other is the conformational selection model (Bosshard 2001), whereby the binding happens by selecting the partners with the conformation in the bound form from the preliminary equilibrium ensemble in a free state. In reality, a protein can bind with its partners through either one or a combination of the two scenarios. To classify the binding mechanism in protein association, it is critical to find out how the protein realizes its function. Recent advances in the structural biology techniques, such as single-molecule and nuclear magnetic resonance (NMR) measurements, have provided useful experimental approaches to characterize the nature of the conformational dynamics in binding. However, the current experiments, which are limited to spatial and temporal scales in the measurements, cannot easily represent a clear and global illustration of how the conformational dynamics proceeds during the binding. As a powerful complementary to the experiments, molecular simulation can provide an accurate, high-resolution picture of protein dynamics as a function of time and therefore can address the specific issues in protein-protein interactions more easily and in more detail than the current available experimental measurements.

In spite of many advantages of molecular simulation in understanding the protein dynamics at atomic or molecular levels, there are two well-known factors limiting its usage. First, due to the large number of degrees of freedom and the rugged energy landscapes of the realistic biological system, simulating with a physiological timescale is challenging as one aims to reconstruct the long timescale behavior of the dynamics with finite computation resources. To resolve this issue, the coarse-grained strategy, in which groups of atoms in a protein can be treated as a pseudobead rather than explicitly represented, is a practical way to accelerate the simulation process in the computer. The coarse-grained pseudobeads interact with others via an effective force field, in which the parameters are inspired by an atomistic force field or theoretical considerations such as structure-based modeling. These types of simplified models with various parametrization methodologies have proved to be a powerful tool for the analysis and comparison with the experimental results and have made noticeable progress in characterization of protein dynamics over long timescales. Second, the parameters of a force field and methodologies in molecular simulations should be accurate to provide a realistic description of the biological system. For this reason, the atomistic force field with rigorous parameterizations from quantum-level calculations or experimental measures should be utilized. However, the more precise and realistic description for the biological system is used in molecular simulations, the more expensive the cost of computational resources will be. Therefore, in practice these two factors are in competition and there seems to be a trade-off between accurate descriptions of the system and computational cost. As a result, with regard to the combination of the coarse-grained strategy and atomistic model with an empirical force field, a so-called multiscale approach is necessary. A coarse-grained model for long time simulations can generally provide qualitative insights to the nature of the transition state ensemble, free-energy landscapes, kinetic rates, and association mechanism from a large number of binding/unbinding transitions by collecting statistics. To achieve a more reliable description of the system, empirical force field models at the atomic level need to employ various computational sampling methodologies, aiming to obtain the thermodynamic properties, for example, by accelerating the sampling in conformational space (Beutler and van Gunsteren 1994; Chou and Carlacci 1991; Hamelberg et al. 2004; Liwo et al. 2008; Okamoto and Sugita 1999) or searching for the most probable pathways through optimizing the kinetic path (Cárdenas and Elber 2003; Kuczera et al. 2009). Different methodologies focus on different aspects of the properties of the system and their combinations with a coarse-grained model will therefore provide a full physical and chemical description on the thermodynamics and kinetics of the system at multiple length and timescales. This chapter will present the basics of a multiscale approach with a coarse-grained structure-based model and atomistic empirical force field model, as well as its applications to explore the conformational dynamics in binding of disordered and ordered proteins with their partners (Figure 9.1).

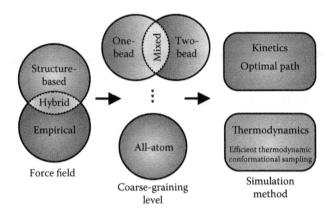

Figure 9.1   Schematic representations of the multiscale approaches.

METHODS AND MODELS

Proteins within a living cell often realize their functions by interacting with their partners, such as enzyme substrates, a peptide segment within a protein and deoxyribonucleic acid (DNA). These interacting events occur at different temporal scales ranging from tens of nanoseconds to seconds. In addition, the relevant protein machines have different sizes spanning several orders of magnitude with a lower limit of nanometers (Tozzini 2010a). In order to understand the process of biological recognition, researchers have developed a number of modeling approaches, each focusing on a particular temporal or spatial range. Up to now, there has been a lack of a universal approach that is suitable to apply to all time and space scales. This demands a combination of different approaches rather than a single technique in a process involving multiscale events. The application of such a combination is also referred to a multiscale strategy (Tozzini 2010b). In fact, the multiscale simulation has proved to be a necessary and powerful tool that can be widely used to explore a wide range of biological behaviors.

Multiscale Force Fields

In principle, given long enough computation time and sufficiently accurate force fields, molecular dynamics (MD) simulation not only can provide a straightforward observation of the action of a protein in solution, but also obtain its macroscopic thermodynamic characterizations. The core of the MD simulation is about the force fields. In a force field, there are two important ingredients. One is the potential energy function used to describe the structure-energy relationship of a protein. The other is the parameters of the potential energy function.

Current force fields can generally be divided into two categories: structure-based (SB) and empirical force field.

SB Force Fields

SB force fields are often built from an atomistic structure from the experimental measurements and thus are highly dependent on the input structures. According to the potential function form, current SB force fields can be categorized into two types: anharmonic and harmonic. The harmonic SB model is also denoted as an elastic network model, which can be also considered as a harmonic version of the former.

In a typical version of elastic network models (Atilgan et al. 2001), the modeled protein is represented by a network of coarse-grained beads that are linked by elastic springs (using a harmonic potential) with a uniform spring constant. The equilibrium positions of all springs are taken from the input structure, which is assumed to be a minimum on the energy landscape of the protein. Despite its simplicity, the elastic network model has been successfully applied in the prediction of the functionally related motions of many proteins. Such models are often limited in extracting harmonic motions around a local energy minimum due to the harmonic potential in the force field (Ma 2005). In fact, large-scale conformational change of proteins is often anharmonic due to the presence of energy barriers.

To simulate the large-scale conformational change, anharmonic potential is required. In a typical version of unharmonic SB models, each amino acid is described by a single bead on a polymer chain located on the C_α position. Note that throughout this chapter, the SB force fields will in general refer to the anharmonic form. The interaction potential energy function is given by the expression:

$$E_{SB}(\Gamma_0) = E_{bonded}(\Gamma_0) + E_{nonbonded}(\Gamma_0)$$

$$= \sum_{bonds} K_r(r-r_0)^2 + \sum_{angles} K_\theta(\theta-\theta_0)^2$$

$$+ \sum_{dihedral} K_\phi^{(n)}[1-\cos(n\times(\phi-\phi_0))]$$

$$+ V_{atrractive}(\gamma^{native}) + V_{repulsive}(\gamma^{nonnative})$$

where the interaction potential energy is a function of a reference structure Γ_0 and is divided into bond stretching, angle bending, torsion, and nonbonded interactions. r, θ, and ϕ are the virtual bond length, bond angle, and torsion angles, respectively. Parameters K_r, K_θ, K_ϕ are used to scale the relative contribution of each term to total potential energy. The nonbonded term contains an attractive potential $V_{attractive}$, which is a function of native contacts and a repulsive potential $V_{repulsive}$, which is a function of nonnative contacts. The native contacts can be derived from a cutoff algorithm, contacts of structural units (CSU) (Sobolev et al. 1999) or shadow contact map (Noel et al. 2010). According to the energy landscape theory (Brooks et al. 2001; Bryngelson and Wolynes 1989; Bryngelson et al. 1995; Leopold et al. 1992), the success of the anharmonic SB force fields is because of the resulting perfectly funneled energy landscape that determines the folding and binding mechanism (Chu et al. 2013; Levy et al. 2004; Wang et al. 2012a). However, both harmonic and anharmonic SB models have a strong bias toward reference structure(s), making them lack transferability (Thorpe et al. 2011).

Empirical Force Fields

Empirical force fields are usually derived from experimental data or accurate quantum mechanical calculations and thus do not require *a priori* structural knowledge. The basic functional form contains two terms: a bonded term describing the covalent links between the basic simulation units (atoms or beads), and a nonbonded term related to the long-range electrostatic and van der Waals forces. Despite the fact that current empirical force fields have some discrepancies among each other, generally they can be written as

$$E_{empirical} = E_{bonded} + E_{nonbonded}$$

$$= (E_{bond} + E_{angle} + E_{dihedral}) + (E_{elec} + E_{VDW})$$

The bonded term E_{bonded} consists of three terms: two-body bonded potential E_{bond}, three-body angle potential E_{angle}, and four-body proper and improper dihedral potential $E_{dihedral}$. The nonbonded term $E_{nonbonded}$ contains electrostatic term E_{elec} and van der Waals term E_{VDW}. Additional terms such as hydrophobic term and hydrogen bond term are also possible in some empirical force fields. The popular empirical force fields include AMBER (Case et al. 2005), CHARMM (Brooks et al. 2009), GROMOS (Christen et al. 2005), OPLS (Jorgensen et al. 1996), and their variants (Lindorff-Larsen et al. 2010).

Hybrid Force Fields

In contrast to SB force fields, empirical force fields have transferability but are computationally expensive. The multiscale idea led to the appearance of hybrid force fields in which empirical potential and SB potential are combined (Chen et al. 2012; Sutto et al. 2011). The introduction of the empirical term in the hybrid force fields makes it transferable due to no bias to a particular conformation while at the same time naturally preserves the atomistic description of the protein structure, whereas the introduction of the SB term can accelerate the sampling of rare events, such as large-scale conformational transitions, due to a smooth and funneled energy landscape. Recently, we successfully applied the hybrid force field in the studies of the coupled folding and binding of an IDP (Wang et al. 2013a). In our atomistic hybrid model, the empirical term is used for local interactions while the SB term for nonlocal interactions. Additionally, an electrostatic term was introduced to describe the nonnative long-range interactions. The hybrid Hamiltonian is given by the expression:

$$U_{hybrid} = U_{empirical} + U_{SB} + U_{charge}$$

$$= U_{empirical} + U_{attraction} + U_{repulsive} + U_{charge}$$

The protein-partner system is modeled in atomic representation with the exception of hydrogen atoms. The local empirical term $U_{empirical}$ maintains the geometry and the correct backbone and side chain rotamers of the protein without any biasing. An attraction term ($U_{attraction}$) in U_{SB} can provide the secondary and tertiary bias to folding and binding, and a repulsive term $U_{repulsive}$ provides the excluded volume.

Considering that the biophysical properties of IDPs have relatively large differences from traditional ordered proteins, the energetic parameters of the SB part in the hybrid model often are required to be calibrated to match with the experimental observations (e.g., structural fluctuations, binding affinity, and residual secondary structure).

Multiscale Structural Representation

Besides focusing on the force field, the multiscale strategy is naturally applied to build the basic simulation units of a protein system at many different structural levels. The more refined the basic units are, the more expensive the computation is, due to more degrees of freedom. Thus, various reduced models with different levels of coarse-graining have been worked out.

One-Bead Coarse-Grained Structure-Based Model

It seems natural to coarse-grain a protein into a residue-based model because the amino acid residue is the building block of a protein structure. The combination of a residue-level coarse-grained topological structure and SB force field is the most popular model in the studies of protein folding. For instance, the Clementi-Onuchic model (Clementi et al. 2000) is a typical C_α SB model (SBM). Its interaction potential energy is the same as E_{SB} (described in the section titled "SB Force Fields"). Note that a 10–12 Lennard-Jones type potential is often used for the $V_{attractive}$ term in the C_α SBM. As a perfectly funneled model, such C_α SBM not only has a huge speed advantage but also can capture the folding/binding mechanism without the requirement of fine-tuning parameters. Its ability to reproduce experimental observations on the folding and binding process led to the commonly accepted notion that protein topology determines the folding and binding mechanisms (Chu et al. 2013; Levy et al. 2004; Wang et al. 2012a).

Two-Bead Coarse-Grained SBM

One-bead coarse-graining seems to be over simplified in some cases in which side chains play a major role in the protein properties, such as thermostability, substrate recognition, and pH dependence (Hyeon et al. 2009). To address the importance of side chains in the conformational dynamics of proteins, an additional bead is introduced in the original one-bead model to represent each side chain except glycine. This improvement resulted in a number of two-bead coarse-grained SBMs (Cheung et al. 2003; Oliveira et al. 2008). Such two-bead models not only allow the consideration of the excluded volume effect of the side chains lacking in the original one-bead models, but also can decouple the nonbonded interactions into backbone-backbone, backbone-side chain, and side chain-side chain types. In this way, amino acid specificity can be reasonably built into the model by hydrogen bonds (Cheung et al. 2003), hydrophobic interactions, and electrostatic forces (Chu et al. 2012).

In our model, one bead (C_A) representing the backbone is located on the C_α atom. Another bead (C_B) representing the side chain is located at the center of mass or the farthest heavy atom of the side chain, depending on its residue characteristic. For the charged residues (include Asp, Glu, Lys, Arg, and sometimes His at low pH), the most distant heavy atoms (among N, C, and O) from the C_α atom were used to model the C_B beads. For the noncharged residues, the positions of C_B beads were placed at the center of mass of the side chain atoms. Because of the introduction of C_B beads, the dihedral term of the bonded potential is slightly different from that in the C_α SBM.

$$U_{backbone}(\Gamma_X) = U_{bond} + U_{angle} + \varepsilon_p {}^* U_{dihedral}$$

$$U_{dihedral}(\Gamma_X) = \sum_{dihedrals} K_\phi [(1 - \cos(\phi - \phi_0)) + 0.5(1 - \cos(3(\phi - \phi_0)))]$$

In contrast to the one-bead SB model, an energetic parameter ε_p was introduced to balance the energetic contribution of dihedral term and native contact terms. From the differences between C_A and C_B beads, different types of dihedral angles are considered. We used different energetic parameters depending on the type of the involved atoms. Improper dihedral angles are described by the same equation as a proper term, keeping the chirality in the protein model. It is worth noting that the increase of the resolution of the protein structure will inevitably lead to the increase of the level of complexity of these models and the increase of parameters that are required to be carefully tuned and calibrated.

Mixed One-Bead and Two-Bead Models

Sometimes the important role of a side chain is reflected mainly on the local regions rather than the global structure. A typical example is protein allostery induced by the binding of small ligands in which most of the functionally related residues are involved near the binding sites, whereas other residues serve as a scaffold with a less functional role. Thus, additional structural details are only required locally. There have been several mixed one-bead and two-bead models developed as an application of a multiscale strategy. For instance, the mixed model was recently used by us for exploring the dynamic functional landscapes of maltose-binding protein (MBP) and adenylate kinase (ADK) in the presence of ligands (Wang et al. 2012c, 2013b). The computational cost is significantly reduced compared to the all-atom simulation and the results are consistent with the experiments.

Atomistic SB or Hybrid Models

With the advance of all-atom SBMs, an SBM is no longer the abbreviation of C_α or a coarse-grained SBM. At present, an atomistic SBM can be easily built by a web server, namely structure-based models for biomolecules (SMOG) (Noel et al. 2010). Whitford et al. further extended the application of atomistic SBM to ribonucleic acid (RNA)/DNA (Whitford et al. 2010). Besides SMOG, there is an alternate way to build an atomic model without the loss of the important features of the SB model (funneled energy landscape and fast dynamic transitions). This type of atomistic model is achieved by integrating the all-atom empirical model and SB potential (Pogorelov and Luthey-Schulten 2004; Sutto et al. 2011), as mentioned in the section titled "Hybrid Force Fields." The atomistic details are perfectly inherited from the empirical model (e.g., AMBER and CHARMM force fields). Although the atomistic SBMs and hybrid models have the same limitation as coarse-grained SBMs that require the information on native structures of modeled proteins, they still have sufficient sampling advantage in contrast to the atomistic empirical models. In addition, these atomistic SB and hybrid models can completely decouple the energetic contributions and geometric contributions on protein dynamics. This feature is of great help to analyze and quantify the underlying energy landscape (Chu et al. 2013).

Multiscale Simulation Methods

Besides the efforts on the force fields and structural modeling, multiscale strategy has also been applied in the combination of different simulation methods.

Recently, we developed a multiscale strategy by combining SB molecular dynamics simulations at the residue level with an optimal path calculation at the atomic level. Such method has been applied to explore the coupled folding and binding of an intrinsically disordered protein inhibitor IA3 to its target enzyme YPrA (Wang et al. 2011). The optimal paths, also denoted as the steepest descent paths, are those routes linking the initial state and the final state on the energy landscape with the highest possibility (Henkelman et al. 2000). The optimal path calculation has been widely used to describe chemical reactions (Olender and Elber 1996; Yang et al. 2009). Given an initial conformation x_i and a final conformation x_f, optimal path calculation will minimize the target function T:

$$T = \sum_{j=1}^{N-1} \sqrt{H_s + \left(\frac{\partial U}{\partial x_j}\right)^2} \; |x_{j+1} - x_j| + \lambda \sum_j (\Delta l_{j,j+1} - <\Delta l>)^2$$

The target function T is minimized by conjugate gradient local minimization. $<\Delta l> = \frac{1}{N-1} \sum_{j=1}^{N-1} \Delta l_{j,j+1}$, where $\Delta l_{j,j+1}$ is the step length and λ is the strength of a penalty function that restrains the step length to the average length $<\Delta l>$. For further details see Majek et al. (2008). The optimal paths are dependent on the choice of initial and end structures. In general, the end structure is the protein-partner complexed structure that is available from experimental measurement or molecular modeling while the initial structure can be randomly chosen from an ensemble of conformations to obtain statistically robust results.

The optimal path calculation is based on an empirical force field whose energy function is a combination of the AMBER and OPLS force fields (Elber et al. 1995). However, in contrast to the classical empirical MD simulation, its computational cost is several orders of magnitude less. This is mostly contributed by the fact that the kinetic path has been optimized to minimize the energy, which significantly reduces the conformational space for the protein to search during the transitions between the initial and end structures.

APPLICATIONS

Binding-Induced Folding of Intrinsically Disordered Protein Inhibitor IA3 to YPrA

We have developed a multiscale approach by combining the SBM at residue level (Levy et al. 2004) and stochastic path method at atomic level (Cárdenas and Elber 2003; Wang et al. 2005) to explore the underlying mechanism of binding of intrinsically disordered proteinase inhibitor IA3 to its target enzyme

YPrA (Wang et al. 2011). Our results showed a clear induced-fit mechanism (Koshland 1958) for IA3 binding to YPrA consistent with the kinetic experiment (Narayanan et al. 2008) and also uncovered the important roles of non-native interactions in the initial stage of this association.

The one-bead coarse-grained SB simulations were performed under constant temperature with varying dissociative configurations. The temperature was chosen to be lower than the binding transition temperature to ensure the stability of the enzyme. To enhance the sampling, an additional harmonic biasing potential was introduced to the Hamiltonian during the thermodynamic simulations. By projecting the free-energy landscapes to Q_{fIA3}, which measures the folding similarity of IA3 to its native structure, and R_{com}, which measures binding degree of IA3 to the target enzyme (Figure 9.2a), we found that the binding and folding of IA3 are decoupled. At the beginning, there is no folding occurring as the system approaches the binding transition states. And after the binding transition states, the binding and folding become strongly coupled. These features lead to a typical binding prior to folding mechanism. In addition, the binding transition states captured by the two-dimensional (2-D) free-energy profiles are found to be characterized by many nonnative contacts with only a few native contacts (Figure 9.2b). The interfacial contacts in the binding transition state are mostly formed at the surface of the active site groove of YPrA, corresponding to the flap region, which is highly fluctuating by root-mean-square fluctuation (RMSF) analysis (Figure 9.2c and d). Since the nonnative interactions in our SBM are only represented as repulsive terms, the flexibility in YPrA is supposed to facilitate the recognition by serving as an entropic factor to shrink the search space through nonspecific nonnative contacts (Zheng et al. 2013). Nevertheless, the entire roles of the nonnative interactions should be explored at the atomic level, using the empirical force field, explicitly including nonnative interactions.

In order to monitor the binding and folding of IA3 to YPrA in full atomic description, the optimal kinetic paths were quantified to capture the most probable pathways in association. Since the pathways are strongly dependent on the choice of the beginning structures, we set up several parallel simulations with different initial configurations including disordered, unfolded IA3 and uninhibited, folded YPrA (Figure 9.2e). The three typical kinetic pathways all showed a binding-induced folding mechanism for IA3 binding to YPrA (Figure 9.2f through h). Interestingly, there is no native interfacial contacts in the initial stage of the binding and therefore the nonnative interactions are likely to be the dominant forces to facilitate the recognition. The folding of IA3 seems to happen only after the native interfacial contacts formed, consistent with results from the coarse-grained model. In addition, we also observed the notable backtracking behavior for IA3, which corresponds to the partial folding/unfolding

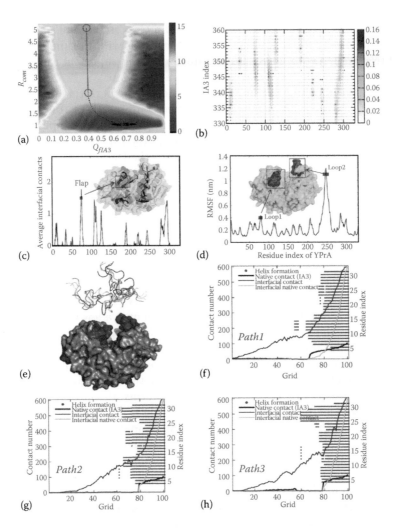

Figure 9.2 Binding-induced folding of IA3 to YPrA. (a) Unbaised free-energy land-scapes as a function of Q_{fIA3} and R_{com} from coarse-grained SBM. Q_{fIA3} is the native contact fraction of IA3 and R_{com} is the center of mass distance between IA3 and YPrA. (b) Binding contact map based on cutoff algorithm, which can take account of nonnative contacts in the transition state. Native contacts are indicated by red squares. (c) Distribution of average number of interfacial contacts. The contacting residues in YPrA are colored blue. The residues at the surface of active site groove of YPrA formed the most contacts and are colored red. (d) RMSF of the C_α atoms of each residue in YPrA from coarse-grained SBM. (e) The initial structures calculated in atomic model. (f–h) The structural evolution in binding-folding transitions for the three typical binding pathways in atomic model. (Adapted from Wang, J. et al., *PLoS Comput Biol* 7 (4), e1001118, 2011. With permission.)

transitions of IA3 at free states during the association (Capraro et al. 2008). The backtracking, formed by unstable native contacts, can be only captured by the empirical force field at the atomic level and is often regarded as the result of the topological frustrations (Gosavi et al. 2006; Hills Jr and Brooks III 2008; Hills et al. 2010). Here, we proposed that it is a way of reducing the entropy in the energy landscapes by stabilizing the protein energetically to facilitate the association. Therefore, we concluded that the nonnative and unstable native interactions can both play a role in protein folding and recognition (Zheng et al. 2012). In this work, we developed a multiscale approach, including the thermodynamic landscapes and kinetic paths to investigate the binding of IA3 to YPrA. This provides a novel way to investigate the function of IDPs.

Coupled Folding and Binding of Intrinsically Disordered Histone Chaperone Chz1 and Histone H2A.Z-H2B

In chromatin, the histone chaperones facilitate the assembly and disassembly of nucleosome by associating and dissociating from the corresponding histone proteins (Das et al. 2010; Hondele and Ladurner 2011; Park and Luger 2008; Zhou et al. 2011). The histones are found to be highly positive, and usually associate with their charged partners, such as DNA and histone chaperones, through electrostatic interactions (Korolev et al. 2007). Here, we investigate the process of an intrinsically disordered histone chaperone Chz1 binding to its targeted histone H2A.Z-H2B using two-bead SB model (Figure 9.3a), in which the Debye-Hückel model is implemented for describing the electrostatic interactions (Chu et al. 2012). Our results uncover an ionic-strength-controlled binding/folding mechanism in the association, in which the interchain electrostatic interactions facilitate the recognition by a fly-casting mechanism and the intrachain electrostatic interactions slows the kinetics by local collapse in Chz1.

To investigate the binding mechanism, we performed the thermodynamic simulations started from varying unbound configurations at physiological salt concentration conditions. The three-dimensional (3-D) free-energy landscapes as a function of Q_i, Q_{Ni}, and Q_{Ci} (representing the similarity of binding between the whole, N-helix, and C-helix of Chz1 and histone relative to the native bounded state) provided a clear illustration that Chz1 passes through two parallel binding pathways to the targeted H2A.Z-H2B (Figure 9.3b). There are two binding intermediates, one for each pathway, with different populations: I_N–84% and I_C–16%. The distinct distributions of the two intermediates are mainly due to the fact that the N-terminal region of Chz1 with abundant negatively charged residues can form many more electrostatic interactions with a positively charged region in H2A.Z-H2B than the C-terminal region does. The long-range electrostatic interactions have been found to be the driving forces for the binding process through the fly-casting mechanism (Shoemaker et al.

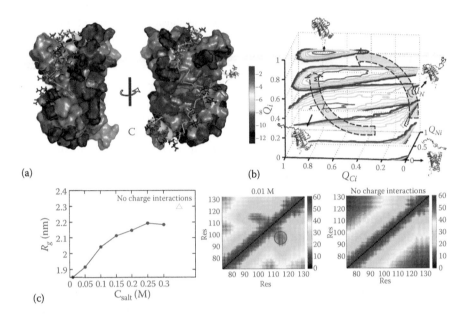

Figure 9.3 Binding-folding of histone chaperone Chz1 to histone H2A.Z-H2B. (a) Structure of the Chz1-H2A.Z-H2B complex. Residue Lys and Arg are colored blue, representing positively charged residue in coarse-grained SBM; residue Glu and Asp are colored red, representing negatively charged residue in coarse-grained SBM. Histones are in the surface model. Chz1 is shown in cartoon representation and the charged residues in Chz1 is shown in stick representation. (b) Parallel binding pathways. The free-energy landscape profile is plotted as a function of Q_i, Q_{Ni}, and Q_{Ci}, which corresponds to the fraction of native binding contacts between the whole, N-helix, and C-helix of Chz1 and histones. (c) Side chain distance of residue pairs in Chz1 and radius of gyration R_g of Chz1 at dissociative states at low salt concentrations and no charged interactions case.
(*Continued*)

2000). After the initial recognition, the short-range electrostatic interactions can stabilize the partly formed intermediates to evolve to the bound complex. Therefore, we found that the interchain electrostatic interactions facilitated the association of Chz1 to H2A.Z-H2B, which is consistent with the experimental measurements (Hansen et al. 2009).

The intrachain electrostatic interactions in IDPs are found to be able to modulate the conformational space by changing the residual distance and the radius of gyration (Haran and England 2010; Muller-Spath et al. 2010; Pappu et al. 2010). The resulting changes in chain dimensions are likely to affect the function of IDPs in the association process. In our studies, we also found the collapsed structured formed by the intrachain electrostatic interactions in Chz1 at low salt concentrations (Figure 9.3c). To quantify the role of electrostatic

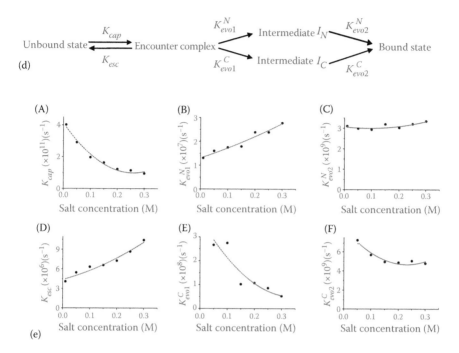

(e)

Figure 9.3 (Continued) Binding-folding of histone chaperone Chz1 to histone H2A.Z-H2B. (d) The divided binding process. The binding process is dissected into four steps along two parallel binding pathways. (e) The binding rates are modulated by the salt concentrations. (A, D) The rate K_{cap} and K_{esc} are shared by the two parallel pathways. The evolving rates in (B, C) I_N binding pathway and (E, F) I_C binding pathway show different behaviors as the salt concentration changes. (Adapted from Chu, X. et al., *PLoS Comput Biol* 8 (7), e1002608, 2012. With permission.)

interactions in both the interchain and intrachain, we performed the kinetic simulations at different salt concentrations. In general, the association process can be dissected into encountering, escaping, evolving to the intermediate states, and forming the bound states (Figure 9.3d). In our work, the rates for each step in the two parallel binding pathways were calculated (Figure 9.3e). Interestingly, we found that it is not always straightforward to conclude that the rates are increased from decreasing the salt concentrations. At the first encounter and escape step shared by the two parallel pathways, the electrostatic interactions facilitated the recognition by increasing the capturing rate and decreasing the escaping rate. However, the role of electrostatic interactions in the following two parallel pathways became distinct. Specifically, for the I_C pathway, the evolving process corresponded to binding of the highly charged N-terminal region and Chz motif, so K^C_{evo1} and K^C_{evo2} increased as electrostatic interactions increased, while for the I_N pathway, K^N_{evo1} decreased with increasing electrostatic interactions. This is due to the fact that there is a local collapsed structure formed in Chz1 at low salt concentrations. To evolve to the intermediate I_N, Chz1 has to unravel the collapsed region and this consumes time. Lastly, the electrostatic interactions do not change K^N_{evo2} very much because this step corresponds to the binding of the C-terminal region, which lacks the charged interactions at the interfaces. Our coarse-grained SB model implemented by the ionic-strength-controlled electrostatic interactions provides a new understanding of the roles of the interchain and intrachain electrostatic interactions in IDPs' binding.

Folding and Binding of Multiple Domains within a Multidomain Protein

We will explore a specific case of folding and binding in which the binding partner is not other proteins or small substrates, but another domain from the same protein. In fact, the proteins consisting of multiple domains are quite common and their population can be up to 80% in eukaryotic proteins (Batey et al. 2008; Han et al. 2007). However, our current knowledge of protein folding is predominantly derived from the studies of small or single-domain proteins. In order to investigate whether the folding principles can be extended for larger multidomain proteins, we explored a four-domain protein, namely DNA polymerase IV (DPO4, as shown in Figure 9.4a), via a combinational strategy consisting of a coarse-grained SB model and its sequence-flavored variant. The combinational strategy allows us to test the robustness of the different simulation models and investigate the sequence dependency in the folding of DPO4. As the most thoroughly studied member of the Y-family DNA polymerases (Wong et al. 2008), DPO4 is an excellent model system as a multiple-domain protein and plays an important role in biological function.

We performed thermodynamic as well as kinetic simulations of DPO4 folding. The free-energy landscape was projected into the folding order parameters for all four domains in DPO4, as shown in Figure 9.4b. It shows that all free-energy minima are located at the corner of the 2-D free-energy surfaces corresponding to complete folding or unfolding states. Remarkably, there is no direct route along the diagonal lines, indicating that all four domains in DPO4 fold independently without any coupling. Despite this, it is difficult to determine a unique folding order of four domains. For DPO4, we can conclude that the folding of domains is highly sequential. In other words, the folding of DPO4 proceeds via a mechanism in which individual domains fold independently first and then bind together. It is worth noting that the pathway of domain

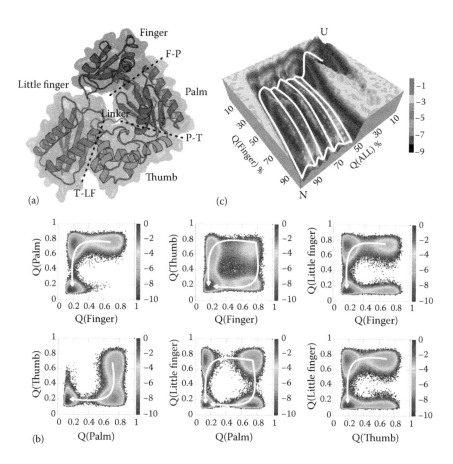

Figure 9.4 Folding and binding of multiple protein domains. (a) Structure of DPO4. (b) Sequential folding mechanism. The free-energy profiles are plotted as a function of the folding similarity to a native structure of four domains. There is no pathway along the diagonal lines in these surfaces, indicating that the four domains in DPO4 fold independently. (c) Diverse parallel folding pathways. The free-energy surface is plotted as a function of total native contacts (Q[ALL]) and native contacts in the F domain (Q[Finger]). There are six routes bridging the unfolded (U) and native folded (N) states, corresponding to different parallel folding pathways. (Reprinted with permission from Wang et al. 2012b. Copyright 2012 American Chemical Society.)

binding is highly diversified, as shown in Figure 9.4c. It shows the 2-D free-energy surface as a function of a folding order parameter of the whole DPO4 (Q[ALL]) and a folding order parameter of the F domain (Q[Finger]). There are six routes bridging the unfolded (U) and native folded (N) states, indicating different parallel folding pathways.

Above all, our results indicate that DPO4 folds by a stepwise assembly process along multiple parallel pathways. In order to provide a theoretical explanation for the question about how the multidomain proteins solve the protein folding problem, we further propose that multidomain proteins fold rapidly into their functional conformations through a divide-and-conquer strategy. That is, a multidomain protein folds its domains one by one. Subsequently, these folded domains bind each other to form a large multidomain protein. In this way, Levinthal's paradox (Levinthal 1969) for searching specific conformation structures can be resolved because the degrees of freedom for multidomain protein folding is in the form of polynomials rather than exponentials (Wang et al. 2012b).

Functional Landscape of ADK Modulated by Its Natural Substrates

It is a subject of great interest to explore the relationship between protein allostery and ligand binding. ADK has been considered as an excellent allosteric model system. This allosteric system involves large-scale domain arrangement, which is necessary to its catalytic function. By catalyzing the reversible phosphoryl transfer reaction, $Mg^{2+}ATP$ + adenosine monophosphate (AMP) $\rightarrow$ $Mg^{2+}ADP$ + ADP, ADK plays an important role in maintaining the energy balance within the cell (Tan et al. 2009). ADK is a multidomain protein that contains a core domain (CORE), an ATP-binding domain (LID) and a nucleoside monophosphate binding domain (NMP). To explore the interplay between functional dynamics of ADK and the binding of its natural substrates (AMP and ATP), we applied a multibasin two-bead SB model in which substrates are explicitly modeled (Wang et al. 2013b). We have to emphasize that such explicit consideration of substrates is important to the investigations of the functional landscape responding to substrate binding. We employed a combination of four models of ADK: in its apo (substrate-free) form, in the presence of AMP/ATP, and in complex with both substrates.

Our models predicted two intermediate states (Figure 9.5a). One is denoted as I_N in which the NMP domain is open while the LID domain is closed. The other is I_L in which the LID domain is open while the NMP domain is closed. The two intermediates are on the pathway between the open and closed basins, resulting in two major parallel kinetic pathways. The probability distributions of LID-CORE distance with and without substrates derived from both fluorescence resonance energy transfer (FRET) experiment (Hanson et al. 2007) and our simulation are summarized in Figure 9.5b. It shows that the theoretical results are in good agreement with the experimental findings, including not only the increase of a closed peak, but also the shift of an open peak in the

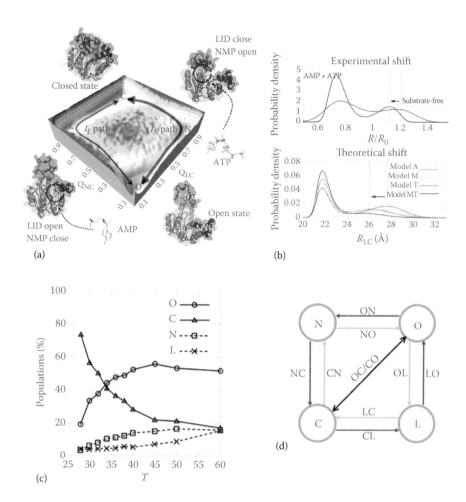

(a)

Figure 9.5   Functional landscape of ADK modulated by its natural substrates. (a) The free-energy landscape. The two intermediates connecting the open and closed basins yield two major parallel kinetic pathways: one through the intermediate state I_N with the LID domain closed and the NMP domain open (I_N pathway), and another through the intermediate state I_L with the NMP domain closed and the LID domain open (I_L pathway). (b) The probability distributions of LID-CORE distance with and without substrates summarized from FRET experiment and our simulation. (c) Pathway populations as a function of temperature. (d) A schematic diagram of four-state conformational transition model. The open and closed states as well as the two intermediates I_N and I_L are labeled as O, C, N, and L, respectively.

(*Continued*)

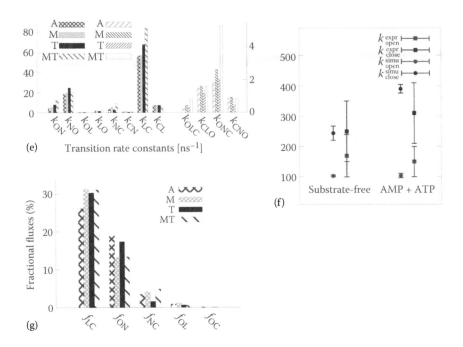

Figure 9.5 (Continued) Functional landscape of ADK modulated by its natural substrates. (e–g) Rate and flux analysis. The results strongly support that the motion of the NMP domain is the rate-limiting step for enzyme turnover. (Reprinted with permission from Wang et al. 2013b, 84–95. Copyright 2013 American Chemical Society.)

presence of substrates. This further confirms that the shift of the open peak originates from the change of conformational distribution of ADK induced by interacting with AMP and ATP. The conformational redistribution cannot be exactly captured by a single-order parameter such as the LID-CORE distance. However, the peak of the distribution of a single-order parameter may shift as a result of conformational redistribution. This implies that a complex multistate landscape, such as that of ADK, is probably insufficiently characterized by only one-dimensional (1-D) free-energy profiles.

At a simulated temperature that is comparable with an experimental temperature at which ADK is about seven-fold less active than its maximal activity (Wolf-Watz et al. 2004), ADK changes its conformation mostly along the I_N pathway, while the I_L pathway becomes dominant near an optimal temperature (Figure 9.5c). Therefore, we proposed that the I_L pathway is more dominant by entropic effect and the I_N pathway is more dominant by enthalpic effect. We also found that the binding of AMP favors the I_L pathway and the binding of ATP favors the I_N pathway.

After the experimental investigation in 2004 of the predominant role of the conformational dynamics of two lids in the catalytic efficiency of ADK (Wolf-Watz et al. 2004), it is still under debate which lid's motion is the rate-limiting step of the conformational cycle of ADK. Based on the thermodynamic and kinetic simulation results, we constructed a four-state kinetic transition model (Figure 9.5d). From the four-state model, we analyzed the pathway fluxes. The results support that the motion of NMP lid is the rate-limiting step for as for enzyme turnover (Figure 9.5e–g). Moreover, the double-substrate model allows us to validate the bi-substrate (Bi) bi-products (Bi) reaction mechanism. Our models support a random Bi-Bi mechanism.

CONCLUSIONS

Recent increase of computational power has significantly improved our capability for exploring the protein dynamics by simulations (Lindorff-Larsen et al. 2011; Shaw et al. 2009, 2010). Nevertheless, for most of the biologically interesting and functioning processes involving folding, recognition, conformational changes, and allosteric motions, the timescale for accurate atomic level simulation is still out of reach. In contrast to the conventional MD empirical force field based all-atom models which are hard to be implemented by brute-force simulations, simplified structure-based models or their hybrid models at the atomic level, have the advantage of the reduction of the number of degrees of freedom and the improvement of the efficiency of conformational sampling. They can be implemented to accelerate the observation of important biological process on a computer. In practice, the level of the simplification of the coarse-grained model does not only just depend on the available computer power but also relies on the concerned biological processes, for which the simplification is expected to be able to capture the properties of the system in comparison with experiments. Although the coarse-grained model has made dramatic advances in the understanding of conformational dynamics in protein binding, the deficient representation without atomic details cannot give a full description on the physical and chemical behaviors of proteins. In such cases, numerous sampling techniques have been implemented into the all-atom molecular dynamics. However, due to the limited computational resources, specific simulation methodologies are always worked out for addressing the specific problems. An all-atom model often cannot take consideration of both thermodynamic and kinetic simultaneously in the description of protein behaviors. Therefore, the multiscale simulation, which combines the simplified coarse-grained models and all-atom models, is applied to the investigation of the conformational dynamics in protein binding and multidomain protein folding. With long-time simulation and accurate all-atom description, the multiscale simulations, with the advantages of both efficiency in coarse-grained model and accuracy of all-atom model, have provided the successful classifications of the folding/binding mechanisms in our studies. At present, the developments of the multiscale simulations on force field, simplified representation, and experimental fitting parametrization are rapidly progressing. This approach provides a novel way to characterize the nature of protein dynamics and will likely become a powerful tool to investigate the complexity of biological process with multiple spatial and temporal scales in the future.

ACKNOWLEDGMENTS

Y.W. and X.C. acknowledge support from the National Science Foundation of China (Grants 21190040, 11174105, and 91227114) and 973 project of China (2009CB930100 and 2010CB933600). J.W. thanks the National Science Foundation for support.

REFERENCES

Atilgan, A. R., S. R. Durell, R. L. Jernigan, M. C. Demirel, O. Keskin, and I. Bahar (2001). Anisotropy of fluctuation dynamics of proteins with an elastic network model. *Biophys J* 80 (1), 505–515.

Batey, S., A. A. Nickson, and J. Clarke (2008). Studying the folding of multidomain proteins. *HFSP J* 2 (6), 365–377.

Beutler, T. C., and W. F. van Gunsteren (1994). The computation of a potential of mean force: Choice of the biasing potential in the umbrella sampling technique. *J Chem Phys* 100, 1492.

Bosshard, H. R. (2001). Molecular recognition by induced fit: How fit is the concept? *News Physiol Sci* 16, 171–173.

Brooks, B. R., C. Brooks, 3rd, A. Mackerell, Jr., L. Nilsson, R. J. Petrella, B. Roux, Y. Won, G. Archontis, C. Bartels, S. Boresch et al. (2009). CHARMM: The biomolecular simulation program. *J Comput Chem* 30 (10), 1545–1614.

Brooks, C. L., J. N. Onuchic, and D. J. Wales (2001). Statistical thermodynamics—Taking a walk on a landscape. *Science* 293 (5530), 612–613.

Bryngelson, J. D., J. N. Onuchic, N. D. Socci, and P. G. Wolynes (1995). Funnels, pathways, and the energy landscape of protein-folding—A synthesis. *Proteins* 21 (3), 167–195.

Bryngelson, J. D., and P. G. Wolynes (1989). Intermediates and barrier crossing in a random energy-model (with applications to protein folding). *J Phys Chem* 93 (19), 6902–6915.

Capraro, D. T., M. Roy, J. N. Onuchic, and P. A. Jennings (2008). Backtracking on the folding landscape of the β-trefoil protein interleukin-1β? *Proc Natl Acad Sci U S A* 105 (39), 14844–14848.

Cárdenas, A. E., and R. Elber (2003). Kinetics of cytochrome c folding: Atomically detailed simulations. *Proteins: Struct Funct Bioinform* 51 (2), 245–257.

Case, D. A., T. E. Cheatham, 3rd, T. Darden, H. Gohlke, R. Luo, K. M. Merz, Jr., A. Onufriev, C. Simmerling, B. Wang, and R. J. Woods (2005). The amber biomolecular simulation programs. *J Comput Chem* 26 (16), 1668–1688.

Chen, K., J. Eargle, J. Lai, H. Kim, S. Abeysirigunawardena, M. Mayerle, S. Woodson, T. Ha, and Z. Luthey-Schulten (2012). Assembly of the five-way junction in the ribosomal small subunit using hybrid MD-Go simulations. *J Phys Chem B* 116 (23), 6819–6831.

Cheung, M. S., J. M. Finke, B. Callahan, and J. N. Onuchic (2003). Exploring the interplay between topology and secondary structural formation in the protein folding problem. *J Phys Chem B* 107 (40), 11193–11200.

Chou, K.-C., and L. Carlacci (1991). Simulated annealing approach to the study of protein structures. *Protein Eng* 4 (6), 661–667.

Christen, M., P. H. Hnenberger, D. Bakowies, R. Baron, R. Brgi, D. P. Geerke, T. N. Heinz, M. A. Kastenholz, V. Krutler, C. Oostenbrink et al. (2005). The GROMOS software for biomolecular simulation: GROMOS05. *J Comput Chem* 26 (16), 1719–1751.

Chu, X., L. Gan, E. Wang, and J. Wang (2013). Quantifying the topography of the intrinsic energy landscape of flexible biomolecular recognition. *Proc Natl Acad Sci U S A* 110 (26), E2342–E2351.

Chu, X., Y. Wang, L. Gan, Y. Bai, W. Han, E. Wang, and J. Wang (2012). Importance of electrostatic interactions in the association of intrinsically disordered histone chaperone Chz1 and histone H2A.Z-H2B. *PLoS Comput Biol* 8 (7), e1002608.

Clementi, C., H. Nymeyer, and J. N. Onuchic (2000). Topological and energetic factors: What determines the structural details of the transition state ensemble and "enroute" intermediates for protein folding? An investigation for small globular proteins. *J Mol Biol* 298 (5), 937–953.

Das, C., J. K. Tyler, and M. E. A. Churchill (2010). The histone shuffle: Histone chaperones in an energetic dance. *Trends Biochem Sci* 35 (9), 476–489.

Dunker, A. K., J. D. Lawson, C. J. Brown, R. M. Williams, P. Romero, J. S. Oh, C. J. Oldfield, A. M. Campen, C. R. Ratliff, K. W. Hipps et al. (2001). Intrinsically disordered protein. *J Mol Graph Model* 19 (1), 26–59.

Dyson, H. J., and P. E. Wright (2002). Coupling of folding and binding for unstructured proteins. *Curr Opin Struct Biol* 12 (1), 54–60.

Dyson, H. J., and P. E. Wright (2005). Intrinsically unstructured proteins and their functions. *Nat Rev Mol Cell Biol* 6 (3), 197–208.

Elber, R., A. Roitberg, C. Simmerling, R. Goldstein, H. Y. Li, G. Verkhivker, C. Keasar, J. Zhang, and A. Ulitsky (1995). MOIL—A program for simulations of macromolecules. *Comput Phys Commun* 91 (1–3), 159–189.

Fischer, E. (1894). Einfluss der Configuration auf die Wirkung der Enzyme. *Ber Dtsch Chem Ges* 27 (3), 2984–2993.

Gosavi, S., L. L. Chavez, P. A. Jennings, and J. N. Onuchic (2006). Topological frustration and the folding of interleukin-1β. *J Mol Biol* 357 (3), 986–996.

Hamelberg, D., J. Mongan, and J. A. McCammon (2004). Accelerated molecular dynamics: A promising and efficient simulation method for biomolecules. *J Chem Phys* 120, 11919.

Han, J.-H., S. Batey, A. A. Nickson, S. A. Teichmann, and J. Clarke (2007). The folding and evolution of multidomain proteins. *Nat Rev Mol Cell Bio* 8 (4), 319–330.

Hansen, D. F., Z. Zhou, H. Q. Fen, L. M. M. Jenkins, Y. W. Bai, and L. E. Kay (2009). Binding kinetics of histone chaperone Chz1 and variant histone H2A.Z-H2B by relaxation dispersion NMR spectroscopy. *J Mol Biol* 387 (1), 1–9.

Hanson, J. A., K. Duderstadt, L. P. Watkins, S. Bhattacharyya, J. Brokaw, J. W. Chu, and H. Yang (2007). Illuminating the mechanistic roles of enzyme conformational dynamics. *Proc Natl Acad Sci U S A* 104 (46), 18055–18060.

Haran, G., and J. L. England (2010). To fold or expand—A charged question. *Proc Natl Acad Sci U S A* 107 (33), 14519–14520.

Henkelman, G., B. P. Uberuaga, and H. Jonsson (2000). A climbing image nudged elastic band method for finding saddle points and minimum energy paths. *J Chem Phys* 113 (22), 9901–9904.

Hills Jr., R. D., and C. L. Brooks, 3rd (2008). Subdomain competition, cooperativity, and topological frustration in the folding of CheY. *J Mol Biol* 382 (2), 485–495.

Hills, R. D., S. V. Kathuria, L. A. Wallace, I. J. Day, C. L. Brooks, and C. R. Matthews (2010). Topological frustration in beta alpha-repeat proteins: Sequence diversity modulates the conserved folding mechanisms of alpha/beta/alpha sandwich proteins. *J Mol Biol* 398 (2), 332–350.

Hondele, M., and A. G. Ladurner (2011). The chaperone-histone partnership: For the greater good of histone traffic and chromatin plasticity. *Curr Opin Struct Biol* 21 (6), 698–708.

Hyeon, C., P. A. Jennings, J. A. Adams, and J. N. Onuchic (2009). Ligand-induced global transitions in the catalytic domain of protein kinase a. *Proc Natl Acad Sci U S A* 106 (9), 3023–3028.

Jorgensen, W. L., D. S. Maxwell, and J. Tirado-Rives (1996). Development and testing of the OPLS all-atom force field on conformational energetics and properties of organic liquids. *J Am Chem Soc* 118 (45), 11225–11236.

Korolev, N., O. V. Vorontsova, and L. Nordenskiold (2007). Physicochemical analysis of electrostatic foundation for DNA-protein interactions in chromatin transformations. *Prog Biophys Mol Biol* 95 (1–3), 23–49.

Koshland, D. E. (1958). Application of a theory of enzyme specificity to protein synthesis. *Proc Natl Acad Sci U S A* 44 (2), 98–104.

Kuczera, K., G. S. Jas, and R. Elber (2009). Kinetics of helix unfolding: Molecular dynamics simulations with milestoning? *J Phys Chem A* 113 (26), 7461–7473.

Leopold, P. E., M. Montal, and J. N. Onuchic (1992). Protein folding funnels—A kinetic approach to the sequence structure relationship. *Proc Natl Acad Sci U S A* 89 (18), 8721–8725.

Levinthal, C. (1969). Proceedings in Mossbauer spectroscopy in biological systems. In *Mossbauer Spectroscopy in Biological Systems: Proceedings of a meeting held at Allerton House, Monticello, Illinois*, 22–24.

Levy, Y., P. G. Wolynes, and J. N. Onuchic (2004). Protein topology determines binding mechanism. *Proc Natl Acad Sci U S A* 101 (2), 511–516.

Lindorff-Larsen, K., S. Piana, R. O. Dror, and D. E. Shaw (2011). How fast-folding proteins fold. *Science* 334 (6055), 517–520.

Lindorff-Larsen, K., S. Piana, K. Palmo, P. Maragakis, J. L. Klepeis, R. O. Dror, and D. E. Shaw (2010). Improved side-chain torsion potentials for the Amber ff99SB protein force field. *Proteins* 78 (8), 1950–1958.

Liwo, A., C. Czaplewski, S. Ołdziej, and H. A. Scheraga (2008). Computational techniques for efficient conformational sampling of proteins. *Curr Opin Struct Biol* 18 (2), 134–139.

Ma, J. (2005). Usefulness and limitations of normal mode analysis in modeling dynamics of biomolecular complexes. *Structure* 13 (3), 373–380.

Majek, P., R. Elber, and H. Weinstein (2008). Pathways of conformational transitions in proteins. In *Coarse-Graining of Condensed Phase and Biomolecular Systems*, Voth G. A., ed. Boca Raton, FL: CRC Press, 185–203.

Muller-Spath, S., A. Soranno, V. Hirschfeld, H. Hofmann, S. Ruegger, L. Reymond, D. Nettels, and B. Schuler (2010). Charge interactions can dominate the dimensions of intrinsically disordered proteins. *Proc Natl Acad Sci U S A* 107 (33), 14609–14614.

Narayanan, R., O. K. Ganesh, A. S. Edison, and S. J. Hagen (2008). Kinetics of folding and binding of an intrinsically disordered protein: The inhibitor of yeast aspartic proteinase YPrA. *J Am Chem Soc* 130 (34), 11477–11485.

Noel, J. K., P. C. Whitford, K. Y. Sanbonmatsu, and J. N. Onuchic (2010). SMOG@ctbp: Simplified deployment of structure-based models in GROMACS. *Nucleic Acids Res* 38 (Web Server Issue), W657–W661.

Okamoto, Y., and Y. Sugita (1999). Replica-exchange molecular dynamics method for protein folding. *Chem Phys Lett* 314 (1–2), 141–151.

Olender, R., and R. Elber (1996). Calculation of classical trajectories with a very large time step: Formalism and numerical examples. *J Chem Phys* 105 (20), 9299–9315.

Oliveira, L. C., A. Schug, and J. N. Onuchic (2008). Geometrical features of the protein folding mechanism are a robust property of the energy landscape: A detailed investigation of several reduced models. *J Phys Chem B* 112 (19), 6131–6136.

Papoian, G. A. (2008). Proteins with weakly funneled energy landscapes challenge the classical structure–function paradigm. *Proc Natl Acad Sci U S A* 105 (38), 14237–14238.

Pappu, R. V., A. H. Mao, S. L. Crick, A. Vitalis, and C. L. Chicoine (2010). Net charge per residue modulates conformational ensembles of intrinsically disordered proteins. *Proc Natl Acad Sci U S A* 107 (18), 8183–8188.

Park, Y. J., and K. Luger (2008). Histone chaperones in nucleosome eviction and histone exchange. *Curr Opin Struc Biol* 18 (3), 282–289.

Pogorelov, T. V., and Z. Luthey-Schulten (2004). Variations in the fast folding rates of the lambda-repressor: A hybrid molecular dynamics study. *Biophys J* 87 (1), 207–214.

Shaw, D. E., R. O. Dror, J. K. Salmon, J. Grossman, K. M. Mackenzie, J. A. Bank, C. Young, M. M. Deneroff, B. Batson, K. J. Bowers et al. (2009). Millisecond-scale molecular dynamics simulations on Anton. In *High Performance Computing Networking, Storage and Analysis, Proceedings of the Conference on*, 1–11, IEEE.

Shaw, D. E., P. Maragakis, K. Lindorff-Larsen, S. Piana, R. O. Dror, M. P. Eastwood, J. A. Bank, J. M. Jumper, J. K. Salmon, Y. Shan et al. (2010). Atomic-level characterization of the structural dynamics of proteins. *Science* 330 (6002), 341–346.

Shoemaker, B. A., J. J. Portman, and P. G. Wolynes (2000). Speeding molecular recognition by using the folding funnel: The fly-casting mechanism. *Proc Natl Acad Sci U S A* 97 (16), 8868–8873.

Sobolev, V., A. Sorokine, J. Prilusky, E. E. Abola, and M. Edelman (1999). Automated analysis of interatomic contacts in proteins. *Bioinformatics* 15 (4), 327–332.

Sutto, L., I. Mereu, and F. L. Gervasio (2011). A hybrid all-atom structure-based model for protein folding and large scale conformational transitions. *J Chem Theory Comput* 7 (12), 4208–4217.

Tan, Y.-W., J. A. Hanson, and H. Yang (2009). Direct Mg2+ binding activates adenylate kinase from *Escherichia coli*. *J Biol Chem* 284 (5), 3306–3313.

Thorpe, I. F., D. P. Goldenberg, and G. A. Voth (2011). Exploration of transferability in multiscale coarse-grained peptide models. *J Phys Chem B* 115 (41), 11911–11926.

Tozzini, V. (2010a). Minimalist models for proteins: A comparative analysis. *Q Rev Biophys* 43 (3), 333–371.

Tozzini, V. (2010b). Multiscale modeling of proteins. *Acc Chem Res* 43 (2), 220–230.

Wang, J., R. J. Oliveira, X. Chu, P. C. Whitford, J. Chahine, W. Han, E. Wang, J. N. Onuchic, and V. B. P. Leite (2012a). Topography of funneled landscapes determines the thermodynamics and kinetics of protein folding. *Proc Natl Acad Sci U S A* 109 (39), 15763–15768.

Wang, J., Y. Wang, X. Chu, S. J. Hagen, W. Han, and E. Wang (2011). Multi-scaled explorations of binding-induced folding of intrinsically disordered protein inhibitor IA3 to its target enzyme. *PLoS Comput Biol* 7 (4), e1001118.

Wang, J., K. Zhang, H. Lu, and E. Wang (2005). Quantifying kinetic paths of protein folding. *Biophys J* 89 (3), 1612–1620.

Wang, Y., X. Chu, S. Longhi, P. Roche, W. Han, E. Wang, and J. Wang (2013a). Intrinsically disordered ensembles and coupled folding and binding of a molecular recognition element in measles virus nucleoprotein. *Proc Natl Acad Sci U S A* 110 (40), E3743–E3752.

Wang, Y., X. Chu, Z. Suo, E. Wang, and J. Wang (2012b). Multidomain protein solves the folding problem by multifunnel combined landscape: Theoretical investigation of a Y-family DNA polymerase. *J Am Chem Soc* 134 (33), 13755–13764.

Wang, Y., L. Gan, E. Wang, and J. Wang (2013b). Exploring the dynamic functional landscape of adenylate kinase modulated by substrates. *J Chem Theory Comput* 9 (1), 84–95.

Wang, Y., C. Tang, E. Wang, and J. Wang (2012c). Exploration of multi-state conformational dynamics and underlying global functional landscape of maltose binding protein. *PLoS Comput Biol* 8 (4), e1002471.

Whitford, P. C., P. Geggier, R. B. Altman, S. C. Blanchard, J. N. Onuchic, and K. Y. Sanbonmatsu (2010). Accommodation of aminoacyl-tRNA into the ribosome involves reversible excursions along multiple pathways. *RNA* 16 (6), 1196–1204.

Wolf-Watz, M., V. Thai, K. Henzler-Wildman, G. Hadjipavlou, E. Z. Eisenmesser, and D. Kern (2004). Linkage between dynamics and catalysis in a thermophilic-mesophilic enzyme pair. *Nat Struct Mol Biol* 11 (10), 945–949.

Wong, J. H., K. A. Fiala, Z. Suo, and H. Ling (2008). Snapshots of a Y-family DNA polymerase in replication: Substrate-induced conformational transitions and implications for fidelity of Dpo4. *J Mol Biol* 379 (2), 317–330.

Wright, P. E., and H. J. Dyson (1999). Intrinsically unstructured proteins: Re-assessing the protein structure-function paradigm. *J Mol Biol* 293 (2), 321–331.

Yang, Z., P. Majek, and I. Bahar (2009). Allosteric transitions of supramolecular systems explored by network models: Application to chaperonin GroEL. *PLoS Comput Biol* 5 (4), e1000360.

Zheng, W., N. P. Schafer, A. Davtyan, G. A. Papoian, and P. G. Wolynes (2012). Predictive energy landscapes for protein–protein association. *Proc Natl Acad Sci U S A* 109 (47), 19244–19249.

Zheng, W., N. P. Schafer, and P. G. Wolynes (2013). Frustration in the energy landscapes of multidomain protein misfolding. *Proc Natl Acad Sci U S A* 110 (5), 1680–1685.

Zhou, Z., H. Q. Feng, B. R. Zhou, R. Ghirlando, K. F. Hu, A. Zwolak, L. M. M. Jenkins, H. Xiao, N. Tjandra, C. Wu et al. (2011). Structural basis for recognition of centromere histone variant CenH3 by the chaperone Scm3. *Nature* 472 (7342), 234–237.

Chapter 10

Coarse-Grained Simulation of Intrinsically Disordered Proteins

David de Sancho, Christopher M. Baker, and Robert B. Best

CONTENTS

INTRODUCTION

The important role played by intrinsically disordered proteins (IDPs) in protein–protein interaction networks and cellular signaling is increasingly being recognized [1–3]. In these interactions, an intrinsically disordered polypeptide (often an intrinsically disordered region of a multidomain protein) binds to another macromolecule, be it a folded protein, another intrinsically disordered protein, or a nucleic acid. Frequently, but not always, the disordered region undergoes a local folding transition coupled to binding [2]. A rich variety of possible binding scenarios is thus generated (in comparison with the simpler case of two folded proteins associating) and consequently many interesting questions arise regarding the binding mechanism, the bound state, and possible advantages of disordered regions versus folded proteins in fulfilling a given role in an interaction network [4]. While bioinformatics provides powerful tools for hypothesis testing by correlation [5], it does not provide direct physical insight into the origin of any observed effects.

Physics-based models do not have this deficiency. Fully atomistic simulations of coupled folding-binding with explicit solvent in principle provide the highest accuracy feasible, albeit requiring large computational resources in addition to enhanced sampling methods in order to sample binding. However, there are a number of reasons why atomistic simulations may not be the best choice for studying folding-binding transitions at the time of writing. The first is that the energy functions, or force fields, themselves have some limitations, which may be particularly detrimental to an accurate treatment of unfolded or disordered proteins. For example, unfolded states using current protein force fields and water models are too collapsed and structured in comparison to any experimental measure [6], and nonspecific protein–protein association appears to be too favorable [7]. Since this deficiency could clearly bias any binding mechanism obtained in simulation, use of such force fields for studying coupled folding and binding may be questionable. The second issue relates to the sort of questions that are being asked. If a set of binding events at atomistic detail is needed, certainly all atom simulations are the only feasible approach. However, for a great number of interesting questions, this level of detail is unnecessary and may even make it harder to identify the effects that are really important. Lastly, the computational cost for studying binding events, even with enhanced sampling, is still very high. This limits the possibility of varying the protein sequence or other conditions in order to test hypotheses. Coarse-grained simulation models, in which the number of configurational degrees of freedom as well as the complexity of the energy function are reduced, can help to overcome some of these deficiencies. They make it easier to tune both sequence and interactions, and, in some cases, to make accurate predictions of binding mechanism. Their major drawback is their limited predictive value for specific cases, particularly if a structure of the bound complex is not known. Nonetheless, there are many generic and specific questions that can be more straightforwardly investigated with the aid of coarse-grained models.

In this chapter, we will give an overview of the types of coarse-grained models that may be useful for investigating intrinsic structure formation as well as coupled folding-binding of IDPs. These allow us to answer questions such as, What is the extent of structure formation in the unbound state? What is the mechanism by which a given IDP binds its target—is it required to adopt its folded structure before binding? Does being unstructured confer any advantages with regard to binding rate, and if so, how? Why are IDPs selected by evolution to fulfill certain functions in protein interaction networks—does this happen by chance or does being unstructured confer a clear functional advantage in these cases? All of these questions can in principle be addressed using coarse-grained models, although the level of coarse-graining needed will vary from case to case.

TYPES OF COARSE-GRAINED MODELS SUITABLE FOR IDPs

In this section, we give a broad overview of the possible types of coarse-grained models that can be used to describe IDPs and their binding; in the following we focus in on coarse-grained modeling of protein–protein and protein–deoxyribonucleic acid (DNA) binding using variations on Gō-like models using one or several beads per residue, which have been particularly successful for this purpose.

The most detailed of the coarse-grained models suitable for describing IDPs and their interactions are those in which the protein is still treated at the atomistic level, while the solvent degrees of freedom are integrated out and replaced by an approximate solvation free-energy functional. This already constitutes a very large saving, since water molecules constitute the majority of atoms in explicit solvent simulations. Some of the most successful types of implicit solvent model are the generalized Born models [8], or approximations to these [9], and Gaussian solvation models such as effective energy function (EEF1) [10,11] and self-assembly of biomolecules studied by an implicit, novel, and tunable Hamiltonian (ABSINTH) [12]. The former models attempt to reproduce the solvation free energy that would be obtained from a continuum electrostatics calculation by means of computationally cheaper approximations. The latter models assign the experimental solvation free energies to each amino acid, which are then reduced by the Gaussian solvation term upon close approach of the residues. Implicit solvent models bring the benefit of enhanced speed to the simulation while retaining atomistic detail. The interested reader is referred to Chapter 6 for an overview of the ABSINTH implicit solvent model [12], which has been the most successful in terms of reproducing the dimensions of unfolded and disordered proteins [13]. Importantly, such models retain sufficient detail so that if the implicit solvent is good enough, they can still be used in a predictive manner.

The next level of detail in coarse-grained models eliminates some of the protein degrees of freedom, typically leaving only one, or a few, interaction centers per residue. The large increase in speed afforded by such models is one of their principle advantages, and for this reason these models have found frequent use in describing the binding of intrinsically disordered proteins. The lack of detail means that either complex orientational potentials must be included in an attempt to restore predictive power [14] (but with additional computational cost), or further assumptions must be made about the nature of the energy landscape. The most frequent assumption is that protein sequences have been designed to fold or bind to a given native structure and that nonnative interactions, which might "frustrate" this process, have been attenuated by evolutionary sequence optimization: known as the principle of minimal frustration [15]. This assumption leads to the Gō-like family of models [16], in which nonnative contacts are neglected or treated only as excluded volume, while native contacts are considered attractive. Some degree of sequence-based nonnative interaction may of course be restored to the model later, as we discuss in the following, in order to assess its role. Apart from speed, this level of coarsegraining permits changes in energy function, which are a powerful tool for investigating the effects which may be important for the binding mechanism. Due to their extensive use in characterizing the binding of IDPs, we have focused on this class of model in this chapter.

The most coarse-grained models are lattice models. Here, each residue (or a group of residues) in the protein is treated as a single bead that can only occupy sites on a lattice, which is usually cubic. Put forward here is the notion that the chain folds to a structure that corresponds to a specific protein or protein complex: clearly such models can only be used to investigate generic questions regarding IDPs. These questions might include general questions such as the relative merit of being intrinsically disordered (versus folded) for playing a specific role in a protein interaction network or for creating certain types of interaction network. However, while they played an important part in addressing questions related to evolution and sequence design in the early years of protein folding [17,18], their application to intrinsically disordered proteins has been relatively limited [19–21], and so we do not discuss them further here. However, we certainly believe they may have an important role in future developments.

COARSE-GRAINED MODELING OF PROTEIN–PROTEIN INTERACTIONS

Except for a few studies with on-lattice simulations, coarse-grained models for IDP binding to other protein are very often native-centric (i.e., Gō) models [22,23]. This type of model has been a natural choice given, on one hand, the

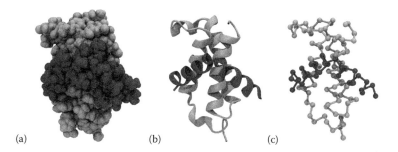

(a)   (b)   (c)

Figure 10.1   Different representations of the experimental structure of pKID (dark gray) bound to its partner KIX (light gray; PDB: 1kdx). (a) Atomic detail, (b) cartoon representation of the secondary structure, and (c) ball and stick representation of the C^{α} atoms.

complexity of IDPs, and on the other, the previous successes of these models in protein folding. As discussed in the section "Types of Coarse-Grained Models Suitable for IDPs," the motivation for such models is not just their computational convenience, but is also based on the understanding that naturally occurring sequences have been designed to fold/bind to a specific structure, and consequently the role of nonnative contacts is reduced.

The systems that have thus far been studied typically consist of at least two proteins exceeding 100 amino acids in total (109 for the most studied example, the complex formed by pKID-KIX, Figure 10.1a). This immediately translates into having to integrate the equations of motion for tens to hundreds of thousands of atoms if we consider the surrounding water molecules. For this reason explicit and even implicit solvent simulations, which are sometimes attempted for the disordered protein alone [24–27] and only rarely for the binding process [28], would be impractical. On the other hand, Gō models have played an instrumental role for understanding general principles and making semiquantitative predictions on protein folding [22,23]. Originally developed for independently folding proteins, Gō models have also been used to explore the topology of energy landscapes for protein–protein binding in a wide catalog of proteins [29]. It is not surprising then that this family of coarse-grained models has seen such widespread usage in the study of IDPs.

Standard Gō Models

In most of the work carried out on IDPs, previously established protein folding Gō models have been utilized, the most important being due to Kaya and Chan [30], Karanicolas and Brooks [31], and Clementi, Nymeyer and Onuchic [32]. For the last two, either a set of scripts (http://mmtsb.org/) [33] or a web server (http://smog-server.org/) [34] are available that will generate parameter files from a Protein Data Bank (PDB) file. These coarse-grained models share a simple description of the protein chain with just a single bead per amino acid residue positioned in the C$^\alpha$ atom (see Figure 10.1c), although an all-atom version exists for the Clementi, Nymeyer and Onuchic model. Also, in all these models the water degrees of freedom are averaged out and their contribution is implicitly defined in the potential energy function.

The potential is defined as a sum of bonded and nonbonded terms

$$V = V_{bonds} + V_{stretches} + V_{torsions} + V_{nonbonded} \tag{10.1}$$

In this expression, the first two terms are harmonic potentials with the minimum in the value corresponding to the native state, and are not dissimilar in form from those used in standard all-atom force fields. The torsional term is also defined to have the angle in the experimental structure as an energy minimum, or alternatively using statistical propensities derived from the PDB. Using statistical potentials can reduce the bias for forming a native-like structure in the unfolded state. Finally, nonbonded terms are defined using a Lennard-Jones like functional form. Either a vanilla homogeneous potential for every pair of residues that form a native contact or a colored potential dependent of the chemistry of the amino acid residues involved (i.e., the Miyazawa-Jernigan potential [35]) is used to define the energetics of contact formation. The specific nonbonded interaction function used is often slightly different from the standard 12-6 form. For example, 12-10 (Equation 10.2) or 12-10-6 (Equation 10.3) functional forms, or inverted Gaussian functions have been successfully used.

$$V(r_{ij}) = \varepsilon_{ij} \left\{ 5 \left(\frac{\sigma}{r_{ij}} \right)^{12} - 6 \left(\frac{\sigma}{r_{ij}} \right)^{10} \right\} \tag{10.2}$$

$$V(r_{ij}) = \varepsilon_{ij} \left\{ 13 \left(\frac{\sigma}{r_{ij}} \right)^{12} - 18 \left(\frac{\sigma}{r_{ij}} \right)^{10} + 4 \left(\frac{\sigma}{r_{ij}} \right)^{6} \right\} \tag{10.3}$$

The common feature of these potentials that they are of shorter range than 12-6 Lennard-Jones potentials. Physically, this can be motivated because the atomistic interactions underlying the coarse particle interaction are more short-range relative to the coarse particle size. It also has a beneficial effect on the cooperativity of the models: by reducing the available volume in which each contact can be formed, an increased entropic penalty for native contact formation is introduced. The 12-10-6 potential used by Karanicolas and Brooks [31] also has repulsive part at intermediate distances, which has been suggested to mimic a solvent desolvation penalty.

Modifications for the Study of IDPs

The IDP community has recognized that a few adaptations were needed in order to be able to apply these coarse-grained models to disordered proteins. First, the unfolded state description of Gō models is usually poor, with considerable amounts of preformed helical structure. Second, nonnative interactions are usually neglected in their description. Third, electrostatic interactions, which modulate the properties of the unfolded ensemble [36], are ignored more frequently than not. These three deficiencies of protein folding Gō models do in fact go hand in hand when it comes to determining the properties of IDPs [37]. Different groups have introduced modifications in preexisting Gō models for the study of IDP binding.

Local Structural Propensities

In conventional Gō models, the torsion angle potential biases the local structural preference of the protein in the unfolded state to be native-like. Even for sequence-based statistical torsion potentials, such as in the Karanicolas/Brooks model, there may be some bias because the database from which these potentials are derived may not be representative of true intrinsic preferences. The best-known example is the tendency of statistical potentials derived from folded proteins to favor alpha-helical structure. In order to correct for this bias, De Sancho and Best introduced a uniform correction term in the torsion angle so that less alpha-helical structure was retained in the unfolded state [38]. Alternatively Ganguly and Chen have tuned the local preferences indirectly by weakening local contacts [39]. The former approach has the advantage of directly tackling the torsional preferences. However, care must be taken so that the correct helical geometry is preserved in the bound state. It is desirable that new schemes, including a more appropriate description of hydrogen bonds, are developed to properly reproduce torsional propensities in the bound and unbound states.

Nonnative and Electrostatic Interactions

Different approaches have been devised in order to model the nonnative contribution, which is usually considered only for interprotein contacts. In a simple version, a hydrophobic/polar (HP) description of amino acids was used to allow for certain nonnative interactions to be favorable (i.e., those between hydrophobic residues) [40]. The most sophisticated approach consisted of modeling the nonnative interactions using the Kim-Hummer transferable potential that was carefully calibrated for weak binding complexes ($K_d > 1$ μM) using second virial coefficients and binding affinities [41]. A Debye-Hückel description of the electrostatic interactions that uses the amino acid residue charge and the dielectric constant of water is also included in the model. A similar approach to model electrostatics has been used in recent studies by Chen [42], Wang [43], and their coworkers.

Insights into Binding Mechanisms from Coarse-Grained Models

The coupled folding and binding of IDPs has received much attention in the last few years. However, only a few experimental studies have focused in the actual mechanism of the folding-binding reaction [44–46]. Much of the work carried out with Gō models has focused on the study of this transition and how details in the model would explain some of the experimental signatures that are supposed to provide functional advantages to IDPs. The most important of these features is a speed-up in the binding due to an increased capture radius, termed "fly casting," a mechanism first proposed by Wolynes and coworkers [47]. Using an analytical model for a free-energy surface, Wolynes and coworkers showed that unfolded configurations could experience attraction at much larger distances from their partners than folded ones, therefore facilitating binding via native interactions.

Explicit chain coarse-grained simulations have further investigated the kinetic advantage of IDPs. In a seminal study, Hummer and coworkers first found for the pKID-KIX complex (see Figure 10.1) that modifying the amount of structure in the unbound pKID only slightly reduced the on-rate [48]. Huang and Liu interrogated this issue further [49]. They found that the speed-up in binding was not due to an increased capture radius, but rather to a decrease in the free-energy barriers when they increased the chain flexibility. Also, Huang and Liu were the first to suggest that binding could be facilitated by nonnative interactions [40]. Two of us later studied the effects of nonnative interactions in a study of HIF1α-CBP where a hierarchy of models was considered [38]. An important conclusion of this work was that nonnative interactions could dramatically lower free-energy barriers, accelerating the binding process to diffusion-limited rates, as experimentally observed [46]. The nonspecific binding between the IDP and its target would be greatly facilitated by nonnative interactions, and then in a relatively slow search process, the specific binding site would be identified. This can be seen clearly in Figure 10.2, where binding is initiated by formation of a nonnative encounter complex followed by subsequent development of native-like intermolecular contacts. This is a second mechanism (nonnative steering [46]) by which binding rates can be enhanced, distinct from fly-casting. Fly-casting is a phenomenon also observed in pure

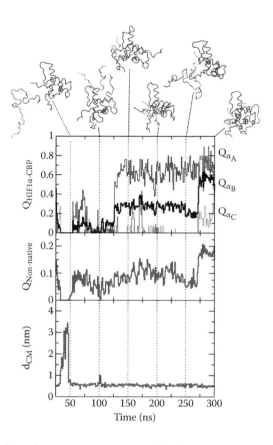

Figure 10.2   Effect of nonnative contacts on binding mechanism of HIF1α to CBP. Top panel: Fraction of native contacts formed by each of the three helices (A, B and C) in HIF1α. Center panel: Fraction of residues making nonnative contacts. Lower panel: Center of mass difference between the two proteins. (From De Sancho, D. & Best, R. B. Modulation of an IDP binding mechanism and rates by helix propensity and non-native interactions: Association of HIF1α with CBP. *Molecular Biosystems* 8, 256–267 [2012]. Reproduced by permission of The Royal Society of Chemistry.)

Gō-models, where it was originally proposed [47], and involves an increase capture radius due to disordered regions. The effect of nonnative interactions is to enhance binding by increasing the probability that two proteins will remain associated after initial contact and hence the probability of native binding. This effect, which has been treated theoretically in a different context by Zhou and Szabo [50], must however be balanced with the need to avoid nonnative interactions being too strong in a crowded cellular environment [51].

Finally, the effect of electrostatics in the binding process has also been studied. Ganguly et al. proposed that electrostatic steering could be a prevalent mechanism for binding [42], a hypothesis supported by the higher charge densities in IDPs [52]. Wang and coworkers also found support for the modulation of rates and mechanisms by electrostatic interactions, paying special attention to the effects of different salt concentrations [43]. All these observations highlight the importance of carefully modeling additional contributions in coarse-grained models of IDP binding.

Conformational Selection and Induced Fit

An outstanding question that coarse-grained models have tried to address is whether a conformational selection scenario or induced fit better describe the folding and binding transition [53]. In their work on pKID-KIX, Hummer and coworkers found evidence for an induced fit type of transition. By studying transition states for the binding reaction they found that structure only formed after binding [48]. Similarly, De Sancho and Best found highly unstructured transition states or HIF1α binding to CBP [38]. On the contrary, in a study of NCBD binding to its partner ACTR, Chen and coworkers found support for conformational selection. They observed that preformed elements of helical structure form a "mini folding core" that allows rapid binding of the IDP to its partner [54]. However, the final picture appears more akin to the extended conformational selection scenario [53], where a range of selection and arrangement processes occur. A more recent study by Knott and Best used a double Gō model to treat the binding of NCBD to both ACTR and IRF3 [55]. This is motivated by the fact that a single Gō model cannot capture binding to both proteins and experimental evidence for transient population of native structure in the unbound state, which is suggestive of conformational selection. In the double Gō model, the energy surface of the individual Gō models are combined in such a way so as to preserve the two native states while allowing transitions between the two energy surfaces. This is accomplished by an exponential averaging of the two energy functions (equivalent to adding their partition functions) [56].

$$e^{-\beta_{mix}U(x)} = e^{-\beta_{mix}U_A(x)} + e^{-\beta_{mix}U_B(x)}$$

where $U_A(x)$ and $U_B(x)$ are the energy surfaces of the individual Gō models and $U(x)$ is the combined surface; β_{mix} is an inverse mixing temperature that controls the smoothness of the switching between the two surfaces. This model also enabled a novel criterion for testing for conformational selection; namely, the relative lifetimes of the binding-incompetent state in the absence and presence of the binding partner. Only for true conformational selection will these lifetimes be the same. This criterion led to the conclusion that NCBD binds both IRF3 and ACTR via an induced fit mechanism [55].

While it is not completely clear from the above studies whether and induced fit or conformational selection mechanism is favored in the binding of IDPs to their targets, an induced fit description has been favored in most studies to date—and of course the exact mechanism will depend on the nature of the IDP.

Beyond Coarse-Grained Models

A particularly promising approach is the combination of coarse-grained models with experimental data, statistical mechanics models, and more detailed simulations. Multiscale approaches with different layers of detail have started to appear in the literature. For example Wang et al. have recently studied the binding-induced folding of IA3 to its target YPrA combining topology based modeling and an atomistic stochastic path method [57]. From the former the authors obtain a potential of mean force for the whole binding transition while the later allows for determining optimal kinetic paths for binding from different starting points. Naganathan and Orozco combined the variable barrier model for the analysis of differential scanning calorimetry thermograms [58] with a Gō model for binding and atomistic molecular dynamics simulations to provide an integrative view of NCBD, which they combined with data from small-angle X-ray scattering, NMR, and circular dichroism [59]. After the extensive usage of Gō models for exploring general principles, this type of hybrid, quantitative approach will probably set the route to follow in future applications of coarse-grained models to IDPs.

COARSE-GRAINED MODELING OF PROTEIN–NUCLEIC ACID INTERACTIONS

Intrinsic Disorder in Nucleic Acid Binding

Since the early 1990s, it has been recognized that some proteins undergo folding transitions when they bind to DNA [60], but it is only relatively recently that it has become apparent just how common intrinsic disorder is in DNA binding proteins. Intrinsically disordered regions of proteins can be identified on the basis of amino acid composition [61], and such an approach has been used [62] to identify that more than 80% of transcription factors—proteins that exist for the purpose of binding DNA in a sequence-specific fashion [63]—possess extended disordered regions. When compared to the overall 30% of eukaryotic proteins that are believed to possess intrinsically disordered regions [64], it is clear that nature must find some particular advantages to employing intrinsic disorder in DNA binding. In fact, disordered regions in nucleic acid binding proteins can be broadly grouped into two classes: those that fold into well-defined structures upon binding and those that remain disordered in the bound complex [1]. Intrinsically disordered domains that fold upon binding to specific DNA regions are obviously biologically functional even when the precise reasons why biology employs intrinsic disorder are unclear. What has been less obvious, but is now becoming apparent, is that intrinsically disordered domains distant from the binding region (and which may never fold) are also important for DNA binding. Fuxreiter et al. [1] have suggested that there are four possible mechanisms by which these disordered regions can affect DNA binding: by modulating the structure of the binding domain, by modulating the flexibility of the binding domain, by competing with DNA to bind to the DNA binding domain, or by tethering structured binding domains together. Simulating protein–DNA binding is always a challenging and computationally demanding problem given that it requires the presence of two large, independent biomolecules. But if we wish to examine the effect of additional domains—domains that do not participate directly in the binding process—then the system becomes larger and even more challenging to simulate. At present, coarse-graining offers perhaps the only route to study such systems on biologically meaningful timescales.

Coarse-Grained Models of DNA

The first stage in developing a coarse-grained model for studying protein–DNA interactions must be the development of a coarse-grained model providing an acceptable representation of DNA. The simplest nucleic acid analogs of Cα-based protein models include one coarse-grained bead per nucleotide (Figure 10.3). Doi et al. [40] developed a model of DNA in which the one bead was placed at the site of the P atom and interactions were determined using a potential that included terms to represent bond, angle, base-pairing, aromatic stacking, and electrostatic interactions. With such an approach, they were able to reproduce experimental melting temperatures and the salt dependence of the persistence length [65]. Recognizing the importance of electrostatic interactions, Savelyev and Papoian have developed a one-bead-per-residue coarse-grained model that allows them to run simulations with the inclusion of explicit ions [66]. Their approach was built on a model that could more accurately be described as a two-beads-per-base-pair model [67], with beads representing base-paired nucleotides connected by bonds (meaning that the model cannot

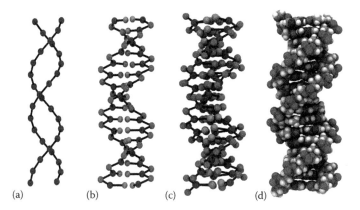

Figure 10.3   Coarse-grained models of DNA. (a) One bead per nucleotide. (b) Three beads per nucleotide. (c) Six beads per nucleotide. (d) An all-atom representation of DNA.

describe DNA melting). They then added additional terms to their potential function to represent ion-ion and DNA-ion interactions [67].

Such one-bead-per-nucleotide approaches are appealing because the resulting models are extremely simple and computationally efficient. But they can also suffer from the drawback of being too simple. A one-bead-per-nucleotide model requires each bead to represent at least 30 atoms, which results in a much coarser description of the biomolecule than found in protein Cα models, where each bead represents between 7 (GLY) and 24 (TRP) atoms. Inevitably, therefore, coarse-graining a nucleic acid with one bead per nucleotide results in the loss of more information than does coarse-graining a protein with one bead per residue; this problem is aggravated by the orientationally specific nature of base stacking and Watson-Crick base pairing via hydrogen bonds. For this reason, coarse-grained nucleic acid models employing more than one bead per residue have also been developed. The most popular scheme is to employ three beads per residue, with one bead representing each of the base, sugar, and phosphate groups (Figure 10.3). One such model was introduced by Knotts et al. [68] with a potential function including the typical bond, angle, dihedral, excluded volume, and electrostatic terms, as well as DNA-specific base pairing and stacking terms; stacking is treated using a Gō-model style native contact scheme [69]. Each of the DNA bases is also represented using a different type of bead, allowing the model to capture sequence-dependent properties. A subsequent variation of this model introduced several new features [70], most significantly a solvent-induced contribution to the potential function, which provides an implicit representation of many body affects associated with the arrangement of water during the reversible denaturation of DNA. An alternative version of the three-bead model, which relies purely on physics-based interactions, has been introduced by Morriss-Andrews et al. [71]. The unique feature of this model is that it explicitly accounts for the anisotropy of the nucleic acid bases. Rather than representing these groups by spherical particles, the model uses squashed elliptical (or "pancake" [72]) beads. This allows the model to capture a number of DNA properties that cannot be seen using isotropic coarse-grained models, including a proper description of base stacking and helicoidal parameters such as tip, roll, or propeller twist.

Yet more detailed coarse-grained models of DNA are also available. Dans et al. have developed a model with six interaction sites per nucleotide (Figure 10.1) [73]. Apart from being intrinsically more detailed than a one- or three-bead model, this representation has the advantage of allowing straightforward mapping from the coarse-grained representation to an all atom representation. De Mille et al. [74] have combined the three-site model of Knotts et al. [68] with coarse-grained models of water and ions [75].

Coarse-Grained Models of IDP–DNA Complexes

In principle, any one of the above models could be combined with an appropriate coarse-grained protein model and used to study protein–DNA interactions. As yet, however, only a small number of researcher groups have attempted such studies. Foremost among those have been Levy and coworkers, who have used coarse-grained models to study protein–DNA interactions involving both well-structured [76] and intrinsically disordered [77] proteins. Here, we will focus our discussion on their work targeting intrinsically disordered proteins.

Coupled Folding and Binding in Protein–Nucleic Acid Complexes

To study the role of the fly-casting mechanism [47] in protein–DNA binding, Levy et al. [78] performed coarse-grained simulations of the specific complex formed when the Ets domain of the transcription factor SAP-1 binds to DNA [79]. This protein is not actually an IDP as it is structured in the absence of its binding partner, but the authors studied its folding in the presence and absence of DNA, allowing them to draw conclusions that DNA can influence the folding of a protein. They also introduced the coarse-grained model that they would later adapt for the study of IDP-DNA interactions. In this model, both the protein and the DNA were represented by a single bead per residue, with nonbonded interactions described by a "native topology based model (Gō model) supplemented by electrostatic interactions." All DNA residues possessed a negative charge, and charges for protein residues were assigned as 0 or ± 1 according to their protonation state at physiological pH. The DNA molecule was then held rigid throughout the simulation with the protein constrained to remain within a sphere of 40 Å centered at the DNA's center of mass. The protein itself was flexible, with its structure determined through bond, angle, and dihedral potentials. Overall, therefore, the potential function included six distinct terms: bond; angle; dihedral; a 10-12 Lennard-Jones term to describe native interactions, a repulsive term to describe nonnative interactions; and a Coulomb term, which is shown in Equation 10.4:

$$V_{Coulomb} = C \sum_{i>j} \frac{q_i q_j}{\varepsilon_r r_{ij}^2} \qquad (10.4)$$

where q_i and q_j are the charges on the beads, r_{ij} is the distance between them and ε_r is the relative dielectric of the solvent. C is a constant with value 322 kcal mol^{-1} Å^2 e^{-2}. Only phosphate beads on the nucleic acid and charged residues on the protein are assigned their net charge and charges of zero are assigned to other beads. Note that the dielectric of the solvent is included, but an additional distance-dependent factor is also used (usually known as a distance-dependent dielectric), presumably to mimic the effect of charge screening in some way.

The authors' objective was to understand the role that electrostatic interactions play in influencing the binding process, and to this end they systematically varied the dielectric constant. They found that decreasing the dielectric constant (and thereby increasing the electrostatic interactions between the protein and the DNA) results in the destabilization of the protein, which encourages the protein to bind to the DNA in a nonspecific fashion before folding to form the specific complex while attached to the DNA. In the absence of any electrostatic interactions (an infinite dielectric constant), the protein only bound to DNA when it was fully folded.

Intrinsically Disordered Regions Not Directly Involved in Sequence-Specific Binding

The same group has used similar approaches to study the role in DNA binding of intrinsically disordered regions that occur outside the binding domain. Toth-Petroczy et al. [80] studied the impact of disordered tails [81] on DNA binding using an approach similar to that described in the previous section, but with a few notable differences. First, the DNA molecule was modeled using three beads per nucleotide, representing the base, sugar, and phosphate groups, respectively, with the phosphate bead carrying the negative charge. Second, the electrostatic term in Equation 10.4 was replaced by a term representing electrostatic interactions via a Debye-Huckel model, shown in Equation 10.5:

$$V_{Coulomb} = CB(\kappa) \sum_{i>j} \frac{q_i q_j}{\varepsilon r_{ij}} \exp[-\kappa r_{ij}] \qquad (10.5)$$

Symbols have the same meaning as in Equation 10.1, except that now C has units of 322 kcal mol^{-1} Å^2 e^{-1}, κ is an ionic strength-dependent screening factor, and $B(\kappa)$ is a salt-dependent coefficient that depends on κ and the ionic radius. This approach uses the Debye-Hückel theory to account for that fact that in a real ionic solution, each ion will, on average, be surrounded by oppositely charged counter-ions, and that this surrounding shell of counter-ions effectively reduces the electrostatic interactions of a given ion by screening its charge.

With this model, the authors studied the binding of two homeodomains (nucleic acid binding domains commonly found in transcription factors [82]) to DNA. The homeodomains were simulated both with and without their disordered N-terminal tails, and the authors found that the N-tails influence both the kinetics and thermodynamics of binding: their primary role was to anchor the protein to the DNA (by nonspecific electrostatic interactions), facilitating the formation of specific interactions between the homeodomain and DNA.

Another approach by the same group has been to use coarse-grained models to study the role of intrinsically disordered regions in the nonspecific binding of DNA, and more specifically, to identify how proteins use intrinsically disordered regions to search DNA sequences. For studying nonspecific protein–DNA interactions, native (Gō-type) contacts cannot be used to describe interactions between proteins and DNA, but otherwise the model employed is identical to that described in the previous section.

There are believed to be four distinct mechanisms by which proteins can search DNA [83]: sliding (the protein moves along the DNA backbone), hopping (the protein dissociates from the DNA and reassociates nearby), three-dimensional search (the protein dissociates from the DNA and reassociates in an unrelated location), and intersegment transfer (the protein transfers from one DNA chain to another). Vuzman et al. [84] looked again at the role of homeodomain N-tails and found that the presence of these tails increases the propensity for sliding while decreasing the rate of protein diffusion. Most interestingly, however, when the authors placed the proteins in a box with two parallel DNA molecules, they observed that the disordered tails are vital components of an intersegment transfer mechanism that they term the monkey bar mechanism (Figure 10.4). Reminiscent of a child progressing along monkey bars, the mechanism requires the transfer to take place through "a bridged intermediate in which the tail is located on one DNA, while the recognition helix is on the second DNA" [84]. Working on the same systems, Vuzman and Levy have also shown that the frequency with which intersegment transfer occurs is determined by the number and distribution of charged residues in the N-tails. Interestingly, a bioinformatics-based analysis reveals that the charge composition and distribution in the N-tails of 1384 wild-type homeodomains is in line with what the authors' simulations identify as the optimum for achieving intersegment transfer.

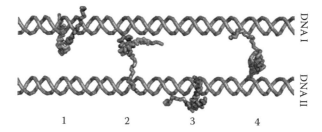

OUTLOOK

We have discussed how highly simplified coarse-grained models containing the essential features describing the binding of intrinsically disordered proteins can be used to address a number of specific questions regarding the binding mechanism of IDPs and the advantages that IDPs may have over structured proteins. Until now, it has often been the case that only limited information on binding mechanism was available experimentally, and so most computational studies have focused on addressing fairly generic questions, although in the context of specific protein systems. An exciting development in recent years has been the emergence of more residue-specific information on binding; for example, information from NMR relaxation-dispersion experiments [45] and mutational analysis, including Φ-value analysis [85–87], and the effects of environmental changes such as temperature and ionic screening on association [88].

REFERENCES

1. Fuxreiter, M., Simon, I. & Bondos, S. Dynamic protein-DNA recognition: Beyond what can be seen. *Trends Biochem. Sci.* **36**, 415–423 (2011).
2. Wright, P. E. & Dyson, H. J. Linking folding and binding. *Curr. Opin. Struct. Biol.* **19**, 31–38 (2009).
3. Dunker, A. K., Silman, I., Uversky, V. N. & Sussman, J. L. Function and structure of inherently disordered proteins. *Curr. Opin. Struct. Biol.* **18**, 756–764 (2008).
4. Fuxreiter, M. Fuzziness: Linking regulation to protein dynamics. *Mol. Biosyst.* **8**, 168–177 (2012).
5. Buljan, M. et al. Tissue-specific splicing of disordered segments that embed binding motifs rewires protein interaction networks. *Mol. Cell.* **46**, 871–883 (2012).

6. Piana, S., Klepeis, J. L. & Shaw, D. E. Assessing the accuracy of physical models used in protein-folding simulations: Quantitative evidence from long molecular dynamics simulations. *Curr. Opin. Struct. Biol.* **24**, 98–105 (2014).

7. Petrov, D. & Zagrovic, B. Are current atomistic force fields accurate enough to study proteins in crowded environments? *PLoS Comput. Biol.* **10**, e1003638 (2014).

8. Feig, M. et al. Performance comparison of generalized Born and Poisson methods in the calculation of electrostatic solvation energies for protein structures. *J. Comp. Chem.* **25**, 265–282 (2004).

9. Haberthür, U. & Caflisch, A. FACTS: Fast analytical continuum treatment of solvation. *J. Comput. Chem.* **29**, 701–705 (2007).

10. Lazaridis, T. & Karplus, M. Effective energy function for proteins in solution. *Proteins* **35**, 133–152 (1999).

11. Bottaro, S., Lindorff-Larsen, K. & Best, R. B. Variational optimization of an all-atom implicit solvent force field to match explicit solvent simulation data. *J. Chem. Theory Comput.* **9**, 5641–5652 (2013).

12. Vitalis, A. & Pappu, R. V. ABSINTH: A new continuum solvation model for simulations of polypeptides in aqueous solutions. *J. Comput. Chem.* **30**, 673–699 (2008).

13. Wuttke, R. et al. Temperature-dependent solvation modulates the dimensions of disordered proteins. *Proc. Natl. Acad. Sci. U. S. A.* **111**, 5213–5218 (2014).

14. Buchete, N.-V., Straub, J. E. & Thirumalai, D. Development of novel statistical potentials for protein fold recognition. *Curr. Opin. Struct. Biol.* **14**, 225–232 (2004).

15. Wolynes, P. G., Onuchic, J. N. & Thirumalai, D. Navigating the folding routes. *Science* **267**, 1619–1620 (1995).

16. Ueda, Y., Taketomi, H. & Gō, N. Studies on protein folding, unfolding and fluctuations by computer simulation. I. The effects of specific amino acid sequence represented by specific inter-unit interactions. *Int. J. Pept. Protein Res.* **7**, 445–459 (1975).

17. Gutin, A. M., Abkevich, V. I. & Shakhnovich, E. I. Evolution-like selection of fast-folding model proteins. *Proc. Natl. Acad. Sci. U.S.A.* **92**, 1282–1286 (1995).

18. Šali, A., Shakhnovich, E. I. & Karplus, M. How does a protein fold? *Nature* **369**, 248–251 (1994).

19. Shirai, N. C. & Kikuchi, M. Structural flexibility of intrinsically disordered proteins induces stepwise target recognition. *J. Chem. Phys.* **139**, 225103 (2013).

20. Bhattacherjee, A. & Wallin, S. Coupled folding-binding in a hydrophobic/polar protein model: Impact of synergistic folding and disordered flanks. *Biophys. J.* **102**, 569–578 (2012).

21. Abeln, S. & Frenkel, D. Disordered flanks prevent peptide aggregation. *PLoS Comput. Biol.* **4**, e1000241 (2008).

22. Hills, R. & Brooks, C. Insights from coarse-grained Go models for protein folding and dynamics. *Int. J. Mol. Sci.* **10**, 889–905 (2009).

23. Chan, H. S., Zhang, Z., Wallin, S. & Liu, Z. Cooperativity, local-nonlocal coupling, and nonnative interactions: Principles of protein folding from coarse-grained models. *Annu. Rev. Phys. Chem.* **62**, 301–326 (2011).

24. Knott, M. & Best, R. B. A preformed binding interface in the unbound ensemble of an intrinsically disordered protein: Evidence from molecular simulations. *PLoS Comput. Biol.* **8**, e1002605 (2012).

25. Ganguly, D. & Chen, J. Atomistic details of the disordered states of KID and pKID. Implications in coupled binding and folding. *J. Am. Chem. Soc.* **131**, 5214–5223 (2009).
26. Zhang, W., Ganguly, D. & Chen, J. Residual structures, conformational fluctuations, and electrostatic interactions in the synergistic folding of two intrinsically disordered proteins. *PLoS Comput. Biol.* **8**, e1002353 (2012).
27. Lindorff-Larsen, K., Trbovic, N., Maragakis, P., Piana, S. & Shaw, D. E. Structure and dynamics of an unfolded protein examined by molecular dynamics simulation. *J. Am. Chem. Soc.* **134**, 3787–3791 (2012).
28. Chen, J. Intrinsically disordered p53 extreme C-terminus binds to S100B(ββ) through "fly-casting." *J. Am. Chem. Soc.* **131**, 2088–2089 (2009).
29. Levy, Y., Wolynes, P. G. & Onuchic, J. N. Protein topology determines binding mechanism. *Proc. Natl. Acad. Sci. U. S. A.* **101**, 511–516 (2004).
30. Kaya, H. S. & Chan, H. S. Solvation effects and driving forces for protein thermodynamic and kinetic cooperativity: How adequate is native-centric topological modeling? *J. Mol. Biol.* **326**, 911–931 (2003).
31. Karanicolas, J. & Brooks, C. L. The origins of asymmetry in the folding transition states of protein L and protein G. *Protein Sci.* **11**, 2351–2361 (2002).
32. Clementi, C., Nymeyer, H. & Onuchic, J. N. Topological and energetic factors: What determines the structural details of the transition state ensemble and "en-route" intermediates for protein folding? An investigation for small globular proteins. *J. Mol. Biol.* **298**, 937–953 (2000).
33. Feig, M., Karanicolas, J. & Brooks, C. L., III, MMTSB Tool Set: Enhanced sampling and multiscale modeling methods for applications in structural biology. *J. Mol. Graph. Model.* **22**, 377–395 (2004).
34. Noel, J. K., Whitford, P. C., Sanbonmatsu, K. Y. & Onuchic, J. N. SMOG@ctbp: Simplified deployment of structure-based models in GROMACS. *Nucleic Acids Res.* **38**, W657–W661 (2010).
35. Miyazawa, S. & Jernigan, R. L. Residue–residue potentials with a favorable contact pair term and an unfavorable high packing density term, for simulation and threading. *J. Mol. Biol.* **256**, 623–644 (1996).
36. Mao, A. H., Crick, S. L., Vitalis, A., Chicoine, C. L. & Pappu, R. V. Net charge per residue modulates conformational ensembles of intrinsically disordered proteins. *Proc. Natl. Acad. Sci. U. S. A.* **107**, 8183–8188 (2010).
37. Müller-Späth, S. et al. Charge interactions can dominate the dimensions of intrinsically disordered proteins. *Proc. Natl. Acad. Sci. U. S. A.* **107**, 14609–14614 (2010).
38. De Sancho, D. & Best, R. B. Modulation of an IDP binding mechanism and rates by helix propensity and non-native interactions: Association of HIF1α with CBP. *Mol. Biosyst.* **8**, 256–267 (2012).
39. Ganguly, D. & Chen, J. Topology-based modeling of intrinsically disordered proteins: Balancing intrinsic folding and intermolecular interactions. *Proteins* **79**, 1251–1266 (2011).
40. Huang, Y. & Liu, Z. Nonnative interactions in coupled folding and binding processes of intrinsically disordered proteins. *PLoS One* **5**, e15375 (2010).
41. Kim, Y. C. & Hummer, G. Coarse-grained models for simulations of multiprotein complexes: Application to ubiquitin binding. *J. Mol. Biol.* **375**, 1416–1433 (2008).

42. Ganguly, D. et al. Electrostatically accelerated coupled binding and folding of intrinsically disordered proteins. *J. Mol. Biol.* **422**, 674–684 (2012).

43. Chu, X. et al. Importance of electrostatic interactions in the association of intrinsically disordered histone chaperone Chz1 and histone H2A.Z-H2B. *PLoS Comput. Biol.* **8**, e1002608 (2012).

44. Rogers, J. M., Steward, A. & Clarke, J. Folding and binding of an intrinsically disordered protein: Fast, but not "diffusion-limited." *J. Am. Chem. Soc.* **135**, 1415–1422 (2013).

45. Sugase, K., Dyson, H. J. & Wright, P. E. Mechanism of coupled folding and binding of an intrinsically disordered protein. *Nature* **447**, 1021–1025 (2007).

46. Sugase, K., Lansing, J. C., Dyson, H. J. & Wright, P. E. Tailoring relaxation dispersion experiments for fast-associating protein complexes. *J. Am. Chem. Soc.* **129**, 13406–13407 (2007).

47. Shoemaker, B. A., Portman, J. J. & Wolynes, P. G. Speeding molecular recognition by using the folding funnel: The fly-casting mechanism. *Proc. Natl. Acad. Sci. U. S. A.* **97**, 8868–8873 (2000).

48. Turjanski, A. G., Gutkind, J. S., Best, R. B. & Hummer, G. Binding-induced folding of a natively unstructured transcription factor. *PLoS Comput. Biol.* **4**, e1000060 (2008).

49. Huang, Y. & Liu, Z. Kinetic advantage of intrinsically disordered proteins in coupled folding-binding process: A critical assessment of the "fly-casting" mechanism. *J. Mol. Biol.* **393**, 1143–1159 (2009).

50. Zhou, H.-X. & Szabo, A. Enhancement of association rates by nonspecific binding to DNA and cell membranes. *Phys. Rev. Lett.* **93**, 178101 (2004).

51. Jacobs, W. M. & Frenkel, D. Predicting phase behaviour in multicomponent mixtures. *J. Chem. Phys.* **139**, 024108 (2013).

52. Uversky, V. N., Gillespie, J. R. & Fink, A. L. Why are "natively unfolded" proteins unstructured under physiologic conditions? *Proteins* **41**, 415–427 (2000).

53. Csermely, P., Palotai, R. & Nussinov, R. Induced fit, conformational selection and independent dynamic segments: An extended view of binding events. *Trends Biochem. Sci.* **35**, 539–546 (2010).

54. Ganguly, D., Zhang, W. & Chen, J. Synergistic folding of two intrinsically disordered proteins: Searching for conformational selection. *Mol. Biosyst.* **8**, 198–209 (2012).

55. Knott, M. & Best, R. B. Discriminating binding mechanisms of an intrinsically disordered protein via a multi-state coarse-grained model. *J. Chem. Phys.* **140**, 175102 (2014).

56. Best, R. B., Chen, Y.–G. & Hummer, G. Slow protein conformational dynamics from multiple experimental structures: The helix/sheet transition of Arc repressor. *Structure* **13**, 1755–1763 (2005).

57. Wang, J. et al. Multi-scaled explorations of binding-induced folding of intrinsically disordered protein inhibitor IA3 to its target enzyme. *PLoS Comput. Biol.* **7**, e1001118 (2011).

58. Muñoz, V. & Sanchez-Ruiz, J. M. Exploring protein-folding ensembles: A variable-barrier model for the analysis of equilibrium unfolding experiments. *Proc. Natl. Acad. Sci. U. S. A.* **101**, 17646–17651 (2004).

59. Naganathan, A. N. & Orozco, M. The native ensemble and folding of a protein molten-globule: Functional consequence of downhill folding. *J. Am. Chem. Soc.* **133**, 12154–12161 (2011).

60. Spolar, R. S. & Record, M. T., Jr. Coupling of local folding to site-specific binding of proteins to DNA. *Science* **263**, 777–784 (1994).

61. Dyson, H. J. & Wright, P. E. Coupling of folding and binding for unstructured proteins. *Curr. Opin. Struct. Biol.* **12**, 54–60 (2002).

62. Liu, J. et al. Intrinsic disorder in transcription factors. *Biochemistry* **45**, 6873–6888 (2006).

63. Latchman, D. S. Transcription factors: An overview. *Int. J. Biochem. Cell Biol.* **29**, 1305–1312 (1997).

64. Dunker, A. K. et al. Intrinsically disordered protein. *J. Mol. Graph. Model.* **19**, 26–59 (2001).

65. Lu, Y., Weers, B. & Stellwagen, N. C. DNA persistence length revisited. *Biopolymers* **61**, 261–275 (2002).

66. Savelyev, A. & Papoian, G. A. Chemically accurate coarse graining of double-stranded DNA. *Proc. Natl. Acad. Sci. U. S. A.* **107**, 20340–20345 (2010).

67. Savelyev, A. & Papoian, G. A. Molecular renormalization group coarse-graining of polymer chains: Application to double-stranded DNA. *Biophys. J.* **96**, 4044–4052 (2009).

68. Knotts, T. A., IV, Rathore, N., Schwartz, D. C. & de Pablo, J. J. A coarse grain model for DNA. *J. Chem. Phys.* **126**, 084901 (2007).

69. Hoang, T. X. & Cieplak, M. Molecular dynamics of folding of secondary structures in Go-type models of proteins. *J. Chem. Phys.* **112**, 6851–6862 (2000).

70. Sambriski, E. J., Schwartz, D. C. & de Pablo, J. J. A mesoscale model of DNA and its renaturation. *Biophys. J.* **96**, 1675–1690 (2009).

71. Morriss-Andrews, A., Rottler, J. & Plotkin, S. S. A systematically coarse-grained model for DNA and its predictions for persistence length, stacking, twist and chirality. *J. Chem. Phys.* **132**, 035105 (2010).

72. Patoyan, D. A., Savelyev, A. & Papoian, G. A. Recent successes in coarse-grained modeling of DNA. *WIRES Comput. Mol. Sci.* **3**, 69–83 (2013).

73. Dans, P. D., Zeida, A., Machado, M. R. & Pantano, S. A coarse grained model for atomic-detailed DNA simulations with explicit electrostatics. *J. Chem. Theory Comput.* **6**, 1711–1725 (2010).

74. DeMille, R. C., Cheatham, T. E., III & Molinero, V. A coarse-grained model of DNA with explicit solvation by water and ions. *J. Phys. Chem. B* **115**, 132–142 (2011).

75. DeMille, R. C. & Molinero, V. Coarse-grained ions without charges: Reproducing the solvation structure of NaCl in water using short-ranged potentials. *J. Chem. Phys.* **131**, 034107 (2009).

76. Givaty, O. & Levy, Y. Protein sliding along DNA: Dynamics and structural characterization. *J. Mol. Biol.* **385**, 1087–1097 (2009).

77. Vuzman, D. & Levy, Y. Intrinsically disordered regions as affinity tuners in protein-DNA interactions. *Mol. Biosyst.* **8**, 47–57 (2012).

78. Levy, Y., Onuchic, J. N. & Wolynes, P. G. Fly-casting in protein-DNA binding: Frustration between protein folding and electrostatics facilitates target recognition. *J. Am. Chem. Soc.* **129**, 738–739 (2007).

79. Mo, Y., Vaessen, B., Johnston, K. & Marmorstein, R. Structures of SAP-1 bound to DNA targets from the E74 and c-fos promoters: Insights into DNA sequence discrimination by Ets Proteins. *Mol. Cell.* **2**, 201–212 (1998).

80. Toth-Petroczy, A., Simon, I., Fuxreiter, M. & Levy, Y. Disordered tails of homeo-domains facilitate DNA recognition by providing a trade-off between folding and specific binding. *J. Am. Chem. Soc.* **131**, 15084–15085 (2009).

81. Crane-Robinson, C., Dragan, A. I. & Privalov, P. L. The extended arms of DNA-binding domains: A tale of tails. *Trends Biochem. Sci.* **31**, 547–552 (2007).

82. Gehring, W. Exploring the homeobox. *Gene* **135**, 215–221 (1993).

83. Berg, O. G., Winter, R. B. & von Hippel, P. H. Diffusion-driven mechanisms of protein translocation on nucleic-acids: 1. Models and theory. *Biochemistry* **20**, 6929–6948 (1981).

84. Vuzman, D. & Levy, Y. DNA search efficiency is modulated by charge composition and distribution in the intrinsically disordered tail. *Proc. Natl. Acad. Sci. U. S. A.* **107**, 21004–21009 (2010).

85. Bachmann, A., Wildemann, D., Praetorius, F., Fischer, G. & Kiefhaber, T. Mapping backbone and side-chain interactions in the transition state of a coupled protein folding and binding reaction. *Proc. Natl. Acad. Sci. U. S. A.* **108**, 3952–3957 (2011).

86. Hill, S. A., Kwa, L. G., Shammas, S. L., Lee, J. C. & Clarke, J. Mechanism of assembly of the non-covalent spectrin tetramerization domain from intrinsically disordered proteins. *J. Mol. Biol.* **426**, 21–35 (2014).

87. Rogers, J. M., Wong, C. T. & Clarke, J. Coupled folding and binding of the disordered protein PUMA does not require particular residual structure. *J. Am. Chem. Soc.* **136**, 5197–5200 (2014).

88. Shammas, S. L., Travis, A. J. & Clarke, J. Remarkably fast coupled folding and binding of the intrinsically disordered transactivation domain of cMyb to CBP KIX. *J. Phys. Chem. B* **117**, 13346–13356 (2013).

Chapter 11

Natural and Directed Evolution of Intrinsically Disordered Proteins

Tali H. Reingewertz and Eric J. Sundberg

CONTENTS

OVERVIEW

An increase in the proportion of intrinsically disordered proteins (IDPs) concomitant with an increase in organism complexity (Tompa, Dosztanyi et al. 2006; Ward, Sodhi et al. 2004) suggests an important role for disorder in evolution. Uncovering details regarding the origin and changes that occurred in the evolution of IDPs may shed light on the evolution of more ordered proteins and the rise of complex protein functions in humans. Moreover, such information will provide insight to the role of disorder, IDP functionality, and their mechanisms of action at the molecular level. Such information for IDPs, however, is difficult to obtain using traditional biochemical and biophysical methods. Directed evolution can mimic such combinatorial processes that occur in natural evolution. Pairing directed evolution with detailed analyses of the biophysical properties of IDPs, therefore, could bring new understanding to protein disorder and evolution, as well as indicate novel strategies for engineering proteins for biotechnological and therapeutic purposes. In this chapter, we provide an overview of recent research concerning the natural evolution of IDPs, as well as advances in the use of directed evolution methods to gain insight to IDP function.

NATURAL EVOLUTION OF INTRINSIC DISORDER IN PROTEINS

Tracing the origin of disorder on the evolutionary timescale can be addressed by following the conservation of intrinsically disordered regions (IDRs) of proteins, in addition to that of proteins that are entirely intrinsically disordered (IDPs). Evolutionary conservation can be expressed in the preservation of sequence, function, and structure (or lack thereof, in the case of IDPs), all of which are correlated. The unique set of traits that distinguish IDPs from structured proteins are vital to their functions and molecular mechanisms of action and may contribute to the understanding of their evolution. Here, we assess these different characteristics as they pertain to evolution and highlight differences between IDPs and more ordered proteins.

Defining Intrinsic Disorder in Proteins

To map the evolution of protein disorder we need to begin by orienting IDPs among the larger spectrum of all proteins, ordered, partially ordered, and disordered. Numerous definitions of IDPs exist in the literature. Some refer to IDPs as unique and distinguish a defined group of proteins with a specific set of traits and characteristics that are not exhibited by ordered proteins (Brown, Johnson et al. 2010; Dyson and Wright 2005; Wright and Dyson 1999). Within this larger group of IDPs/IDRs exist a wide variety of proteins that depend on the length of the IDR (e.g., long disordered regions vs. short loops), the functionality of the IDP (if known), and the degree of foldedness of the IDP (well-structured vs. unstructured loops) (Schlessinger, Liu et al. 2007). Other definitions place IDPs at one extreme of protein flexibility within the entire spectrum of order/disorder (Schlessinger, Schaefer et al. 2011; Vucetic, Brown et al. 2003). Defining IDPs as a distinct group from that of structured protein challenges a central dogma of structural biology that a protein's structure is necessary for its biological function (Wright and Dyson 1999) while placing IDPs and protein disorder within the spectrum of protein structure (albeit at an extreme end) is essentially consistent with the structure-function framework (Schlessinger, Schaefer et al. 2011). What all definitions share is that in the unbound state, IDPs do not fold into a defined and stable tertiary structure but are instead composed of an ensemble of conformations interchanging dynamically.

The structure of any protein can be expressed in terms of the degree of fluctuation in the average positions of backbone atoms, where ordered proteins undergo small fluctuations and disordered proteins exhibit large fluctuations (Brown, Johnson et al. 2010). An alternative definition describes the average time that a protein spends in one major conformation relative to all others (Palmer and Massi 2006). From an energetic standpoint, IDPs do not have a single global minimum in conformational space, but instead adopt several structural states separated by low-energy barriers (Tompa 2010). The lack of unity in the definition of protein disorder poses a barrier to tracing their evolutionary origins. However, regardless of definition, IDPs can be viewed as a distinct phenomenon in structural biology.

Conformational Diversity and Dynamic Transition among Conformations

Conformational diversity and flexibility are considered essential traits of IDPs, giving rise to some of their unique advantages/characteristics, such as their abilities to interact with multiple binding partners through the same region or at the same time, to use increased extended interaction interfaces in the extended conformation along with fine-tuning of these interactions through conformational changes (Dyson and Wright 2002). A special evolutionary advantage is conferred upon IDPs by their conformational diversity, which provides increased access to new functions related to new/different conformations (James and Tawfik 2003).

The conformational diversity of IDPs can be expressed as the ability to adopt different conformations in the unbound state, the flexibility/unstructured nature of some IDPs in the free state, and/or the ability to adopt different conformations in the bound state. Coupled folding and binding events present important implications for research on protein folding in general and the evolution of particular protein folds. At the functional level, it can teach us about misfolded proteins in disease, such as neurodegenerative disorders and the amyloidosis, or the role by which chaperones help proteins fold successfully (and in many cases are disordered themselves). At the evolutionary level, it has been suggested that conformational diversity may support the hypothesis of coevolution of protein fold and function, as a new fold can be selected from a new evolved function that serves as a selection pressure (James and Tawfik 2003).

The conformational diversity of IDPs is highly dependent on factors in the local environment that influence the dynamic transitions between conformations. This enables tight control of conformational transitions but at the same time may result in the ability of IDPs to better adapt to environmental changes. The composition of the solution in the cell is dynamic, varying in solutes, pH, ionic strength, and the concentration and proximity of binding partners, which can result in the existence of temporally constrained specific conformations (Johansson, Gudmundsson et al. 1998; Uversky 2002). Moreover, posttranslational modifications and proteolysis/degradation, both particularly frequent in IDPs, can change the conformation and/or select between different functions of the protein, as well as distinct regulation modes (Dyson and Wright 2005).

The link between conformational diversity and adaptation is exemplified by viruses that utilize intrinsic disorder throughout nearly every stage of their life cycles. Many viral proteins (especially those from ribonucleic acid [RNA] viruses) exhibit intrinsic disorder that is critical for their functions (Xue, Williams et al. 2011). Most viral proteins are enriched in short disordered regions, are loosely packed, and include a high fraction of polar residues and residues that are not involved in secondary structure (Tokuriki, Oldfield et al. 2009). Viruses must survive extreme changes in environment, ranging from the different organelles of their host cells to the extracellular environment, and often must adapt to entirely new host cells (Tokuriki, Oldfield et al. 2009; Xue, Williams et al. 2011). Viral proteins typically interact with numerous binding partners, either viral or host macromolecules, including deoxyribonucleic acid (DNA), RNA, proteins, and membranes. It is the intrinsically disordered nature of many viral proteins that enables viruses to adjust to rapid changes in their biological and physical environments (Tokuriki, Oldfield et al. 2009). Moreover, viruses exhibit a high level of mutation and adaptation compared to bacteria and eukaryotes (Drake, Charlesworth et al. 1998). In combination with multiple overlapping reading frames that results from the need to keep the viral genome compact (Reanney 1982), these mutations may affect more than one protein at a time. Intrinsic disorder may also provide a platform for enabling this type of adaptation.

On the evolutionary timescale, viruses evolved early, as they share homologous features regardless of the origin of their infected host cell (e.g., bacteria, archaea, and eukarya). Such homology does not exist between viral and cellular proteins (Forterre 2006). This may be a result of their need to rapidly and constantly adapt to changing environments that cells are not required to do. The HIV-1 Vif protein is an example of a viral protein that is believed to have evolved as a result of the need to counteract a defense mechanism presented by the host cell (Strebel, Daugherty et al. 1987). A major component of the human innate immune response to human immunodeficiency virus type 1 (HIV-1) is the APOBEC-3G protein, which inserts dC→dU mutations during viral DNA synthesis from RNA. Vif inhibits the action of APOBEC-3G through several distinct mechanisms that involve interactions with multiple binding partners (Lake, Carr et al. 2003). Many of these interactions are mediated through the C-terminal domain (CTD, residues 141–192) of Vif, which has been shown to be intrinsically disordered in the unbound state (Reingewertz, Benyamini et al. 2009) while undergoing conformational changes in the bound state (Bergeron, Huthoff et al. 2010; Reingewertz, Benyamini et al. 2009). In addition to direct binding to APOBEC-3G, Vif uses its CTD to interact with numerous other binding partners, including three components of E3 ubiquitination complex devised to target APOBEC-3G for proteasomal degradation; it binds the viral reverse transcriptase and Gag protein, interacts with the inner side of cells membrane, self-oligomerizes to undergo posttranslational modifications (i.e., phosphorylation) and protease cleavage in that region (summarized in Reingewertz, Shalev et al. 2011). The disordered nature of the Vif CTD presents a viable solution for its need to exhibit multiple functional and binding abilities that require conformational diversity within the constraints of the viral genome length and as a product of the selection pressure imposed by APOBEC-3G on viral replication.

Complex Interactions and Functional Patterns: Promiscuity and Moonlighting

Like all proteins, many of the functions of IDPs are accomplished via binding of other macromolecules. One of the most important outcomes of conformational diversity is the expanded complexity of the available modes of molecular recognition, which, in turn, leads to increased functional diversity. Several types of interactions/functions for IDPs have been defined in the literature. These types of complex molecular recognition strategies are described in many ordered proteins and not only IDPs, including enzymes and other structural or regulatory proteins (Copley 2003; James and Tawfik 2003). In *moonlighting*, an IDP performs distinct functions using different or overlapping binding surfaces and involving interactions with different ligands (Copley 2003; Tompa, Szasz et al. 2005). Although the extent of each function may be different, usually all functions in moonlighting have a known *in vivo* role (Jeffery 1999). Another complex interaction mechanism is *cross reactivity*, in which the IDP has an extended binding diversity (i.e., it exerts a similar function by binding to closely related biological targets). For example, the cyclin-dependent kinase (CDK) inhibitor p21 binds to and inhibits several CDKs through an IDR in a coupled binding and folding event (Kriwacki, Hengst et al. 1996). *Promiscuity* refers to an IDP performing different functions using the same binding interface or active site and through interactions with different ligands (Copley 2003; Cumberworth, Lamour et al. 2013). In this case, the same region exerts alternative binding modes that proceed through different mechanisms using the same binding region (for example, the Vif CTD; see the section titled "Conformational Diversity and Dynamic Transition among Conformations"). The lack of a defined ordered conformation and the ability to adopt different conformations in the unbound state facilitate such complex interaction modes in IDPs and endows IDPs with the conformational flexibility that is often needed in each of the bound states.

The evolutionary importance of such complex interaction mechanisms is that they endow a single IDP with the ability to perform different, or even opposing, functions, which could serve as a selection criterion in the evolution of new protein functions and possibly new structures. James and Tawfik refer to two of these important traits of IDPs, conformational diversity and promiscuity, as evolvability traits that enable proteins to rapidly evolve new functions (James and Tawfik 2003). Furthermore, they suggest that these characteristics form the basis of the coevolution of protein fold and function as opposed to the alternative scenario in which fold evolved before function—a scenario that lacks any selective advantage/driving force (James and Tawfik 2003). IDPs that undergo gene duplication, mutation, and selection can easily evolve into new proteins with expanded functions, either through natural evolution or directed evolution. Moreover, the conformational flexibility of IDPs by itself may be used to create or enhance promiscuity, thus introducing new functions from which to select. A schematic representation of the factors involved in the evolution of IDPs is presented in Figure 11.1. Understanding how to control these molecular recognition mechanisms of IDPs will lead to their increased use as engineered proteins for therapeutic and technological purposes.

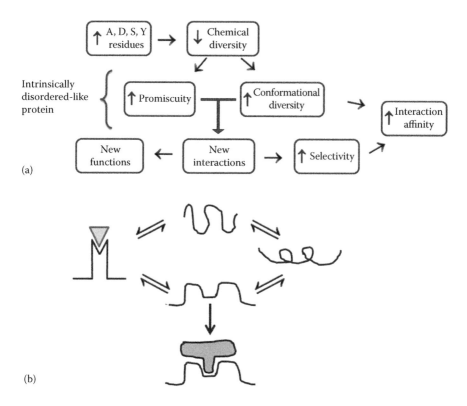

(a)

(b)

Figure 11.1 Causality of factors governing changes in directed evolution. (a) A flow-chart describing the changes in traits and factors that contribute to engineering of new interaction and functions. Decreasing the chemical diversity of a given sequence by increasing the content of A, D, S, Y residues increases the conformational diversity of the proteins and hence its functional promiscuity. These traits that are found in IDPs can give rise to the creation of a new interaction that may result in a new function that can be further improved. (b) Schematic representation of the conformational changes that may lead to the formation of new interactions and functions.

Sequence Characteristics

The disordered nature of IDPs, similar to the structured nature of globular proteins, is presented in their primary amino acid sequence (Dunker, Lawson et al. 2001; Lise and Jones 2005; Romero, Obradovic et al. 2001; Uversky, Gillespie et al. 2000). In general, IDP sequences are of relatively low complexity with low variation and frequent repetition of residues. IDP sequences also typically have a high content of small and/or polar/charged amino acid residues (E, K, R, G, Q, S, P, and A) as opposed to the bulky hydrophobic and aromatic residues (I, L,

V, W, F, Y, and C) that are preferentially found in protein cores. Based on these characteristics, numerous computational algorithms have been developed that accurately predict regions of intrinsic protein disorder (Ferron, Longhi et al. 2006; Radivojac, Iakoucheva et al. 2007). As the sequential characteristics of IDPs differ substantially from those of ordered proteins, the differences provide insights to the evolutionary processes that IDPs have undergone.

Repeats and Low-Complexity Regions

Approximately half of the human genome is composed of repetitive elements and although they appear most frequently in noncoding regions they are also found in genes encoding proteins (Heringa 1998; Marcotte, Pellegrini et al. 1999; Karlin, Brocchieri et al. 2002). Repeats are three times more frequent in eukaryotic than in prokaryotic proteins (Heringa 1998) and are much more prevalent in IDPs (39%) than in all Swiss-Prot (14%), yeast (18%), or human (28%) proteins (Tompa 2003). Expansion of internal repeat regions, followed by positive selection, has been suggested to be one of the mechanisms of increasing the genetic information with increased complexity of organisms, as part of the evolution of the IDPs (Tompa 2003).

The lower sequence complexity of IDPs, which indicates low variation and frequent repetition of residues, may have stemmed from the lack of structural constraints aimed to form a tightly packed core (Romero, Obradovic et al. 2001). Thus, IDPs are free to contain fewer residues, which is manifested in lower complexity and increased repeats. However, low complexity is not analogous to randomness. Accordingly, *in silico* sequences assembled randomly result in a lower abundance of long IDPs than is observed in nature. Moreover, random mutagenesis of long intrinsically disordered regions *in silico* resulted in reduced intrinsic disorder. This data indicates a possible evolutionary pressure toward intrinsic protein disorder (Schaefer, Schlessinger et al. 2010; Schlessinger, Schaefer et al. 2011).

Low sequence complexity does impinge, however, on the ability to compare IDP sequences to those of other IDPs and folded proteins, as sequence alignment methods typically exclude regions with low sequence complexity. Thus, when comparing or predicting disorder from sequence, the fact that the sequence is simple and highly repetitive can result in reduced statistical significance and increased false-positive rates. The shorter the sequence segment is, the higher the probability of such misleading comparisons.

Mutational Analysis

Protein evolution is driven by insertions, deletions, amino acid substitutions, and insertion/deletion of entire protein domains (Benner, Cohen et al. 1993; Grishin 2001; Pascarella and Argos 1992). Mutations that impair a needed function, conformation, or the stability of the protein will generally be selected against. IDPs show a distinct pattern of mutation and substitution relative to structured proteins. Although insertion and deletion (indel) mutations are less frequent than substitution mutations, they can impart significant structural changes and have a potentially larger impact on evolutionary processes (Grishin 2001). It was suggested more than three decades ago that during evolution the majority of protein loops in ancestral structures were targeted for indel mutations (Pascarella and Argos 1992). More recently, several mutational analysis studies revealed that long indels in homologous proteins include disorder promoting residues, and the longer the indel is, the more disordered residues it includes (Light, Sagit et al. 2013). This implies that IDPs can tolerate higher indel mutation rates than can structured proteins. Moreover, improved sequence alignment of disordered regions showed that optimization of gap penalties improves the alignment of disordered proteins (Radivojac, Obradovic et al. 2002).

The prevalence of polar and charged disorder-promoting residues in IDPs and the relative infrequency of bulky hydrophobic and aromatic residues are well established (Lise and Jones 2005; Radivojac, Iakoucheva et al. 2007; Romero, Obradovic et al. 2001). It has been suggested (Brown, Takayama et al. 2002) that these residues have a relatively higher substitution rate, as they tend to be exposed to solvent that is generally less conserved (Goldman, Thorne et al. 1998), suggesting that IDPs may typically have faster evolutionary rates. Relative frequencies of amino acid substitution matrices depend on the pool of proteins examined and is based on the evolutionary pattern of the proteins, which can be confounding. Differences in substitution frequency matrices result from analyzing different kinds of residues derived from distinct structural elements, such as α-helixes, β-sheets, coils, and turns (Goldman, Thorne et al. 1998). These results indicate not only the constraints that structure imposes on evolution but also the importance of the specific fold or secondary structure involved. Calculation of a substitution matrix for disordered proteins has been performed and compared to a substitution matrix of ordered proteins, better demonstrating differences between ordered and disordered proteins (Brown, Johnson et al. 2010). The results show that most amino acids in IDPs are less conserved than their counterparts in ordered proteins and that point mutations in disordered regions are more accepted. This implies that, in general, IDPs have less evolutionary constraints compared to ordered proteins.

Prediction of Function from the Sequence of IDPs

Another way to assess/address the relation between the sequence and function of IDPs is by predicting function from sequence. A link between patterns of intrinsic disorder and protein function has been suggested, implying a level of conservation of the sequence of disordered regions (Lobley, Swindells et al. 2007). When included in function prediction algorithms, disorder patterns improve prediction accuracies for gene ontology (GO) categories related to molecular recognition and biological processes. A recent study used evolutionary conservation of short linear motifs, often found in interaction interfaces of IDPs, as a tool to recognize these motifs and infer their function (Davey, Cowan et al. 2012). These short linear motifs include less than four residues in a stretch of approximately 2–10 residues and often play vital regulatory roles, such as for ligand binding, in modification sites and as target signals. Although these short motifs are under weak evolutionary constraints as they lack the need to conserve a folded structure, they are strongly constrained functionally (Davey, Shields et al. 2009). This trait enables them to recognize the more functionally important and conserved residues versus adjacent nonfunctional residues.

The Evolution of IDPs

Determining the evolutionary origin of intrinsic protein disorder would shed light on the role and mechanisms of action of disordered protein regions. An increase in the disordered fraction of the genome correlated to increased complexity of the species indicates that organisms must have evolved toward at least some degree of intrinsic disorder. However, it is not necessarily the case that disordered protein regions originated from ordered protein regions. Such a scenario would involve gene duplication and mutation of ordered proteins into disordered ones, which is unrealistic as structured proteins have high functional selection constraints that would disfavor such transitions (Tompa 2010). Instead several mechanisms have been suggested:

1. Gene duplication and domain rearrangements may account for the emergence of some disordered domains (Tompa 2010; Tompa, Fuxreiter et al. 2009). Such disordered domains are distinguished from short disordered recognition motifs and have a functional and evolutionary origin (Tompa, Fuxreiter et al. 2009). Several protein families share a common disordered domain, while the rest of the protein can be extremely variable. The shared domains were probably generated by duplication and relocation by gene rearrangement.

2. IDPs and their unique traits of promiscuity and conformational diversity were suggested to facilitate the evolution of new proteins from old ones (see the section titled "Complex Interactions and Functional Patterns: Promiscuity and Moonlighting") (James and Tawfik 2003; Tokuriki and Tawfik 2009a). The ability to select for evolved properties might be favored by itself, thus undergoing evolutionary selection toward promiscuity and conformational diversity may increase the content of disordered regions in the genome. It has been demonstrated how evolvability can be a selectable trait of Darwinian selection

(Earl and Deem 2004). In this case, IDPs may evolve from other IDPs, expanding disorder in the genome.

3. As discussed in the section titled "Conformational Diversity and Dynamic Transition among Conformations," intrinsic disorder is prevalent in the viral proteome and confers several advantages to viruses (Tokuriki, Oldfield et al. 2009; Xue, Williams et al. 2011). Together with the hypothesis that viruses were involved in the evolution of the cell (Claverie 2006), it is possible that viruses have contributed to the emergence of disorder in evolution. It is already accepted that viruses possess an evolutionary role as they can implement horizontal gene transfer between distinct organisms without being the offspring of that organism, thus increasing genetic diversity. For example, during mitochondrial evolution viral proteins replaced bacterial DNA/RNA polymerases and possibly other proteins (Forterre 2006), and influenced bacterial chromosome evolution (Canchaya, Fournous et al. 2003).

An increase in the number of disordered segments in the genome of several organisms might have resulted from an elevated need for disorder, for example, to evolve new and complexed functions. The fact that there had been an increase in intrinsic protein disorder over evolutionary time indicates the importance of these regions, or else their emergence and expansion would not have been favored.

Factors Restricting the Evolution of IDPs

If IDPs have such functional advantages relative to ordered proteins and higher organisms have evolved toward increased fractions of their proteomes represented as IDPs, why have humans not evolved to have an even larger number of IDPs? Perhaps this is because evolution and selection criteria are influenced by biophysical processes related to the encoded protein, such as protein synthesis, folding, and interactions (Wilke and Drummond 2010). Protein stability is a major factor that may determine the probability of a particular protein to evolve. Certainly, ordered proteins are required and/or are preferred for certain molecular mechanisms, but there is a wide range of additional factors that may restrict the evolution of IDPs. Indeed, organisms that evolve to have increased numbers of IDPs and/or more regions of intrinsic disorder within a given protein must deal with certain challenges, such as the fact that overexpression of several IDPs can be toxic to the cell (Uversky, Oldfield et al. 2008), and thus their expression must be tightly regulated. An increased abundance of IDPs in a cell can result in undesirable interactions. Because IDPs are often involved with multiple interactions and processes in the cell, their overabundance could potentially disrupt the exquisite balance in signaling networks. In fact, the intrinsic disorder content of a protein was found to be an important determinant in predicting dosage sensitivity and strongly associated with dosage sensitivity oncogenes (Vavouri, Semple et al. 2009). A reason for this toxicity was suggested to be increased promiscuous interactions at the elevated concentrations resulting from overexpression. Another reason that may cause toxicity to cells in a dose-dependent manner is the accumulation of aggregation-prone proteins, which include IDPs with Q/N-rich stretches associated with β-aggregation (Gsponer and Babu 2012), even though IDPs are usually soluble and may even serve as linkers added to increase protein solubility (Santner, Croy et al. 2012). No matter the reason for their toxicity, the harmful potential of IDPs has likely been one of their restricting factors in evolution.

IDPs are frequently degraded rapidly, either by ubiquitin-dependent or ubiquitin-independent proteasomal degradation, in which a disordered initiation site is sometimes required (Prakash, Tian et al. 2004). IDPs are extremely susceptible to proteolytic degradation *in vitro* as well. Fast degradation is a particularly useful way in which to control the level of a given protein, for example in signal transduction. A correlation between protein half-life and disorder has been suggested and proposed to play a key signaling role for degradation in many cases (Tompa, Prilusky et al. 2008). Interestingly, a protection mechanism was recently suggested in which "nanny" proteins protect newly synthesized IDPs from degradation (Tsvetkov, Reuven et al. 2009). Still, rapid degradation may impose a constraint on the evolution of IDPs.

The level of IDPs inside the cell is tightly regulated on several levels to make sure that IDPs are present in the needed amount at the appropriate time and place. Babu and coworkers (Gsponer, Futschik et al. 2008) have demonstrated evolutionarily conserved tight control of IDPs through comprehensive analysis of whole proteins in the genome of *Saccharomyces cerevisiae* (*S. cerevisiae*) both at the level of production and clearance of most IDPs. The regulation of IDPs in the cell is manifested in myriad ways, such as by degradation at the protein level, control at the level of the mRNA, posttranslational modifications to regulate the activity of the IDP, control of the activity IDPs through specific interactions, and specific cellular localizations.

Are IDPs Evolutionary Conserved or Evolving Rapidly?

It is generally accepted that protein structure is more conserved than sequence (Chothia and Lesk 1986; Doolittle 1981). In the traditional view, evolutionary constraints are mainly driven by functional constraints that lead to a conserved functional structure, which leads to a conserved sequence (Tatusov, Koonin et al. 1997). However, in the event that function is dependent on the disordered nature of the protein, would it still be conserved? Should we expect sequence conservation or evolutionary maintenance of flexibility? Will evolution conserve instead the equilibrium between the different conformations that may exist in a given IDP? Would this result in sequence conservation or not? Would sequence conservation in that case even be a reliable indication for evolutionary conservation?

Structured proteins have structural constraints that correspond to evolutionary restraints. As with any protein, IDPs may have functional constraints, but these may be less restraining than the structural constraints. From the sequence characteristics of IDPs, we see that although the sequence is not necessarily random it can accommodate mutation more easily. The rate of evolutionary change of IDPs is considered to be relatively fast, and thus IDPs could be defined as less conserved. Still, some analyses report short disordered regions as conserved (Chen, Romero et al. 2006a,b) and others even report conservation of long disordered domains (Tompa, Fuxreiter et al. 2009).

IDPs Are Fast Evolving, Thus Not Conserved

The rate of protein evolution may encompass a wide range of related factors. It has been shown that binding surfaces evolve slower than those not involved in binding (Eames and Kortemme 2007), which can be related to the relative solvent accessibility of the surface (Goldman, Thorne et al. 1998). Another interesting correlation showed that proteins with many interaction partners evolve more slowly and hence are more conserved (Fraser, Hirsh et al. 2002). A correlation with mRNA levels implies that highly expressed proteins were generally evolving most slowly (Drummond, Bloom et al. 2005). Drawing conclusions from these correlations about the conservation of IDPs is complicated, however, as IDPs often have multiple binding partners and their mRNA levels are often tightly regulated (Gsponer, Futschik et al. 2008). Such a general conclusion regarding IDPs is still missing.

Most studies that measure evolution rates have used the rate of change of the sequence throughout entire proteins, thus an average of the residues in any given protein region. Recently, the evolution rate of proteins was assessed by analyzing the evolution rate per residue position within a protein and its relation to that of the overall per protein rate (Toth-Petroczy and Tawfik 2011). Three main types of residues were deduced by this analysis: (1) slowly evolving residues that include mainly core residues, (2) quickly evolving residues that correspond to surface residues, and (3) residues evolving very fast, which include residues from disordered regions. Thus, disordered regions may be less conserved even compared to fast-evolving residues on the surface of the proteins. This study also found a correlation between the surface and core mutations. Although the relative rates of surface and core mutations are not constant, sufficient change in the surface will eventually lead to core mutations as well and the more constrained/conserved the surface is, the more conserved its core region will be. This data demonstrates the complexity in attempting to deduce protein properties that correlate with evolutionary rates of proteins. Such properties may not be independent and may be influenced by other factors (Wilke and Drummond 2010).

Other studies also suggest that disordered regions are not conserved. By sampling the evolution rate of proteins that include disordered regions ≥ 30 and ordered regions, it has been shown that disordered regions evolve more rapidly than ordered regions within the same protein (Brown, Takayama et al. 2002). By calculating the substitution matrix of disordered proteins, Brown et al. (Brown, Johnson et al. 2010) demonstrated that disordered residues are less conserved than ordered ones, and thus that IDPs are less conserved, supporting the general notion that the IDPs are less conserved (Brown, Johnson et al. 2011).

Sequence Conservation of IDPs

Early attempts to create a database of conserved predicted disordered regions resulted in numerous short sequences of about 20–30 amino acids (Chen, Romero et al. 2006a, b). Several functions were inferred as associated with conserved disordered domains. In general, conserved predicted disorder was much less common than observed predicted disorder. The evolutionary conservation of short linear motifs in disordered regions was addressed more recently (Davey, Shields et al. 2009), showing that although they are under weak evolutionary constraints, they are under strong functional constraints and thus can be used as a fingerprint for functional motifs (Davey, Cowan et al. 2012).

Conservation of longer disordered domains has been addressed by defining groups of IDRs longer than 20–30 residues as a new type of protein domain (Tompa, Fuxreiter et al. 2009). This study focused on several examples and shows that these domains have conserved functional, evolutionary, and disordered natures. The most common function of these disordered domains is the recognition mechanism. A more structurally oriented approach analyzed loopy proteins, which are regions (>70 AA) that have no or very low content (less than 12%) of regular secondary structure (Liu, Tan et al. 2002). These regions differ from loops found in the protein data bank (PDB) and were grouped into four types, including connecting loops, loopy ends, wrapping loops, and loopy domains. Specific functions and sequence characteristics were assigned to these regions, which are IDPs by the new accepted definition. The study showed that these long disordered loopy regions are evolutionary conserved at least as much as other nonloopy segments in the same proteins.

Other Approaches to Assess Conservation of Disorder

Conservation of disorder with functional conservation of IDPs was demonstrated in spite of low sequence conservation (Toth-Petroczy, Meszaros et al.

2008). Using a unique analysis, evaluation of the conservation of disordered regions was based on three measures: (1) similarity of the predicted disorder profile in different prediction conditions, (2) conservation of amino acid propensity (order versus disorder promoting), and (3) an overlap between disordered and ordered regions. The multiple sequence alignments included low-complexity regions as well. This new approach was demonstrated on three homologues of the Mediator complex of transcription regulation that exhibit low sequence conservation, showing conservation of disorder and function.

In another study, nuclear magnetic resonance (NMR) analysis was used to demonstrate the evolutionary conservation of the dynamic behavior of intrinsically unstructured linker domains in replication protein A homologues that showed no significant sequence similarity (Daughdrill, Narayanaswami et al. 2007). The dynamic behavior had been shown to be under selection even in the disordered regions that are not involved in any interaction. Thus, the constraints are from the flexibility and not binding determinants.

Tracking the phylogenetic tree of two protein families that include IDRs and are composed from paralogues (i.e., distantly related copies) due to gene duplication highlighted another aspect of disorder conservation (Siltberg-Liberles 2011). Computational predictions of the conformational flexibility (i.e., disorder propensity) and secondary structure element propensities of the proteins in each family demonstrate that these propensities change in concert, suggesting that disordered regions are not universally conserved but interconvert with secondary structure conservation.

Concluding Remarks

While some studies argue that IDPs have a certain level of sequence conservation, other studies claim that not only are IDPs not conserved, but that they are evolving rapidly. The truth probably lies somewhere in between. IDPs are diverse in size, function, and flexibility. Disordered domains with definite patterns in evolution result in high levels of conservation. Conversely, short disordered regions aimed at maintaining flexibility and not involved in any function could be less conserved or even rapidly evolving. Thus, conservation of a disordered region may vary as well. Moreover, we potentially require a different scale of conservation when we assess the evolution of IDPs. It is possible that the conversation concerning IDPs lies at the structural and/or functional levels (conservation of flexibility or function) while not being obvious at the sequence level. It might be that we will need to assess disorder conservation relative to other disordered regions and not compare it to ordered proteins. In that case, a new criterion for the differentiation between types of IDPs might be the level of their sequence conservation, which will most probably be related to any functional role they have.

DIRECTED EVOLUTION OF IDPs

Deciphering the mechanisms of action of IDPs at the molecular level has proved challenging using common biochemical and biophysical methods. First, IDPs tend to be less abundant in the cell and their levels are typically tightly regulated (Gsponer, Futschik et al. 2008). IDPs do not lend themselves to crystallization, and in proteins that do crystallize, IDRs most often have scant or missing electron density. A number of other biophysical methods, including circular dichroism (CD) and NMR spectroscopies, and other spectroscopic and hydrodynamic techniques, have been employed historically to analyze IDPs. However, these methods are largely descriptive, characterizing the disordered nature—they are used to catalog the extent of protein disorder in IDPs, the structural changes induced upon binding, and those resulting from distinct environments. Determining the exact mechanism of action of IDPs at the molecular level is not a straightforward task and this challenge calls for new tools, or more likely, a novel combination of techniques.

Directed evolution is a useful method by which to introduce a controlled perturbation in a given system, for instance, by changing the disordered nature of the IDP or modulating a binding interface with single or multiple binding partners. Monitoring the change in the conformation or the interaction would enable going back to the original position of the system before the change, thus gathering information on the disordered state. Combining directed evolution with structural (crystallographic or NMR) and biophysical methods (thermodynamic/kinetic analysis) could prove to be a valuable tool in the case of IDPs when trying to resolve complex interaction patterns driven by conformational diversity.

Directed Evolution: An Overview

Aimed to mimic natural evolution, directed evolution changes a characteristic of a protein by applying a selective pressure in a way that will favor the change and create an improved variant. In natural evolution, mutations occur spontaneously, resulting in variants with favorable properties that would promote the survival of their host organisms. Subsequently, these variants undergo further mutation and selection resulting in evolution over generations, favoring the beneficial mutations. Directed evolution is a lab technique that operates in a similar way, where the protein of interest is subjected to multiple rounds of random/combinatorial mutations and selection according to a selected criterion, all on a much shorter timescale. In a broader respect, directed evolution represents a retrospective way to analyze changes in protein–protein interactions; first, random mutations are created and then their effects analyzed, as opposed to the rational design of mutations based on proposed mechanisms (Hoseki, Okamoto et al. 2003).

Directed evolution was originally employed to create proteins or enzymes with improved stability while maintaining the function of the protein (Arnold 2001). It was thought that increasing the rigidity of the protein will increase its stability (Hoseki, Okamoto et al. 2003). Schreiber et al. (Schreiber, Buckle et al. 1994), however, suggested that the functionality of the protein might represent a much stronger evolutionary pressure than the stability of the protein and concluded that proteins evolve primarily to optimize their function *in vivo* and that there is a lower selective pressure to increase stability above a certain threshold.

The selection criteria can vary on the basis of the desired knowledge we wish to gain or the change we wish to introduce to the protein. These include improved or new functions, increased interaction affinity, extended promiscuity, changes in selectivity/specificity, increased stability, targeted structural changes, and increased rigidity, among others. A dual selection pressure can be applied to a system, as in the case of Levin et al. (Levin, Dym et al. 2009), in which selection for both increased affinity and altered selectivity (i.e., increased affinity to a different ligand and improved selectivity against the original ligand) was made. Interestingly, consecutive attempts of selection for increased affinity and then improved selectivity failed. Only when selecting for both traits at the same time were they able to reduce cross reactivity and thus increase selectivity.

Each experimental design in the directed evolution process includes three major steps, as presented in Figure 11.2 (Bonsor and Sundberg 2011):

1. *Library generation*— randomly mutating a gene of interest in selected residues to create a library comprised of variants of the protein, generated by error-prone polymerase chain reaction (PCR), DNA shuffling, and/or PCR cassette mutagenesis.

2. *Selection*—screening the generated library against the selection criterion. The screening method depends on the selected trait that is being modified, in which the variants with the improved trait are selected. For example, when trying to monitor increased binding affinity, selection is preceded by expressing and displaying the protein using one of several display systems, including phage, yeast, bacterial display, ribosome, messenger RNA (mRNA) or *in vitro* compartmentalization. Subsequently, binding to the target protein(s) is tested. Other selection procedures may be performed *in vivo* or *in vitro* according to the examined character.

3. *Amplification*—enrichment of selected variants either by generating the resulting template by reverse transcription and PCR or growing/infecting the cells with the improved variants.

The selected mutants/variants are then subjected to iterative rounds of mutation and selection until there is no additional improvement of the selected trait.

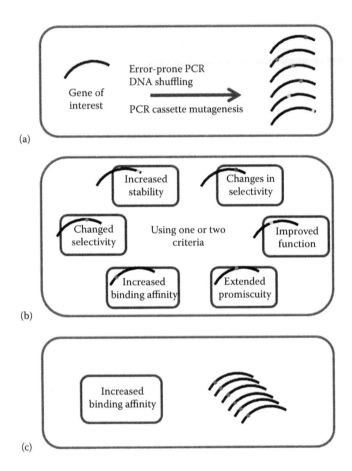

Figure 11.2   Steps in a directed evolution method. The technique is comprised of three steps: (a) library generation, (b) selection, and (c) amplification.

Characterization of resulting variants using biophysical methods may be performed during or after the directed evolution process.

Analysis of Protein–Protein Interactions

Using increased interaction affinity as a selected criterion in the directed evolution of proteins enables a detailed understanding of molecular recognition processes that are important in nearly all aspects of life (Bonsor and Sundberg 2011). Studying the affinity maturation process of an interaction system can provide information concerning the mechanism of the interaction and the molecular rules governing the interaction and may result in protein-based products to be used for therapeutic or technological purposes. Monitoring the changes between the free and bound states concomitant with increased affinity may shed light on the contribution and role of each interacting residue as well as the selectivity and specificity interchange in the interaction. The optimal scenario is to solve and compare the structures of the free and bound species involved in the complex. In the case of IDPs the structure of the unbound and sometimes even the bound state cannot be fully determined.

The binding pattern of IDPs may be complex compared with most structured proteins as presented above (see the section titled "Complex Interactions and Functional Patterns: Promiscuity and Moonlighting"). In the case of an interaction between two well-structured proteins (i.e., by a lock and key mechanism), the interaction is typically formed between specific binding points found in defined conformations. Defining the interaction interface in IDPs is inherently more challenging. In the case of IDPs, the interaction interface may include residues that contribute to a particular interaction, while the same residues, and potentially others as well, might also contribute to a distinct interaction or change the stability of the unbound protein. The exact position of the binding determinants may be critical in some cases, while in other cases the relative proximity of the binding residues to one another might be important. The binding determinants can reside far away from each other, and without the ability of having a detailed structure it may not be possible to anticipate the interactions between them. Energetically, multiple binding sites may contribute to high affinity for the whole protein region with a possible allosteric effect, and thus it might be hard to recognize the contribution of each binding determinant. Moreover, individual residues may contribute to the interaction to a different extent, and thus, changing one interaction may significantly influence other interactions mediated through the same region.

Assessing the role of intrinsically disordered regions in a single protein–protein interaction is also not straightforward. Experimentally calculating the energetic contribution of a disordered loop that was modified by directed evolution may reveal differences not addressed by prediction algorithms used for computational analysis. Improving the predictive ability of computational algorithms to assess the affinity and specificity of protein–protein interactions would be of great value in designing and engineering new proteins. Current prediction methods are based primarily on static interactions and rely predominantly on additive energetic contributions from binding residues. Directed evolution can provide model systems that could improve these algorithms to include intrinsic disorder and energetic cooperativity.

The role of the disordered region might impact the structure and energetics of binding beyond the direct interaction. For example, an IDR may play a conformational role by hindering or positioning the interacting interfaces in the correct orientation needed for complex formation. In that case, mutations outside the binding interface could also influence the affinity of the interaction by modulating a bound conformation of the binding site or an active site, or by changing the complex arrangement needed for the ultimate complex structure. Another challenge imposed by disorder in the study of increased affinity is assessing the relationship between specificity, selectivity, and affinity. Within an interface the forces responsible for driving affinity increases can also drive binding promiscuity, and thus, there may not be a simple, correlative relationship between affinity and specificity in protein–protein interactions involving intrinsic disorder.

Introducing Disorder to Study Natural and Evolutionary Mechanisms

Disordered regions were suggested to be a possible origin/mechanism for protein evolution by presenting the conformational diversity and promiscuity as a starting point in the evolution of new functions and folds (Tompa 2010; Tompa, Fuxreiter et al. 2009) (see point 2 in the section titled "The Evolution of IDPs"). Directed evolution can be employed to mimic natural evolution and enable us to gain insight to the creation of new proteins. Using IDPs as scaffolds for modification may provide an important tool in revealing/validating this possible evolutionary mechanism. Flexibility and dynamism are vital properties of all aspects of proteins including enzymatic and other catalytic activities, protein–protein interactions, and the formation of multicomponent assemblies. Monitoring controlled change in the disordered content of a protein may enable a better understanding of the role of dynamism and disorder in protein mechanism and evolution.

How then should we use directed evolution to study IDPs? Should we increase disorder or stabilize a desired conformation? A reduction in the chemical diversity of a protein can lead to increases in conformational diversity (Koide, Gilbreth et al. 2007) that can drive the creation of novel protein structures and functions (Figure 11.1). A useful way to do so is through the process of directed evolution by restricting the codon usage to four (e.g., Ala, Asp, Ser, and Thr) or even fewer residues (e.g., Ser, Tyr), which has resulted in increased conformational diversity, giving rise to higher interaction affinities as well as new interactions (Fellouse, Li et al. 2005; Fellouse, Wiesmann et al. 2004; Koide, Gilbreth et al. 2007). IDPs already exhibit lower sequence complexity (see the section titled "Repeats and Low-Complexity Regions"), which can be used as is or be further increased by restricting the chemical diversity of the region. Along with using the promiscuity of IDPs we can modify disordered regions to create new functions. Still, we might wish to approach IDPs in directed evolution differently.

Directed evolution of a disordered region could affect IDPs in different ways, including (1) to reduce or increase the flexibility of the disordered region, (2) change some of the conformations in the prebound state, (3) alter the conformations formed upon binding, and (4) drive changes in the disorder region that can influence other more ordered parts of the proteins. Modification of the disorder nature of IDPs may require different experimental tools. Several servers exist that guide library design in the directed evolution process. These servers calculate the evolutionary conservation of each residue in the protein and differentiate between conserved and variable regions. As demonstrated in the section "Are IDPs Evolutionary Conserved or Evolving Rapidly?", assessing conservation of disorder is a very complex task and there is a disagreement regarding the conservation of disorder. Thus, choosing the location for modification may require a new approach.

A major challenge in achieving the desired variant in a directed evolution process is the potential destabilizing effect of any given mutation. It has been suggested that many mutations are never detected, as they negatively affect the stability of the protein (Shoichet, McGovern et al. 2002). Thus, attempts to increase the disordered fraction of a target protein might simply result in destabilized proteins that are less preferentially displayed. A possible solution to this problem may lie in the use of chaperone proteins, which normally rescue misfolded proteins and may be able to rescue some mutant proteins and accelerate directed evolution. For instance, the GroEL/GroES chaperone has been used to rescue about one-third of the adaptive mutant proteins that, without the help of the chaperone, would have been too unstable to be viable (Tokuriki and Tawfik 2009b).

Examples of IDPs Studied Using Protein Engineering/Directed Evolution

Although studying IDPs at the molecular level could have a significant impact on understanding protein function, folding, and evolution, limited research has been performed using molecular modifications combined with biophysical and structural studies of IDPs. Even fewer are examples of using directed evolution to study IDPs. In the following, we present some recent progress in the study of IDPs at the molecular level using directed evolution in disordered regions followed by biophysical/structural analysis.

Increased Rigidity of Protein Domains Results in Increased Protein Stability

Traditionally, directed evolution has been used to create stable mutants of proteins, mainly enzymes, for protein engineering purposes. One way in which to do this is to increase the rigidity of the desired protein, thus restricting the dynamic properties of the protein (i.e., reducing the disordered content). As an example, the selection marker kanamycin nucleotidyltransferase (KNT) was evolved in a directed evolution process to create a variant with 20°C higher thermostability called highly thermostable kanamycin nucleotidyltransferase (HTK) that included 19 mutations (Hoseki, Yano et al. 1999). In order to analyze the effect of each mutation, the evolved variant was mutated back to a wild type one mutation at a time and all resulting mutants were tested for stability (Hoseki, Okamoto et al. 2003). Analysis of the dynamic properties of the mutations was performed by molecular dynamics simulations. Both the wild-type KNT and the evolved variant HTK form dimers, in which the C-terminal domain (CTD) of each molecule interacts while the N-terminal domain (NTD) is mainly exposed to solvent. All of the mutations that exhibited a large effect on the stability of the evolved variant were found to be located in the exposed NTD. Other mutations located in the NTD, which showed lower or negligible effect on stabilization when added individually, appeared to work in a cooperative way, resulting in a large increase in stability when added combinatorially. Cooperativity of additional resides was also found within the CTD. Although the wild type and evolved variant have a similar overall structural rigidity, there was an increase in rigidity within domains that reduced internal fluctuations, along with an increase in interdomain motions. The NTD was shown to fluctuate in a hinge-like motion relative to the CTD while the overall conformation of each domain remained the same. Therefore, the introduced mutations changed the dynamic properties of the internal domains in the protein without significantly changing the overall structure of the protein, thus increasing its rigidity.

Energetic Contribution of a Disordered Region to Affinity Maturation

The role of a disordered region in complex formation was inferred from a detailed structural and thermodynamic analysis of complex formation by affinity matured variants selected by directed evolution (Cho, Swaminathan et al. 2005). A 16-residue disordered disulfide loop of the superantigen (SAG) *Staphylococcus aureus* enterotoxin C3 (SEC3) was subjected to repeated cycles of directed evolution and selected for improved binding to the murine T-cell receptor β chain 8.2 (mVβ8.2) (Andersen, Geisler et al. 2001). Three of the resulting variants having progressive increase in the binding affinity (12 [wild type], 4, 0.57, and 0.1 μM) were selected to represent points in the affinity maturation model constructed to study the role of the disordered region (Cho, Swaminathan et al. 2010). Comparison of the crystal structures of the variants in the free state and in complex with mVβ8.2 indicated increased ordering of the loop and increased intermolecular interactions along with increase in binding affinity. However, the disordered nature of the loop in the unbound state remained the same. The energetic contribution of the conformational changes induced upon binding was estimated by comparing the difference between observed (by isothermal titration calorimetry [ITC]) and calculated (using buried surface areas obtained from the crystal structures) heat capacity changes, with the heat capacities estimated by modeling the folding from the extended conformations. The attributed energetic contribution was assigned as the reason for the increase in binding affinities along the affinity maturation pathway. A comparison of the binding energies calculated from the thermodynamic analysis with estimated binding energies from predictive algorithms implied that major differences resulted from the lack of ability of the computational methods to assess folding transitions of disordered regions. Increased knowledge from experimentally measured energetic contributions of intrinsically disordered regions to protein–protein interactions would be valuable in refining the predictive capacity of such computational methods.

Evolved Disorder Increased Promiscuity

Affibodies are a group of engineered protein binders selected to bind target proteins with high affinity and selectivity (Lofblom, Feldwisch et al. 2010). The *Staphylococcus aureus* protein A (SPA) has five homologous domains (A–E), each of which binds tightly to immunoglobulins (IgGs) mainly through their Fc region. The Z domain is a more stable, engineered analog of the B domain of SPA that folds into a three-helix bundle. Affibodies are Z domain-based proteins engineered to bind a specific target protein of interest by selection from a combinatorial library against the target protein (Nord, Gunneriusson et al. 1997). Having the capability to evolve new interactions with desired proteins, affibodies represent an alternative for antibody recognition and have been widely used for biotechnological and therapeutic applications (Lofblom, Feldwisch et al. 2010).

Z_{SPA-1} is an affibody in which the Z domain-based variant was engineered and selected by binding to the original five-domain SPA protein. The library was generated by randomly mutating 13 residues in the Fc binding surface of the Z domain. The evolved variant was able to bind each of the five domains of wild-type SPA and the Z domain itself with dissociation constants (K_D) in the micromolar range (2–6.5 µM) (Eklund, Axelsson et al. 2002). Structural and biophysical analysis of Z_{SPA-1} indicated that it was intrinsically disordered, described as molten globule (Wahlberg, Lendel et al. 2003). The NMR spectrum of the unbound Z_{SPA-1} showed poor dispersion and broadening of peaks, indicative of an interconversion between multiple poorly packed conformations. Addition of the Z domain resulted with sharp resonances and good peak dispersion, suggesting folding of the Z_{SPA-1} domain upon binding to the Z domain. The NMR structure of the Z_{SPA-1} with the Z-domain showed that the Z_{SPA-1} adopted the typical three-helix bundle, characteristic of the Z-domain.

Z_{SPA-1} was subsequently described as an intermediate that exists in equilibrium between a molten globulelike protein and one completely unfolded in the unbound state (Dincbas-Renqvist, Lendel et al. 2004; Lendel, Dincbas-Renqvist et al. 2004). Detailed thermodynamic- and structure-based energetic calculations were performed, accounting for the structural changes and coupled folding upon binding process. While the energetic calculations suggested the ability to form strong binding, the weaker binding affinity (6 μM) results from a large conformational entropy penalty derived from stabilization of the ordered structure in the bound state (Dincbas-Renqvist, Lendel et al. 2004). Overall, these Z_{SPA-1} affibody experiments demonstrate that directed evolution of an ordered protein into a less ordered one can introduce new interactions with an expanded number of homologous domains of SPA, including the Z domain. In this case, the increase of disorder in the evolved variant hampered tight interaction of the complex, resulting from an entropic penalty associated with folding upon binding.

Directed Evolution of a Binding Motif in an Intrinsically Disordered Region

In the search for an improved binding motif, directed evolution followed by structural and thermodynamic analysis were used to improve the interaction between the hub protein LC8 dynein light chain (DYNLL) with linear peptide segments that included a binding motif found in disordered segments (Rapali, Radnai et al. 2011). Based on the consensus motif: $[D/S]_{-4}K_{-3}X_{-2}[T/V/I]_{-1}Q_0[T/V]_1[D/E]_2$, a library of randomly mutated seven-residue peptide variants was constructed, while keeping the conserved Gln residue at position 0 fixed (for peptides totaling eight residues). A leucine zipper was introduced in order to mimic the natural dimer binding pattern.

The highly improved variant sequence ($V_{-5}D_{-4} K_{-3}S_{-2}T_{-1}Q_0T_1D_2$) resembled the natural consensus sequence. Comparing the original and evolved amino acid preference at each position showed an *in vitro* preference for a Val in position -5 that does not exist in the native motif. ITC binding experiments of the synthesized evolved variants with DYNLL showed a 20-fold increase in affinity attributed to Val_{-5}. Dimerization of the tested peptides improved the binding as well. Comparison of the x-ray crystal structure of DYNLL in complex with the improved dimer peptide and a complex with a peptide that lacked the N-terminal Val_{-5} residue showed that this residue extended the β-sheet and established two backbone hydrogen bonds that were missing in the complex with the peptide lacking the Val. A search for an *in vivo* DYNLL binding partner that contains the phage selected consensus sequence revealed a protein named EML3, which had not previously been identified as a DYNLL binding protein. A recombinant fragment derived from EML3 (residues 8–94) was shown to bind DYNLL with high affinity (K_D = 50 nM) in ITC binding experiments.

SUMMARY

An increase in disorder along with increased complexity of organisms emphasizes the role of disorder as an evolutionary tool used to advance from single-cell prokaryotes to complex eukaryotes. Employing disorder in protein engineering by directed evolution methods might be a valuable tool for the creation of new proteins. Increasing our understanding of the evolutionary origin of IDPs and how such a selective pressure operates would inform our study of nearly all aspects of life. The role of disorder is being increasingly appreciated both in the increasing number of IDPs and IDRs discovered and also by the recognition of how their dynamics play important roles in protein structure, function, and complex formation. Intrinsically disordered proteins provide a unique perspective through which to better understand proteins, such as the origin of function and structure, the process of folding and misfolding, and evolutionary processes leading to increased complexity of living organisms. Continued study of natural protein evolution will enable us to better engineer proteins by directed evolution with new and useful functions.

At the molecular level, the directed evolution of intrinsically disordered proteins will enable the study of the role of disorder/dynamic properties of a particular region in relation to its function and interactions, as well as to evaluate the energetic contributions of a disordered region to the formation of an interaction. IDPs provide model systems to address basic questions in protein recognition and intrinsic disorder. Elucidation of the relationships between numerous biophysical factors related to protein interactions may be addressed by such experiments. Learning about the relationship between specificity, selectivity, and promiscuity will enable a route by which to manipulate these important protein properties.

Directed evolution of IDPs may also be used as a way to design new surfaces with increased or reduced flexibility, in order to improve existing functions/interaction, select between interactions/functions, or even design *de novo* novel functions by creating new interactions. Evolutionary methods may also be utilized to study/increase protein stability by increasing the conformational rigidity of the IDP. The combination of directed evolution, structural, and biophysical methods to analyze and explore the role of IDPs has proved highly informative. Increased understanding regarding the contribution of disordered regions to the function and mechanism of action of a protein can be used at the therapeutic level to counteract diseases that involve IDPs. Such information will contribute to our understanding not only of IDPs, but of all proteins.

REFERENCES

Andersen, P. S., C. Geisler, S. Buus, R. A. Mariuzza and K. Karjalainen (2001). Role of the T cell receptor ligand affinity in T cell activation by bacterial superantigens. *J Biol Chem* **276**(36): 33452–33457.

Arnold, F. H. (2001). Combinatorial and computational challenges for biocatalyst design. *Nature* **409**(6817): 253–257.

Benner, S. A., M. A. Cohen and G. H. Gonnet (1993). Empirical and structural models for insertions and deletions in the divergent evolution of proteins. *J Mol Biol* **229**(4): 1065–1082.

Bergeron, J. R., H. Huthoff, D. A. Veselkov, R. L. Beavil, P. J. Simpson, S. J. Matthews, M. H. Malim and M. R. Sanderson (2010). The SOCS-box of HIV-1 Vif interacts with ElonginBC by induced-folding to recruit its Cul5-containing ubiquitin ligase complex. *PLoS Pathog* **6**(6): e1000925.

Bonsor, D. A. and E. J. Sundberg (2011). Dissecting protein–protein interactions using directed evolution. *Biochemistry* **50**(13): 2394–2402.

Brown, C. J., A. K. Johnson and G. W. Daughdrill (2010). Comparing models of evolution for ordered and disordered proteins. *Mol Biol Evol* **27**(3): 609–621.

Brown, C. J., A. K. Johnson, A. K. Dunker and G. W. Daughdrill (2011). Evolution and disorder. *Curr Opin Struct Biol* **21**(3): 441–446.

Brown, C. J., S. Takayama, A. M. Campen, P. Vise, T. W. Marshall, C. J. Oldfield, C. J. Williams and A. K. Dunker (2002). Evolutionary rate heterogeneity in proteins with long disordered regions. *J Mol Evol* **55**(1): 104–110.

Canchaya, C., G. Fournous, S. Chibani-Chennoufi, M. L. Dillmann and H. Brussow (2003). Phage as agents of lateral gene transfer. *Curr Opin Microbiol* **6**(4): 417–424.

Chen, J. W., P. Romero, V. N. Uversky and A. K. Dunker (2006a). Conservation of intrinsic disorder in protein domains and families: I. A database of conserved predicted disordered regions. *J Proteome Res* **5**(4): 879–887.

Chen, J. W., P. Romero, V. N. Uversky and A. K. Dunker (2006b). Conservation of intrinsic disorder in protein domains and families: II. Functions of conserved disorder. *J Proteome Res* **5**(4): 888–898.

Cho, S., C. P. Swaminathan, D. A. Bonsor, M. C. Kerzic, R. Guan, J. Yang, M. C. Kieke, P. S. Andersen, D. M. Kranz, R. A. Mariuzza and E. J. Sundberg (2010). Assessing energetic contributions to binding from a disordered region in a protein–protein interaction. *Biochemistry* **49**(43): 9256–9268.

Cho, S., C. P. Swaminathan, J. Yang, M. C. Kerzic, R. Guan, M. C. Kieke, D. M. Kranz, R. A. Mariuzza and E. J. Sundberg (2005). Structural basis of affinity maturation and intramolecular cooperativity in a protein–protein interaction. *Structure* **13**(12): 1775–1787.

Chothia, C. and A. M. Lesk (1986). The relation between the divergence of sequence and structure in proteins. *EMBO J* **5**(4): 823–826.

Claverie, J. M. (2006). Viruses take center stage in cellular evolution. *Genome Biol* **7**(6): 110.

Copley, S. D. (2003). Enzymes with extra talents: Moonlighting functions and catalytic promiscuity. *Curr Opin Chem Biol* **7**(2): 265–272.

Cumberworth, A., G. Lamour, M. M. Babu and J. Gsponer (2013). Promiscuity as a functional trait: Intrinsically disordered regions as central players of interactomes. *Biochem J* **454**(3): 361–369.

Daughdrill, G. W., P. Narayanaswami, S. H. Gilmore, A. Belczyk and C. J. Brown (2007). Dynamic behavior of an intrinsically unstructured linker domain is conserved in the face of negligible amino acid sequence conservation. *J Mol Evol* **65**(3): 277–288.

Davey, N. E., J. L. Cowan, D. C. Shields, T. J. Gibson, M. J. Coldwell and R. J. Edwards (2012). SLiMPrints: Conservation-based discovery of functional motif fingerprints in intrinsically disordered protein regions. *Nucleic Acids Res* **40**(21): 10628–10641.

Davey, N. E., D. C. Shields and R. J. Edwards (2009). Masking residues using context-specific evolutionary conservation significantly improves short linear motif discovery. *Bioinformatics* **25**(4): 443–450.

Dincbas-Renqvist, V., C. Lendel, J. Dogan, E. Wahlberg and T. Hard (2004). Thermodynamics of folding, stabilization, and binding in an engineered protein–protein complex. *J Am Chem Soc* **126**(36): 11220–11230.

Doolittle, R. F. (1981). Similar amino acid sequences: Chance or common ancestry? *Science* **214**(4517): 149–159.

Drake, J. W., B. Charlesworth, D. Charlesworth and J. F. Crow (1998). Rates of spontaneous mutation. *Genetics* **148**(4): 1667–1686.

Drummond, D. A., J. D. Bloom, C. Adami, C. O. Wilke and F. H. Arnold (2005). Why highly expressed proteins evolve slowly. *Proc Natl Acad Sci U S A* **102**(40): 14338–14343.

Dunker, A. K., J. D. Lawson, C. J. Brown, R. M. Williams, P. Romero, J. S. Oh, C. J. Oldfield, A. M. Campen, C. M. Ratliff, K. W. Hipps, J. Ausio et al. (2001). Intrinsically disordered protein. *J Mol Graph Model* **19**(1): 26–59.

Dyson, H. J. and P. E. Wright (2002). Coupling of folding and binding for unstructured proteins. *Curr Opin Struct Biol* **12**(1): 54–60.

Dyson, H. J. and P. E. Wright (2005). Intrinsically unstructured proteins and their functions. *Nat Rev Mol Cell Biol* **6**(3): 197–208.

Eames, M. and T. Kortemme (2007). Structural mapping of protein interactions reveals differences in evolutionary pressures correlated to mRNA level and protein abundance. *Structure* **15**(11): 1442–1451.

Earl, D. J. and M. W. Deem (2004). Evolvability is a selectable trait. *Proc Natl Acad Sci U S A* **101**(32): 11531–11536.

Eklund, M., L. Axelsson, M. Uhlen and P. A. Nygren (2002). Anti-idiotypic protein domains selected from protein A-based affibody libraries. *Proteins* **48**(3): 454–462.

Fellouse, F. A., B. Li, D. M. Compaan, A. A. Peden, S. G. Hymowitz and S. S. Sidhu (2005). Molecular recognition by a binary code. *J Mol Biol* **348**(5): 1153–1162.

Fellouse, F. A., C. Wiesmann and S. S. Sidhu (2004). Synthetic antibodies from a four-amino-acid code: A dominant role for tyrosine in antigen recognition. *Proc Natl Acad Sci U S A* **101**(34): 12467–12472.

Ferron, F., S. Longhi, B. Canard and D. Karlin (2006). A practical overview of protein disorder prediction methods. *Proteins* **65**(1): 1–14.

Forterre, P. (2006). The origin of viruses and their possible roles in major evolutionary transitions. *Virus Res* **117**(1): 5–16.

Fraser, H. B., A. E. Hirsh, L. M. Steinmetz, C. Scharfe and M. W. Feldman (2002). Evolutionary rate in the protein interaction network. *Science* **296**(5568): 750–752.

Goldman, N., J. L. Thorne and D. T. Jones (1998). Assessing the impact of secondary structure and solvent accessibility on protein evolution. *Genetics* **149**(1): 445–458.

Grishin, N. V. (2001). Fold change in evolution of protein structures. *J Struct Biol* **134**(2–3): 167–185.

Gsponer, J. and M. M. Babu (2012). Cellular strategies for regulating functional and nonfunctional protein aggregation. *Cell Rep* **2**(5): 1425–1437.

Gsponer, J., M. E. Futschik, S. A. Teichmann and M. M. Babu (2008). Tight regulation of unstructured proteins: From transcript synthesis to protein degradation. *Science* **322**(5906): 1365–1368.

Heringa, J. (1998). Detection of internal repeats: How common are they? *Curr Opin Struct Biol* **8**(3): 338–345.

Hoseki, J., A. Okamoto, N. Takada, A. Suenaga, N. Futatsugi, A. Konagaya, M. Taiji, T. Yano, S. Kuramitsu and H. Kagamiyama (2003). Increased rigidity of domain structures enhances the stability of a mutant enzyme created by directed evolution. *Biochemistry* **42**(49): 14469–14475.

Hoseki, J., T. Yano, Y. Koyama, S. Kuramitsu and H. Kagamiyama (1999). Directed evolution of thermostable kanamycin-resistance gene: A convenient selection marker for Thermus thermophilus. *J Biochem* **126**(5): 951–956.

James, L. C. and D. S. Tawfik (2003). Conformational diversity and protein evolution—a 60-year-old hypothesis revisited. *Trends Biochem Sci* **28**(7): 361–368.

Jeffery, C. J. (1999). Moonlighting proteins. *Trends Biochem Sci* **24**(1): 8–11.

Johansson, J., G. H. Gudmundsson, M. E. Rottenberg, K. D. Berndt and B. Agerberth (1998). Conformation-dependent antibacterial activity of the naturally occurring human peptide LL-37. *J Biol Chem* **273**(6): 3718–3724.

Karlin, S., L. Brocchieri, A. Bergman, J. Mrazek and A. J. Gentles (2002). Amino acid runs in eukaryotic proteomes and disease associations. *Proc Natl Acad Sci U S A* **99**(1): 333–338.

Koide, A., R. N. Gilbreth, K. Esaki, V. Tereshko and S. Koide (2007). High-affinity single-domain binding proteins with a binary-code interface. *Proc Natl Acad Sci U S A* **104**(16): 6632–6637.

Kriwacki, R. W., L. Hengst, L. Tennant, S. I. Reed and P. E. Wright (1996). Structural studies of p21Waf1/Cip1/Sdi1 in the free and Cdk2-bound state: Conformational disorder mediates binding diversity. *Proc Natl Acad Sci U S A* **93**(21): 11504–11509.

Lake, J. A., J. Carr, F. Feng, L. Mundy, C. Burrell and P. Li (2003). The role of Vif during HIV-1 infection: Interaction with novel host cellular factors. *J Clin Virol* **26**(2): 143–152.

Lendel, C., V. Dincbas-Renqvist, A. Flores, E. Wahlberg, J. Dogan, P. A. Nygren and T. Hard (2004). Biophysical characterization of Z(SPA-1)—a phage-display selected binder to protein A. *Protein Sci* **13**(8): 2078–2088.

Levin, K. B., O. Dym, S. Albeck, S. Magdassi, A. H. Keeble, C. Kleanthous and D. S. Tawfik (2009). Following evolutionary paths to protein–protein interactions with high affinity and selectivity. *Nat Struct Mol Biol* **16**(10): 1049–1055.

Light, S., R. Sagit, D. Ekman and A. Elofsson (2013). Long indels are disordered: A study of disorder and indels in homologous eukaryotic proteins. *Biochim Biophys Acta* **1834**(5): 890–897.

Lise, S. and D. T. Jones (2005). Sequence patterns associated with disordered regions in proteins. *Proteins* **58**(1): 144–150.

Liu, J., H. Tan and B. Rost (2002). Loopy proteins appear conserved in evolution. *J Mol Biol* **322**(1): 53–64.

Lobley, A., M. B. Swindells, C. A. Orengo and D. T. Jones (2007). Inferring function using patterns of native disorder in proteins. *PLoS Comput Biol* **3**(8): e162.

Lofblom, J., J. Feldwisch, V. Tolmachev, J. Carlsson, S. Stahl and F. Y. Frejd (2010). Affibody molecules: Engineered proteins for therapeutic, diagnostic and biotechnological applications. *FEBS Lett* **584**(12): 2670–2680.

Marcotte, E. M., M. Pellegrini, T. O. Yeates and D. Eisenberg (1999). A census of protein repeats. *J Mol Biol* **293**(1): 151–160.

Nord, K., E. Gunneriusson, J. Ringdahl, S. Stahl, M. Uhlen and P. A. Nygren (1997). Binding proteins selected from combinatorial libraries of an alpha-helical bacterial receptor domain. *Nat Biotechnol* **15**(8): 772–777.

Palmer, A. G., III and F. Massi (2006). Characterization of the dynamics of biomacromolecules using rotating-frame spin relaxation NMR spectroscopy. *Chem Rev* **106**(5): 1700–1719.

Pascarella, S. and P. Argos (1992). Analysis of insertions/deletions in protein structures. *J Mol Biol* **224**(2): 461–471.

Prakash, S., L. Tian, K. S. Ratliff, R. E. Lehotzky and A. Matouschek (2004). An unstructured initiation site is required for efficient proteasome-mediated degradation. *Nat Struct Mol Biol* **11**(9): 830–837.

Radivojac, P., L. M. Iakoucheva, C. J. Oldfield, Z. Obradovic, V. N. Uversky and A. K. Dunker (2007). Intrinsic disorder and functional proteomics. *Biophys J* **92**(5): 1439–1456.

Radivojac, P., Z. Obradovic, C. J. Brown and A. K. Dunker (2002). Improving sequence alignments for intrinsically disordered proteins. *Pac Symp Biocomput* **7**: 589–600.

Rapali, P., L. Radnai, D. Suveges, V. Harmat, F. Tolgyesi, W. Y. Wahlgren, G. Katona, L. Nyitray and G. Pal (2011). Directed evolution reveals the binding motif preference of the LC8/DYNLL hub protein and predicts large numbers of novel binders in the human proteome. *PLoS One* **6**(4): e18818.

Reanney, D. C. (1982). The evolution of RNA viruses. *Annu Rev Microbiol* **36**: 47–73.

Reingewertz, T. H., H. Benyamini, M. Lebendiker, D. E. Shalev and A. Friedler (2009). The C-terminal domain of the HIV-1 Vif protein is natively unfolded in its unbound state. *Protein Eng Des Sel* **22**(5): 281–287.

Reingewertz, T. H., D. E. Shalev and A. Friedler (2011). Making order in the intrinsically disordered regions of the HIV-1 Vif protein. In *Flexible Viruses: Structural Disorder in Viral Proteins*. V. N. Uversky and S. Longhi, eds. Hoboken, NJ: John Wiley & Sons, 201–221.

Romero, P., Z. Obradovic, X. Li, E. C. Garner, C. J. Brown and A. K. Dunker (2001). Sequence complexity of disordered protein. *Proteins* **42**(1): 38–48.

Santner, A. A., C. H. Croy, F. H. Vasanwala, V. N. Uversky, Y. Y. Van and A. K. Dunker (2012). Sweeping away protein aggregation with entropic bristles: Intrinsically disordered protein fusions enhance soluble expression. *Biochemistry* **51**(37): 7250–7262.

Schaefer, C., A. Schlessinger and B. Rost (2010). Protein secondary structure appears to be robust under in silico evolution while protein disorder appears not to be. *Bioinformatics* **26**(5): 625–631.

Schlessinger, A., J. F. Liu and B. Rost (2007). Natively unstructured loops differ from other loops. *PLoS Comput Biol* **3**(7): 1335–1346.

Schlessinger, A., C. Schaefer, E. Vicedo, M. Schmidberger, M. Punta and B. Rost (2011). Protein disorder—a breakthrough invention of evolution? *Curr Opin Struct Biol* **21**(3): 412–418.

Schreiber, G., A. M. Buckle and A. R. Fersht (1994). Stability and function: Two constraints in the evolution of barstar and other proteins. *Structure* **2**(10): 945–951.

Shoichet, B. K., S. L. McGovern, B. Wei and J. J. Irwin (2002). Lead discovery using molecular docking. *Curr Opin Chem Biol* **6**(4): 439–446.

Siltberg-Liberles, J. (2011). Evolution of structurally disordered proteins promotes neostructuralization. *Mol Biol Evol* **28**(1): 59–62.

Strebel, K., D. Daugherty, K. Clouse, D. Cohen, T. Folks and M. A. Martin (1987). The HIV 'A' (sor) gene product is essential for virus infectivity. *Nature* **328**(6132): 728–730.

Tatusov, R. L., E. V. Koonin and D. J. Lipman (1997). A genomic perspective on protein families. *Science* **278**(5338): 631–637.

Tokuriki, N., C. J. Oldfield, V. N. Uversky, I. N. Berezovsky and D. S. Tawfik (2009). Do viral proteins possess unique biophysical features? *Trends Biochem Sci* **34**(2): 53–59.

Tokuriki, N. and D. S. Tawfik (2009a). Protein dynamism and evolvability. *Science* **324**(5924): 203–207.

Tokuriki, N. and D. S. Tawfik (2009b). Chaperonin overexpression promotes genetic variation and enzyme evolution. *Nature* **459**(7247): 668–673.

Tompa, P. (2003). Intrinsically unstructured proteins evolve by repeat expansion. *BioEssays* **25**(9): 847–855.

Tompa, P. (2010). *Structure and Function of Intrinsically Disordered Proteins*. Boca Raton, FL: CRC Press.

Tompa, P., Z. Dosztanyi and I. Simon (2006). Prevalent structural disorder in E. coli and S. cerevisiae proteomes. *J Proteome Res* **5**(8): 1996–2000.

Tompa, P., M. Fuxreiter, C. J. Oldfield, I. Simon, A. K. Dunker and V. N. Uversky (2009). Close encounters of the third kind: Disordered domains and the interactions of proteins. *BioEssays* **31**(3): 328–335.

Tompa, P., J. Prilusky, I. Silman and J. L. Sussman (2008). Structural disorder serves as a weak signal for intracellular protein degradation. *Proteins* **71**(2): 903–909.

Tompa, P., C. Szasz and L. Buday (2005). Structural disorder throws new light on moonlighting. *Trends Biochem Sci* **30**(9): 484–489.

Toth-Petroczy, A., B. Meszaros, I. Simon, A. K. Dunker, V. N. Uversky and M. Fuxreiter (2008). Assessing conservation of disordered regions in proteins. *Open Proteomics J* **1**: 46–53.

Toth-Petroczy, A. and D. S. Tawfik (2011). Slow protein evolutionary rates are dictated by surface-core association. *Proc Natl Acad Sci U S A* **108**(27): 11151–11156.

Tsvetkov, P., N. Reuven and Y. Shaul (2009). The nanny model for IDPs. *Nat Chem Biol* **5**(11): 778–781.

Uversky, V. N. (2002). What does it mean to be natively unfolded? *Eur J Biochem* **269**(1): 2–12.

Uversky, V. N., J. R. Gillespie and A. L. Fink (2000). Why are natively unfolded proteins unstructured under physiologic conditions? *Proteins* **41**(3): 415–427.

Uversky, V. N., C. J. Oldfield and A. K. Dunker (2008). Intrinsically disordered proteins in human diseases: Introducing the D2 concept. *Annu Rev Biophys* **37**: 215–246.

Vavouri, T., J. I. Semple, R. Garcia-Verdugo and B. Lehner (2009). Intrinsic protein disorder and interaction promiscuity are widely associated with dosage sensitivity. *Cell* **138**(1): 198–208.

Vucetic, S., C. J. Brown, A. K. Dunker and Z. Obradovic (2003). Flavors of protein disorder. *Proteins* **52**(4): 573–584.

Wahlberg, E., C. Lendel, M. Helgstrand, P. Allard, V. Dincbas-Renqvist, A. Hedqvist, H. Berglund, P. A. Nygren and T. Hard (2003). An affibody in complex with a target protein: Structure and coupled folding. *Proc Natl Acad Sci U S A* **100**(6): 3185–3190.

Ward, J. J., J. S. Sodhi, L. J. McGuffin, B. F. Buxton and D. T. Jones (2004). Prediction and functional analysis of native disorder in proteins from the three kingdoms of life. *J Mol Biol* **337**(3): 635–645.

Wilke, C. O. and D. A. Drummond (2010). Signatures of protein biophysics in coding sequence evolution. *Curr Opin Struct Biol* **20**(3): 385–389.

Wright, P. E. and H. J. Dyson (1999). Intrinsically unstructured proteins: Re-assessing the protein structure-function paradigm. *J Mol Biol* **293**(2): 321–331.

Xue, B., R. W. Williams, C. J. Oldfield, G. K.-M. Goh, A. K. Dunker and V. N. Uversky (2011). Do viral proteins possess unique features? In *Flexible Viruses: Structural Disorder in Viral Proteins.* V. N. Uversky and S. Longhi, eds. Hoboken, NJ: John Wiley & Sons, 1–34.

Large-Scale Dynamics

Chapter 12

Discrete Molecular Dynamics

Foundations and Biomolecular Applications

Pedro Sfriso, Agustí Emperador,
Josep Lluis Gelpí, and Modesto Orozco

CONTENTS

DYNAMIC NATURE OF PROTEINS

Proteins are dynamic entities whose conformations change in response to external stimuli [1]. Many experimental studies have demonstrated that protein flexibility is crucial for functioning [2–4], and clear evidence exists that evolution has made an effort to refine not only the structures, but also the flexibility pattern of proteins [5–14]. An analysis of structural databases outlines many examples of proteins displaying structural polymorphism [8,12–15]; that is, cases where, depending on the crystallization conditions or the presence of effectors, the same protein is found in two different conformations (in some cases, separated by more than 20 Å in root-mean-square deviation (RMSD) [16]). Undoubtedly, as structural techniques advance, more examples of dramatic conformational changes in proteins will emerge, showing the magnitude of the biological impact of protein flexibility [12,13].

Experimental characterization of protein flexibility is extremely difficult [17–20] and most of our current knowledge on protein dynamics is derived from molecular simulation techniques [17,21–30]. There is a plethora of theoretical methods for representing protein flexibility, but all of them can be classified based on three independent criteria: (1) the level of resolution in the representation of the protein, (2) the potential energy functional (force field) used to describe molecular interactions, and (3) the sampling method used to generate protein ensembles.

Some simulation methods use atomistic or quasi-atomistic (all heavy atoms) representations of the proteins, typically coupled with similar levels of detail in the representation of the solvent. Other methods (the coarse-grained ones) reduce drastically the huge number of degrees of freedom in the system, modeling the solvent implicitly, and simplifying the protein representation using models with a reduced number of particles. Force fields used in standard atomistic simulations include nonbonded (electrostatic and van der Waals) and bonded (stretching, bending, and torsion) terms, which are carefully calibrated from experimental data and accurate quantum mechanical calculations on small model systems. On the other hand, when coarse-grained resolution is used, the force fields used are simpler (see for example [30–33]) and are directly calibrated using known experimental data (either experimental data or derived from atomistic simulations). The simplest coarse-grained oriented force fields follow Go-like models [34] and assume that the energy related to the distortion of a residue from its equilibrium geometry increases with the square of the distortion distance (see Equation 12.1 and [15,35–40]):

$$E = \sum_{i,j} \delta_{ij} K_{ij} \left(R_{ij} - R_{ij}^0 \right)^2 \tag{12.1}$$

where i and j are residues, δ_{ij} is a delta function equal to 1 when i and j are at less than a given distance, and 0 otherwise, K is a spring constant (linear or distance dependent), R_{ij} stands for interresidue distance, and the superscript 0 refers to the value of R_{ij} in the reference structure.

Once a level of resolution and a force field are selected, sampling of protein movements can be obtained by using a variety of methods. Within the normal mode analysis (NMA) approach movements are assumed to be harmonic along the normal modes obtained by diagonalization of the Hessian (second derivatives) matrix. NMA methods are usually coupled with coarse-grained representations of proteins and elastic network model (ENM) Hamiltonians (see Equation 12.1). More complex force fields are used together with more flexible sampling methods that do not assume harmonicity in the protein movements. Thus, some authors have suggested the use of internal coordinates Monte Carlo methods for sampling of protein movements at a given temperature [41]. However, the most prevalent technique used to represent protein flexibility within the atomistic level relies on the use of classical molecular dynamics.

Within the molecular dynamics (MD) [42,43] framework protein movements are obtained by integration of Newton's equations of motion. Accordingly, forces acting on the protein (and solvent) atoms are obtained by differentiation of the interaction potentials defined in the force field, then deriving atomic accelerations, which are integrated to obtain new atomic velocities and positions. Variants of the basic Newtonian MD have been formulated to guarantee, for example, that collected ensembles maintain a constant temperature, pressure, or volume [42,43]. Thirty-five years after its first use in the biological world, atomistic MD has become the most accurate and universal simulation technique for the study of protein flexibility. Unfortunately, practical use of MD is dramatically limited by the huge gap between the very small time step used in the integration of Newton's equations of motion (femtosecond scale) and the time step where biological processes happen (typically milliseconds to hours). Despite the dramatic improvements in the new MD codes, the accessibility of large supercomputer resources, graphics processing unit (GPU)-based platforms, and even specific-purpose computers, there is in general a 3 to 6 orders of magnitude difference between the MD-accessible and the experimental relevant timescales.

In the following, we will present an inexpensive alternative to MD called discrete molecular dynamics (dMD). Within this approach, protein trajectories are assumed to follow ballistic equations of motion, avoiding the need for integration of Newton's equations of motion. After presenting the basic foundation of the technique, we will comment on some recent applications that illustrate the power and generality of this approach.

BASIC dMD ALGORITHM

dMD assumes that particles move in the ballistic regime, with constant velocity until an event (collision) occurs. dMD potentials are then stepwise, with flat potential regions broken by energy discontinuities at collision distances, forming single or multiple square wells. This allows no computing forces, making the integration of Newton's equations of motion unnecessary. As a consequence, a simulation can jump from collision to collision (independently of the time required for such collision) without requiring a recalculation of energies and forces every femtosecond. Drastic acceleration of the dynamic calculation is then achieved [44].

In the absence of any collision, the particles move linearly with constant velocity. The position of a given particle at the time of the next collision is

$$\vec{r}_i(t+t_c) = \vec{r}_i(t) + \vec{v}_i t_c \tag{12.2}$$

where $\vec{r}_i$ and $\vec{v}_i$ stand for positions and velocities and t_c is the minimum among the collision times t_{ij} between each pair of particles i and j:

$$t_{ij} = \frac{-b_{ij} \pm \sqrt{b_{ij}^2 - v_{ij}^2 \left(r_{ij}^2 - d^2 \right)}}{v_{ij}^2}, \tag{12.3}$$

where r_{ij} is the magnitude of $\vec{r}_{ij} = \vec{r}_j - \vec{r}_i$, v_{ij} is the magnitude of $\vec{v}_{ij} = \vec{v}_j - \vec{v}_i$, $b_{ij} = \vec{r}_{ij} \cdot \vec{v}_{ij}$, and d is the distance corresponding to the wall of the square well.

When two particles collide, there is a transfer of linear momentum into the direction of the vector $\vec{r}_{ij}$:

$$m_i \vec{v}_i = m_i \vec{v}_i' + \Delta \vec{p} \tag{12.4}$$

$$m_j \vec{v}_j + \Delta \vec{p} = m_j \vec{v}_j'$$

where the prime indices denote the variables after the collision.

In order to calculate the change in velocities upon collision the velocity of each particle is projected in the direction of the vector $\vec{r}_{ij}$ and conservation rules are applied:

$$m_i v_i + m_j v_j = m_i v_i' + m_j v_j' \tag{12.5}$$

$$\frac{1}{2} m_i v_i^2 + \frac{1}{2} m_j v_j^2 = \frac{1}{2} m_j v_i'^2 + \frac{1}{2} m_j v_j'^2 + \Delta V, \tag{12.6}$$

where ΔV stands for the depth of the square well defining the inter-atomic potential.

The transferred momentum can be easily determined from

$$\Delta p = \frac{m_i m_j}{m_i + m_j} \left\{ \sqrt{(v_j - v_i)^2 - 2 \frac{m_i + m_j}{m_i m_j} \Delta V} - (v_j - v_i) \right\}, \tag{12.7}$$

Note that the two particles can overcome the potential step as long as

$$\Delta V < \frac{m_1 m_2}{2(m_1 + m_2)}(v_j - v_i)^2 \tag{12.8}$$

Otherwise, if the particles remain in the well, Equation 12.6 reduces to

$$\Delta p = \frac{m_i m_j}{m_i + m_j}\left\{\sqrt{(v_j - v_i)^2} - (v_j - v_i)\right\} \tag{12.9}$$

which, taking the negative solution of the root, leads to

$$\Delta p = \frac{2m_i m_j}{m_i + m_j}(v_i - v_j). \tag{12.10}$$

Once a collision happens and new positions and velocities are computed, the system moves in the ballistic regime until a new collision occurs, irrespective of whether the new collision happens after 1 fs or 1 ms, and resolving this new event is a matter of few operations, which multiplies the time range accessible for dMD.

dMD FORCE FIELDS

An absolute requirement of dMD is the use of discontinuous force fields defined by a combination of square wells. Typically, as in most physical force fields used in standard MD simulations, the interaction potential is composed by bonded and nonbonded terms. The bonded contribution accounts for energy changes related to deviations in bond lengths, angles, and torsions. Within most dMD implementations, stretching (solid black line in Figure 12.1) is represented by a single square well centered at the known equilibrium distance and with infinite walls that guarantee maintenance of covalent structure.

Pseudobonds (dashed lines in Figure 12.1), defined also as single square wells with infinite walls, are used to keep bond angles around equilibrium values as well as to guarantee peptide bond planarity [44]. In all cases the amplitude of the well is adjusted to reproduce observed harmonic oscillations at room temperature (from the analysis of MD simulations [45]).

Torsions around single bonds cannot be generally reproduced by single minimum potentials. In normal continuum potentials these interactions are reproduced by means of Fourier expansions with the dihedral angles used as descriptive coordinates. Within the dMD framework Fourier potentials are approximated by multiple finite well potentials using the 1–4 distance as

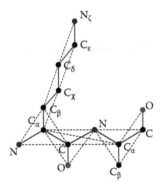

Figure 12.1 Scheme of bonds and pseudobonds used to the Lys-Ala di-peptide in a standard dMD force field.

descriptor. For the nonbonded terms finite step potentials are used. The number of such steps can be chosen arbitrarily, taking into account that a higher number of steps will give a more accurate description of the interaction potential (dashed line in Figure 12.2), but will lead to a higher number of collisions and therefore the simulation will be slower.

Several types of nonbonded potentials can be chosen depending on the granularity of the model used in the simulation. In the case of atomistic simulations, van der Waals and implicit solvation have been included [44,46]. For the implicit solvation term, Ding et al. [46] used a discretization of the EEF1 potential of Lazaridis and Karplus [47]. Hydrogen bonds are defined with pseudobonds. To decide whether the hydrogen bond is formed, Ding et al. [48] used

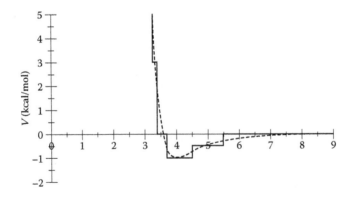

Figure 12.2 Example of fitting of a typical Lennard-Jones potential by multiple step discrete potentials. The x-axis accounts for nonbonded distances (in Å) and the y-axis for energies (in kcal/mol).

the reaction algorithm: upon formation of a hydrogen bond the particle types of the hydrogen-bonding atoms change and the energetic interactions adapt accordingly. If the kinetic energy is enough to overcome the potential energy difference, the hydrogen bond is formed and the pseudobonds defining the correct orientation are placed; otherwise, the atoms undergo an elastic collision (Figure 12.3). In a more recent work [49], a long-range electrostatic term and a sequence-dependent semiempirical potential accounting for secondary structure propensity of each aminoacid were included in the force field.

When coarse-graining the system, atoms are substituted by beads and effective knowledge-based potentials should be used to reproduce either attractive or repulsive interactions between beads [48,50]. Such potentials are parametrized to reproduce the experimental behavior of the known protein system at room temperature.

Dokholyan and coworkers have used simple coarse-grained potentials to study, for example, the folding of the Trp-cage protein, obtaining successful results after optimizing the strength of each of the terms in their potential [50] that included a hydrophobic interaction term, salt bridges, interaction between aromatic residues, aromatic-proline interactions, and hydrogen bonds, with a different strength for backbone-backbone and backbone-side-chain hydrogen bonds (Figure 12.4).

Figure 12.3   Representation of a side chain hydrogen bond between TYR and ASN used in dMD force fields.

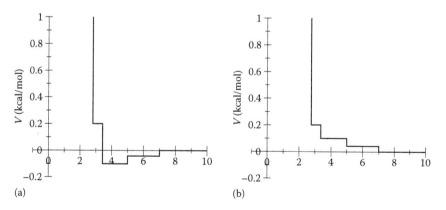

Figure 12.4   Example of attractive (a) and repulsive (b) effective bead-bead multiple-step potentials.

Protein and peptide aggregation is a very slow process, and accordingly, an ideal subject of study for dMD simulations. Thus, Marchut and Hall [51] developed an empirical force field to be used in an intermediate resolution model. The study, designed for investigation of the aggregation of polyglutamine peptides, was provided with a detailed level of description, especially for the side chains, which allowed the formation of hydrogen bonds between them. Four-bead (N, C_α, C, and one bead for side chain) force fields have been largely used to study amyloid aggregation [52,53]. These potentials reproduce hydrogen bonds with pseudobonds using the three backbone beads as a reference (see Figure 12.5), while hydrophobic/hydrophilic and electrostatic interactions are defined only between side chain beads. As with the previous model, the overall strength of hydrophobic and electrostatic interactions were scaled with respect to the hydrogen bond [52] strength to maintain a proper balance of interactions. The relative hydrophobicity of each side chain was evaluated using the empirical hydropaticity scale by Kyte and Doolitle [54].

A lower-resolution model (the one-bead model) has been used to generate protein-folding pathways using structure-based potentials [55]. These are attractive potentials defined only between residues that are in contact in the native conformation. In this model each residue is represented with just one bead, creating polymer-like structures. Similar Go-like potentials were developed in our group to reproduce in an inexpensive way the dynamics in equilibrium of solvated proteins [56]. In the process of translating the ENM-based functional to the discrete framework, square wells with infinite walls were defined between all C_αs within a given cutoff. The amplitude of the wells were adjusted to reproduce the distance-decay of neighbor interactions [15,56]. This ultrasimplified method showed extreme computational efficiency and was able to reproduce with accuracy the equilibrium dynamics of proteins as determined from explicit solvent atomistic MD simulations [25,56].

As illustrated in the previous paragraph, most dMD force fields have been created for the study of proteins, but recently, Dokholyan's group has made a great effort to move dMD into the nucleic acids world, which seemed to be an

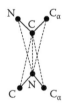

Figure 12.5 Hydrogen bonding in the four-bead model. Pseudobonds are between the acceptor (N) and donor (C) of each residue. Neighboring backbone atoms of the other residue enforce the correct orientation of the hydrogen bond.

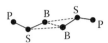

Figure 12.6   Scheme of the hydrogen bonding pattern between two nucleotides in the three-bead model of nucleic acids. Each nucleotide is modeled with a phosphate bead (P), a sugar bead (S), and a nucleobase bead (B). When the hydrogen bond (thick dashed line) is formed the axes S-B in each nucleotide are nearly aligned. This alignment is conserved by means of pseudobonds (thin dashed lines).

exclusive territory for continuum force fields [57,58]. They developed a simplified model where each nucleotide is represented with three beads: one for the phosphate, one for the sugar, and another for the nucleobase [59]. In this model hydrogen-bonding between the nucleobases is assigned with the aforementioned reaction algorithm and imposed geometrical scheme (Figure 12.6). Despite the minimalistic approach, this model allows the definition of electrostatic repulsion between phosphates, stacking, and hydrophobic interactions between nucleobases. dMD folding simulations were coupled to a Metropolis algorithm and experimental tabulated empirical energies of loop formation to properly reproduce the stability of loop arrangements.

This coarse-grained representation of nucleic acids was created to study ribonucleic acid (RNA) structures, and despite its simplicity displayed a good ability to reproduce experimental structures, especially when it was combined with low-resolution experimental structural data. Thus, Gherghe et al. [60] successfully predicted the study of tRNA^{ASP} by combining in dMD simulations secondary structure data (derived from selective 2′-hydroxyl acylation analyzed by primer extension [SHAPE] chemistry), empirical interresidue distances (from intercalation experiments) and force field evaluation. Similarly, Ding et al. [61] added base pair recognition rules and hydroxyl radical probing (HRP) experimental information into the dMD force field, deriving a very powerful simulation protocol (see "RNA Structural Predictions" in application section). The nucleic acids world is clearly an open field where dMD is expected to make a large impact in the near future.

IMPLEMENTATION OF dMD

The dMD protocol has been applied to biomolecular simulations since the 1990s (see "Application of dMD in Biomolecular Problems" section [62–64]). However, most studies found in the literature use in-house codes based on the early descriptions of the method [65], which are adapted ad hoc to solve specific problems of interest. To our knowledge only two codes are publicly available: πDMD from Dokholyan's group, which is accessible for nonprofit research from MoleculesInAction.com, and DISCRETE, a software developed within the Scalalife initiative [66]. Clearly, compared to standard MD simulations, the work done in dMD code engineering is minuscule. This might be justified by two reasons: (1) the small number of groups using dMD compared to MD, and (2) due to the basic formalism of dMD reasonably high simulation rates can be achieved without sophisticated software optimization. Figure 12.7 show simulation times obtained with DISCRETE for a series of protein systems using a quasi-atomistic representation [44]. Considering that dMD runs in a single laptop processor, dMD simulation rates are indeed comparable to equivalent MD running in state-of-the-art supercomputer equipment. The advantage is even clearer when considering that dMD controls the sampled degrees of freedom, which means that dMD simulations multiply by a factor of around 10 the sampling quality of MD for equivalent simulation times.

dMD is an event-driven simulation technique where the calculation of collision times takes most of the CPU time, as it involves, with a naïve approach, the calculation of $N(N-1)/2$ event times (N is the number of particles of the system). In principle, upon every event, collision times corresponding to the colliding particles are recalculated and placed in the appropriate order, making calculation inefficient. To improve searching and updating, a list is usually stored

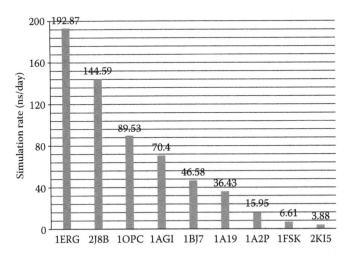

Figure 12.7   DISCRETE benchmark. 1ERG: extracellular region of human CD59 (560 atoms); 2J8B: human CD59 protein (793 atoms); 1OPC: *E. coli* Ompr DNA-binding protein (806 atoms); 1AGI: bovine angiogenin (1024 atoms); 1BJ7: bovine lipocalin allergen (1295 atoms); 1A19: *Bacillus amyloliquefaciens* Barstar (1438 atoms); 1A2P: *Bacillus amyloliquefaciens* Barnase (2624 atoms); 1FSK: major birch pollen allergen (4577 atoms); 2KI5: herpes virus thymidine kinase (4627 atoms). Computations done with an Intel Core2 Duo E6750 at 2.66 GHz. DISCRETE version 0.2.6.1.

using data structures such as binary trees [65]. The availability of a full event list allows also predicting the future positions of the particles [67]. Unfortunately, storing a collision list for a large system can use a significant amount of memory. An alternative approach [68] is to only store the soonest-to-occur event with its partner in a single-event time list for each particle. Since only one event per particle is stored, list management is much faster and the storage needed is much lower. The main issue with the single-event implementation is that more collision times must be recalculated, not only for the colliding particles, but for all particles predicted to collide with them in the future.

Additional improvements over the naïve implementation of dMD can be obtained by the construction of a neighbor list. Recent implementations follow the well-known spatial decomposition used in most simulation algorithms [69]. Simulation space is divided into cells whose sizes are chosen such that considered events are limited to the particles belonging to the given cell and to its neighboring cells. For standard MD the list of particles in a cell is obtained directly for the particle position and can be easily updated after motion integration. However, in the case of dMD, this requires an extra overhead as particles moving at a constant velocity but not colliding can escape from the containing cell. To solve that problem, cell boundary crossings need to be considered as extra events and included in the event list [70]. Our DISCRETE software uses an alternative approach, where neighbor lists are constructed on the basis of the interaction potentials. In this way, the definition of such potentials naturally sets the cutoff distances to select the neighbors for the different particle types.

One of the consequences of the intrinsic simplicity of the dMD algorithm is that parallelization of the code to scale to a significant number of processors is difficult. Data dependencies arise from the evolution of the simulation and are highly stochastic, making it difficult to predict them. Traditional parallelization strategies in MD are based in the spatial decomposition where one or more cells are assigned to each thread. Events can then be processed in parallel assuming a small dependence between cells. However, if events are processed "optimistically," with few communication steps between threads, some kind of rollback procedure should be provided to avoid inconsistencies, as cell dependences always exist. This makes the data management extremely costly for large systems. Khan and Herbordt [71] use a complementary approach called event-based decomposition. On top of a spatial decomposition approach, a unique event list is maintained and shared among threads. The availability of the full event list for all threads allows processing it in parallel while no concurrency issues appear. The authors claim that this approach scales nearly 6× in an eight-core processor and over 9× in a 12-core processor [71]. Recent versions of πdMD from Dokholyan's group have implemented a Khan-Herbordt parallelization scheme [49], reaching good performance in different platforms.

dMD in Web Servers

The fast response of the dMD algorithm has allowed the inclusion of this type of simulation as a back-end engine in a number of web-based tools. Relevant examples are the set of tools oriented to protein and RNA folding offered by the Dokholyan group: iFold [72] simulates folding using a two-bead/residue Go model, and iFoldRNA [73] performs interactive folding of RNA, again using a two-bead/residue model. DISCRETE is available as a stand-alone simulation tool (http://www.scalalife.eu) but can be used also through the general-purpose tool MDWeb [74]. A recent implementation based on DISCRETE, MDdMD [16] provides transition paths between two protein conformations. In MDdMD, the dMD simulation engine is complemented by NMA to assure that trajectories follow realistic paths. Finally DISCRETE, in this case using a one-bead model, is one of the methods providing trajectories for FlexServ [35], a web-based tool that performs a complete analysis of protein flexibility based on essential dynamics. Certainly, as dMD becomes more popular, an increasing number of dMD web applications will be available for the community.

APPLICATION OF dMD IN BIOMOLECULAR PROBLEMS

To our knowledge application of dMD in the biomolecular world was first proposed by the Karplus group in 1996 [62]. Since then, dMD has arisen as a powerful tool to capture the biophysical properties of macromolecules. We will review in the following some of these recent bioapplications of dMD. The reader is directed to Proctor et al. [75] to gain a wider view on the range of applications of the technique in other fields.

Protein Folding

Predicting the three-dimensional (3-D) structure out of the amino acid sequence has been, for decades, the holy grail of structural and computational biology. The use of dMD in the protein-folding problem was first proposed in 1996 by Zhou et al. [62]. Just one year later Zhou and Karplus [63] studied the folding thermodynamics of the three-helix-bundle fragment of *Staphylococcus aureus* protein A by using dMD coupled to a C_α coarse-grained model with Go-like potentials. Despite the extreme simplicity of the model, the computed phase diagram captured the essential expected features like the coil-to-globule transition or the ordered-to-disordered transition in globular structures. This seminal work was followed by Dokholyan et al. [64], who studied the temperature-dependent folding-unfolding transition of a peptide model. Dokholyan et al. used a force field that was a little bit more elaborate, including penalties for nonnative contacts, which was able to reproduce the kinetics of the first collapse, even trapping misfolded intermediates.

Ding et al. presented in 2005 a folding study for Trp-Cage protein [50] using a higher-resolution description where each residue was represented by at least four beads (except Gly), allowing them to consider previously ignored side chain entropy effects. An elaborate physical force field including many terms was used [46,50] (see the section titled "dMD Force Fields"). Using this formalism, Ding et al. found several folded ↔ unfolded transitions describing the native state with a RMSD below 2 Å from experimental structures. The same authors developed the first dMD atomistic force field for proteins [46], which was successfully used (RMSD to target structure from 2 to 6 Å) to predict the folding of six small proteins (20–60 residues) using replica exchange dMD simulations.

More recently dMD has been used to obtain a detailed insight into the kinetics of protein folding by running multiple parallel MD simulations [49,71]. For example, Shirvanyants et al. successfully studied the folding pathways of small (60–120 residues) proteins using a very elaborate all-atom force-field that incorporated secondary structure propensity and was calibrated with seven fast-folder proteins [49]. dMD is still far from being able to reproduce folding of realistic (100–1000 residues) proteins, but it can be an inexpensive, robust method to gain structural information on this complex process.

Protein Structural Refinement and Design

Homology modeling is the most powerful way to predict the structure of proteins with close homologs of known structure. Unfortunately, homology models, even globally correct, often present local errors such as structural distortions or steric clashes. dMD can be use to remove these local errors. Along these lines, Ramachandran et al. developed a tool that robustly minimizes structures based on short dual-temperature dMD simulations [76].

Protein design is another of the as yet elusive objectives in biochemistry and biotechnology. The potential impact of accurate protein rational design is unimaginable, as well as the problems to overcome. dMD is still not mature enough to design completely *de novo* proteins as has been done with some *ab initio* folding strategies [77,78], but the methods seem robust enough to predict the structural and functional impact of site-directed mutagenesis. Thus, Ding et al. [79] used modified Go-like potentials [80] to determine mutations that can move to protein to adopt alternative folds, a first step toward rational protein design.

RNA Structural Predictions

The expanding functionality of RNA molecules has boosted the interest in RNA 3-D structure prediction. Unfortunately, RNA is extremely flexible and folds in very long timescales, making prediction of RNA structure extremely challenging. Using a low-resolution force field (see above) Dokholyan's group obtained significant success in predicting structure from sequence in model systems [59]. An improved force field that included loop penalties and other additional corrections was benchmarked against a dataset of 153 RNA sequences using replica exchange to improve sampling [81]. Derived results were promising for small RNAs (RMSD with experiment typically below 6 Å for RNAs shorter than 50 nucleotides), but less predictive power was found for longer oligos. In order to improve predictive power, low-resolution experimental constraints were introduced into the dMD simulations, allowing Gherghe et al. [60] to obtain a refined 3-D model for tRNA^{ASP} with deviation lower than 4 Å from experiment. Using HRP data as a measure of solvent accessibility, Ding et al. managed to obtain a good 3-D structural prediction for four complicate RNA systems (*Oceanobacillus iheyensis* [*O. iheyensis*] group II intron, RNase P catalytic domain, Lysine riboswitch, and glmS ribozyme) [61].

Protein Aggregation

Protein aggregation is a slow process happening continuously in the cell that can very seriously challenge human health. Different authors have taken advantage of the capability of dMD to study slow diffusion processes and have applied the technique to study protein aggregation related to different pathologies [82–84]. Special efforts have been made in the study of amyloid β-protein (Aβ) monomer aggregation, due to its close relationship with Alzheimer's disease [85–87]. Thus, Peng et al. [88] first demonstrated in 2004 the suitability of dMD simulations to investigate amyloid peptide aggregation using a simple two-bead model for each residue that was capable of reproducing the formation of fibrils in an ensemble of 28 Aβ$_{1-40}$ peptides. Soon after, Urbanc et al. [52,89] successfully reproduced the distinct oligomerization pattern of two Aβ alloforms, Aβ$_{1-40}$ and Aβ$_{1-42}$, providing structural clues to the largely pro-Alzheimer characteristics of the Aβ$_{1-42}$ variant and reproducing well available experimental data on these systems [52,90–92]. Further studies were carried out to address in detail folding ↔ unfolding transitions of fragments of the Aβ peptide with united-atom representation [93] as well as the effect of the presence of short fragments of peptides in the aggregation process [94].

Equilibrium Protein Dynamics

Analysis of the equilibrium dynamics of proteins reveals important information about their intrinsic deformability that usually correlates with functional movements [14]. Simple Go-like dMD simulations were able to reproduce very well the equilibrium dynamics of a wide representative range of proteins [56], with the advantage with respect to ENM that introduction of anharmonic per-turbations is straightforward in dMD. More recently, Emperador et al. intro-duced an all-atom model coupled to a quasi-physical potential to treat protein dynamics around an equilibrium state [95] with atomistic details. dMD derived trajectories using this *ab initio* force field showed good correlations with stan-dard MD simulations in solution for a large range of proteins [44]. In this field, dMD presents many advantages other than sampling efficiency, the most important one being the facility to use hybrid coarse-grained atomistic force fields and the full control over the simulated system.

Conformational Transitions

There is an increasing number of proteins that are experimentally detected in two very distinct conformations, illustrating the existence of huge (as large as 27 Å RMSD) functionally relevant conformational transitions. Obviously, great interest exists in finding potential pathways for such conformational transi-tions, a task that considering the typical timescale of the processes can be extremely difficult for atomistic MD simulations. Sfriso et al. [16] took advan-tage of the remarkable dMD sampling capabilities to track conformational transitions in proteins. The procedure (named MDdMD) couple atomistic dMD simulations with a double-core Monte Carlo method that implements a Maxwell Demon biasing scheme and a second bias procedure that guaran-tees that the predicted transition follows, as much as possible, the essential deformation modes of the protein [15]. The authors tested the method in 94 conformational transitions (derived from 47 pairs of experimental protein con-formations) with a success rate above 90%. MDdMD finds reasonable pathways in the minutes-to-hours wall clock time using a laptop computer even in very difficult cases. Interestingly, the method was able to detect, without any fur-ther bias, known intermediates, sampling in all cases structures with excel-lent protein-like properties [15], something that is not always granted by other morphing methods. Despite its simplicity, dMD seems an excellent approach to trace conformational transition, generating snapshots that can be used as starting points for more accurate explicit solvent atomistic MD simulations (Figures 12.8 and 12.9).

Protein–Protein Docking

The determination of structural models of the complexes of proteins with other macromolecules or small drugs is a requirement in computer-aided drug dis-covery and also a crucial step in systems biology and even network medicine. Proctor et al. [96] presented very recently an approach to discriminate native ligand processes from nonnative ones based on the differential residence time

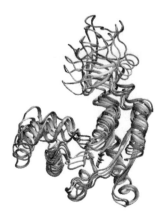

Figure 12.8   Adenylate kinase transition path modeled with MDdMD application. Red (1AKE) and blue (4AKE) structures are initial known states. The morphing algorithm establishes plausible intermediate structures within minutes.

5′NT (1hp1 ⟵⟶ 1oi8 ⟵⟶ 1hpu)

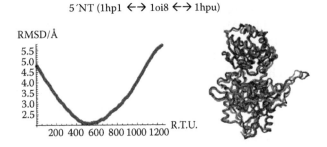

Figure 12.9   Structural comparison of the obtained path for 1hp1 → 1hpu conformational transition respect to experimental intermediate 1oi8. The proposed path samples near experimental structures without any further bias.

of native and nonnative processes in dMD trajectories. Using this approach the authors were able to successfully predict binding sites for six out of eight cases studied. Some applications of dMD have also appeared in the very complex field of protein–protein docking. Thus, Emperador et al. developed a dMD-based framework to introduce protein flexibility at the interface of models of protein complexes [97] derived from standard rigid-docking exhaustive sampling methods. The method took advantage of a multiple granularity scheme across the interacting proteins (low resolution in the core and high resolution in the interacting interface) and was tested in a well-balanced database of 61 complexes for which not only the complex, but also the isolated monomers were structurally known. The approach, which was coupled to pyDock scoring

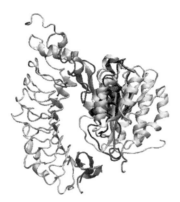

Figure 12.10   Experimental structure of a protein–protein complex (orange and green) and the best-scored model obtained after dMD simulation started from a rigid docking pose. For the simulated structure (red and blue) only the protein–protein interface is displayed.

functional [98], demonstrated significant increase in the quality of the complexes (especially of those whose formation implies monomer flexibility) with respect to state-of-the-art docking protocols, with a modest computational cost, which allows its implementation in the high-throughput regime (Figure 12.10).

CONCLUSIONS

Discrete molecular dynamics is probably one of the simplest molecular simulation techniques. Its basic foundations can be deduced from high-school physics books. dMD allows the expert user a great deal of control over the simulated systems, makes the implementation of knowledge-based potentials trivial, and is computationally highly efficient. Despite is short life, dMD has proved its power in a wide range of cases and certainly will became a default technique for large-scale dynamics calculations in biomolecular systems.

REFERENCES

1. Henzler-Wildman, K. A.; Lei, M.; Thai, V.; Kerns, S. J.; Karplus, M.; Kern, D. A hierarchy of timescales in protein dynamics is linked to enzyme catalysis. *Nature* **2007**, *450*, 913–916.
2. Velazquez-Muriel, J. A.; Rueda, M.; Cuesta, I.; Pascual-Montano, A.; Orozco, M.; Carazo, J.-M. Comparison of molecular dynamics and superfamily spaces of protein domain deformation. *BMC Struct. Biol. [Online]* **2009**, *9*, 6.

3. Bakan, A.; Bahar, I. The intrinsic dynamics of enzymes plays a dominant role in determining the structural changes induced upon inhibitor binding. *Proc. Natl. Acad. Sci. U. S. A.* **2009**, *106*, 14349–14354.
4. Yang, L.; Song, G.; Jernigan, R. L. How well can we understand large-scale protein motions using normal modes of elastic network models? *Biophys. J.* **2007**, *93*, 920–929.
5. Bahar, I.; Chennubhotla, C.; Tobi, D. Intrinsic dynamics of enzymes in the unbound state and relation to allosteric regulation. *Curr. Opin. Struct. Biol.* **2007**, *17*, 633–640.
6. Tobi, D.; Bahar, I. Structural changes involved in protein binding correlate with intrinsic motions of proteins in the unbound state. *Proc. Natl. Acad. Sci. U. S. A.* **2005**, *102*, 18908–18913.
7. Eyal, E.; Dutta, A.; Bahar, I. Cooperative dynamics of proteins unraveled by network models. *WIREs Comput. Mol. Sci.* **2011**, *1*, 426–439.
8. Dobbins, S. E.; Lesk, V. I.; Sternberg, M. J. E. Insights into protein flexibility: The relationship between normal modes and conformational change upon protein–protein docking. *Proc. Natl. Acad. Sci. U. S. A.* **2008**, *105*, 10390–10395.
9. Gerstein, M.; Krebs, W. A database of macromolecular motions. *Nucleic Acids Res.* **1998**, *26*, 4280–4290.
10. Falke, J. J. Enzymology. A moving story. *Science* **2002**, *295*, 1480–1481.
11. Leo-Macias, A.; Lopez-Romero, P.; Lupyan, D.; Zerbino, D.; Ortíz, A. R. An analysis of core deformations in protein superfamilies. *Biophys. J.* **2005**, *88*, 1291–1299.
12. Micheleti, C. Comparing proteins by their internal dynamics: Exploring structure-function relationships beyond static structural alignments. *Phys. Life Rev.* **2013**, *10*, 1–26.
13. Orozco, M. The dynamic view of proteins: Comment on "Comparing proteins to their internal dynamics: Exploring structure-function relationships beyond static structural alignments." *Phys. Life Rev.* **2013**, *10*, 29–30.
14. Stein, A.; Rueda, M.; Panjkovich, A.; Orozco, M.; Aloy, P. A systematic study of the energetics involved in structural changes upon association and connectivity in protein interaction networks. *Structure* **2011**, *19*, 881–889.
15. Orellana, L.; Rueda, M.; Ferrer-Costa, C.; Lopez-Blanco, J. R.; Chacón, P.; Orozco, M. Approaching elastic network models to molecular dynamics flexibility. *J. Chem. Theory Comput.* **2010**, *6*, 2910–2923.
16. Sfriso, P.; Emperador, A.; Orellana, L.; Hospital, A.; Gelpí, J. L.; Orozco, M. Finding conformational transition pathways from discrete molecular dynamics simulations. *J. Chem. Theory Comput.* **2012**, *8*, 4707–4718.
17. Henzler-Wildman, K. A.; Thai, V.; Lei, M.; Ott, M.; Wolf-Watz, M.; Fenn, T.; Pozharski, E.; Wilson, M. A.; Petsko, G. A.; Karplus, M. et al. Intrinsic motions along an enzymatic reaction trajectory. *Nature* **2007**, *450*, 838–844.
18. Lindorff-Larsen, K.; Best, R. B.; Depristo, M. A.; Dobson, C. M.; Vendruscolo, M. Simultaneous determination of protein structure and dynamics. *Nature* **2005**, *433*, 128–132.
19. Ban, D.; Funk, M.; Gulich, R.; Egger, D.; Sabo, T. M.; Walter, K. F. A.; Fenwick, R. B.; Giller, K.; Pichierri, F.; de Groot, B. L. et al. Kinetics of conformational sampling in ubiquitin. *Angew. Chem. Int. Ed.* **2011**, *50*, 11437–11440.

20. Fenwick, R. B.; Esteban-Martin, S.; Richter, B.; Lee, D.; Walter, K. F. A.; Milovanovic, D.; Becker, S.; Lakomek, N. A.; Griesinger, C.; Salvatella, X. Weak long-range correlated motions in a surface patch of ubiquitin involved in molecular recognition. *J. Am. Chem. Soc.* **2011**, *133*, 10336–10339.

21. Kubitzki, M. B.; de Groot, B. L. The atomistic mechanism of conformational transition in adenylate kinase: A TEE-REX molecular dynamics study. *Structure* **2008**, *16*, 1175–1182.

22. Shimamura, T.; Weyand, S.; Beckstein, O.; Rutherford, N. G.; Hadden, J. M.; Sharples, D.; Sansom, M. S. P.; Iwata, S.; Henderson, P. J. F.; Cameron, A. D. Molecular basis of alternating access membrane transport by the sodium-hydantoin transporter Mhp1. *Science* **2010**, *328*, 470–473.

23. Paci, E.; Lindorff-Larsen, K.; Dobson, C. M.; Karplus, M.; Vendruscolo, M. Transition state contact orders correlate with protein folding rates. *J. Mol. Biol.* **2005**, *352*, 495–500.

24. Orozco, M.; Orellana, L.; Hospital, A.; Naganathan, A. N.; Emperador, A.; Carrillo, O.; Gelpí, J. L. Coarse-grained representation of protein flexibility. Foundations, successes, and shortcomings. *Adv. Protein Chem. Struct. Biol.* **2011**, *85*, 183–215.

25. Rueda, M.; Ferrer-Costa, C.; Meyer, T.; Pérez, A.; Camps, J.; Hospital, A.; Gelpí, J. L.; Orozco, M. A consensus view of protein dynamics. *Proc. Natl. Acad. Sci. U. S. A.* **2007**, *104*, 796–801.

26. Karplus, M.; Kuriyan, J. Molecular dynamics and protein function. *Proc. Natl. Acad. Sci. U. S. A.* **2005**, *102*, 6679–6685.

27. Tozzini, V. Coarse-grained models for proteins. *Curr. Opin. Struct. Biol.* **2005**, *15*, 144–150.

28. Bolhuis, P. G.; Chandler, D.; Dellago, C.; Geissler, P. L. Transition path sampling: Throwing ropes over rough mountain passes, in the dark. *Annu. Rev. Phys. Chem.* **2002**, *53*, 291–318.

29. Bolhuis, P. G. Transition-path sampling of beta-hairpin folding. *Proc. Natl. Acad. Sci. U. S. A.* **2003**, *100*, 12129–12134.

30. Wales, D. J. Energy landscapes: Calculating pathways and rates. *Int. Rev. Phys. Chem.* **2006**, *25*, 237–282.

31. Jamroz, M.; Orozco, M.; Kolinski, A.; Kmiecik, S. Consistent view of protein fluctuations from all-atom molecular dynamics and coarse-grained dynamics with knowledge-based force-field. *J. Chem. Theory Comput.* **2013**, *9*, 119–125.

32. Naganathan, A. N.; Orozco, M. The native ensemble and folding of a protein molten-globule: Functional consequence of downhill folding. *J. Am. Chem. Soc.* **2011**, *133*, 12154–12161.

33. García, A. E.; Onuchic, J. N. Folding a protein in a computer: An atomic description of the folding/unfolding of protein A. *Proc. Natl. Acad. Sci. U. S. A.* **2003**, *100*, 13898–13903.

34. Taketomi, H.; Ueda, Y.; Gō, N. Studies on protein folding, unfolding and fluctuations by computer simulation. *Int. J. Pept. Protein Res.* **1975**, *7*, 445–459.

35. Camps, J.; Carrillo, O.; Emperador, A.; Orellana, L.; Hospital, A.; Rueda, M.; Cicin-Sain, D.; D'Abramo, M.; Gelpí, J. L.; Orozco, M. FlexServ: An integrated tool for the analysis of protein flexibility. *Bioinformatics* **2009**, *25*, 1709–1710.

36. Yang, Z.; Májek, P.; Bahar, I. Allosteric transitions of supramolecular systems explored by network models: Application to chaperonin GroEL. *PLoS Comput. Biol.* **2009**, *5*, e1000360.

37. Lezon, T. R.; Sali, A.; Bahar, I. Global motions of the nuclear pore complex: Insights from elastic network models. *PLoS Comput. Biol.* **2009**, *5*, e1000496.

38. Bahar, I.; Lezon, T. R.; Yang, L.-W.; Eyal, E. Global dynamics of proteins: Bridging between structure and function. *Annu. Rev. Biophys.* **2010**, *39*, 23–42.

39. Bahar, I.; Lezon, T. R.; Bakan, A.; Shrivastava, I. H. Normal mode analysis of bio-molecular structures: Functional mechanisms of membrane proteins. *Chem. Rev.* **2010**, *110*, 1463–1497.

40. Lopez-Blanco, J. R.; Garzón, J. I.; Chacón, P. iMod: Multipurpose normal mode analysis in internal coordinates. *Bioinformatics* **2011**, *27*, 2843–2850.

41. Jorgensen, W. L.; Tirado-Rives, J. Molecular modeling of organic and biomolecular systems using BOSS and MCPRO. *J. Comput. Chem.* **2005**, *26*, 1689–1700.

42. McCammon, J. A.; Harvey, S. C. *Dynamics of Proteins and Nucleic Acids*. Cambridge University Press, Cambridge, 1988.

43. Karplus, M.; Petsko, G. A. Molecular dynamics simulations in biology. *Nature* **1990**, *347*, 631–639.

44. Emperador, A.; Meyer, T.; Orozco, M. Protein flexibility from discrete molecular dynamics simulations using quasi-physical potentials. *Proteins* **2010**, *78*, 83–94.

45. Meyer, T.; D'Abramo, M.; Hospital, A.; Rueda, M.; Ferrer-Costa, C.; Pérez, A.; Carrillo, O.; Camps, J.; Fenollosa, C.; Repchevsky, D. et al. MoDEL (Molecular Dynamics Extended Library): A database of atomistic molecular dynamics trajectories. *Structure* **2010**, *18*, 1399–1409.

46. Ding, F.; Tsao, D.; Nie, H.; Dokholyan, N. V. Ab initio folding of proteins with all-atom discrete molecular dynamics. *Structure* **2008**, *16*, 1010–1018.

47. Lazaridis, T.; Karplus, M. Effective energy function for proteins in solution. *Proteins* **1999**, *35*, 133–152.

48. Ding, F.; Borreguero, J.; Buldyrev, S. V.; Stanley, H.; Dokholyan, N. Mechanism for the α-helix to β-hairpin transition. *Proteins* **2003**, *53*, 220–228.

49. Shirvanyants, D.; Ding, F.; Tsao, D.; Ramachandran, S.; Dokholyan, N. V. Discrete molecular dynamics: An efficient and versatile simulation method for fine protein characterization. *J. Phys. Chem. B* **2012**, *116*, 8375–8382.

50. Ding, F.; Buldyrev, S.; Dokholyan, N. Folding Trp-cage to NMR resolution native structure using a coarse-grained protein model. *Biophys. J.* **2005**, *88*, 147–155.

51. Marchut, A. J.; Hall, C. K. Side-chain interactions determine amyloid formation by model polyglutamine peptides in molecular dynamics simulations. *Biophys. J.* **2006**, *90*, 4574–4584.

52. Urbanc, B.; Cruz, L.; Yun, S.; Buldyrev, S. V.; Bitan, G.; Teplow, D. B.; Stanley, H. E. In silico study of amyloid beta-protein folding and oligomerization. *Proc. Natl. Acad. Sci. U. S. A.* **2004**, *101*, 17345–17350.

53. Nguyen, H. D.; Hall, C. K. Molecular dynamics simulations of spontaneous fibril formation by random-coil peptides. *Proc. Natl. Acad. Sci. U. S. A.* **2004**, *101*, 16180–16185.

54. Kyte, J.; Doolittle, R. F. A simple method for displaying the hydropathic character of a protein. *J. Mol. Biol.* **1982**, *157*, 105–132.

55. Zhou, Y.; Karplus, M. Interpreting the folding kinetics of helical proteins. *Nature* **1999**, *401*, 400–403.

56. Emperador, A.; Carrillo, O.; Rueda, M.; Orozco, M. Exploring the suitability of coarse-grained techniques for the representation of protein dynamics. *Biophys. J.* **2008**, *95*, 2127–2138.

57. Pérez, A.; Luque, F. J.; Orozco, M. Frontiers in molecular dynamics simulations of DNA. *Acc. Chem. Res.* **2012**, *45*, 196–205.

58. Orozco, M.; Noy, A.; Pérez, A. Recent advances in the study of nucleic acid flexibility by molecular dynamics. *Curr. Opin. Struct. Biol.* **2008**, *18*, 185–193.

59. Ding, F.; Sharma, S.; Chalasani, P.; Demidov, V. V.; Broude, N. E.; Dokholyan, N. V. Ab initio RNA folding by discrete molecular dynamics: From structure prediction to folding mechanisms. *RNA* **2008**, *14*, 1164–1173.

60. Gherghe, C. M.; Leonard, C. W.; Ding, F.; Dokholyan, N. V.; Weeks, K. M. Native-like RNA tertiary structures using a sequence-encoded cleavage agent and refinement by discrete molecular dynamics. *J. Am. Chem. Soc.* **2009**, *131*, 2541–2546.

61. Ding, F.; Lavender, C. A.; Weeks, K. M.; Dokholyan, N. V. Three-dimensional RNA structure refinement by hydroxyl radical probing. *Nat. Methods* **2012**, *9*, 603–608.

62. Zhou, Y.; Hall, C.; Karplus, M. First-order disorder-to-order transition in an isolated homopolymer model. *Phys. Rev. Lett.* **1996**, *77*, 2822–2825.

63. Zhou, Y.; Karplus, M. Folding thermodynamics of a model three-helix-bundle protein. *Proc. Natl. Acad. Sci. U. S. A.* **1997**, *94*, 14429–14432.

64. Dokholyan, N. V.; Buldyrev, S. V.; Stanley, H. E.; Shakhnovich, E. I. Discrete molecular dynamics studies of the folding of a protein-like model. *Fold. Des.* **1998**, *3*, 577–587.

65. Rapaport, D. C. *The Art of Molecular Dynamics Simulation.* Cambridge University Press, New York, 2004.

66. Apostolov, R.; Axner, L.; Agren, H.; Ayugade, E.; Duta, M.; Gelpí, J. L.; Gimenez, J.; Goni, R.; Hess, B.; Jamitzky, F.; Kranzmuller, D.; Labarta, J.; Laure, E.; Lindahl, E.; Orozco, M.; Peterson, M.; Satzger, H.; Trefethen, A. ScalaLife: Scalable software services for life science. In Proceedings of 9th HealthGrid Conference, **2011**.

67. Erpenbeck, J. J.; Wood, W. W. Molecular dynamics techniques for hard-core systems. In Berne, B. J., ed., *Modern Theoretical Chemistry*, Plenum, New York, **1977**, 6B, 1.

68. Lubachevsky, B. D. How to simulate billiards and similar systems. *J. Comput. Phys.* **1991**, *94*, 255–283.

69. Larsson, P.; Hess, B.; Lindahl, E. Algorithm improvements for molecular dynamics simulations. *WIREs Comput. Mol. Sci.* **2011**, *1*, 93–108.

70. Smith, S. W.; Hall, C. K.; Freeman, B. D. Molecular dynamics for polymeric fluids using discontinuous potentials. *J. Comput. Phys.* **1997**, *134*, 16–30.

71. Khan, M. A.; Herbordt, M. C. Parallel discrete molecular dynamics simulation with speculation and in-order commitment. *J. Comput. Phys.* **2011**, *230*, 6563–6582.

72. Sharma, S.; Ding, F.; Nie, H.; Watson, D.; Unnithan, A.; Lopp, J.; Pozefsky, D.; Dokholyan, N. V. iFold: A platform for interactive folding simulations of proteins. *Bioinformatics* **2006**, *22*, 2693–2694.

73. Sharma, S.; Ding, F.; Dokholyan, N. V. iFoldRNA: Three-dimensional RNA structure prediction and folding. *Bioinformatics* **2008**, *24*, 1951–1952.

74. Hospital, A.; Andrio, P.; Fenollosa, C.; Cicin-Sain, D.; Orozco, M.; Gelpí, J. L. MDWeb and MDMoby: An integrated web-based platform for molecular dynamics simulations. *Bioinformatics* **2012**, *28*, 1278–1279.

75. Proctor, E. A.; Ding, F.; Dokholyan, N. Discrete molecular dynamics. *WIREs Comput. Mol. Sci.* **2011**, *1*, 80–92.

76. Ramachandran, S.; Kota, P.; Ding, F.; Dokholyan, N. V. Automated minimization of steric clashes in protein structures. *Proteins* **2011**, *79*, 261–270.

77. Fleishman, S. J.; Whitehead, T. A.; Ekiert, D. C.; Dreyfus, C.; Corn, J. E.; Strauch, E.-M.; Wilson, I. A.; Baker, D. Computational design of proteins targeting the conserved stem region of influenza hemagglutinin. *Science* **2011**, *332*, 816–821.

78. Azoitei, M. L.; Correia, B. E.; Ban, Y.-E. A.; Carrico, C.; Kalyuzhniy, O.; Chen, L.; Schroeter, A.; Huang, P.-S.; McLellan, J. S.; Kwong, P. D. et al. Computation-guided backbone grafting of a discontinuous motif onto a protein scaffold. *Science* **2011**, *334*, 373–376.

79. Ding, F.; Dokholyan, N. V. Emergence of protein fold families through rational design. *PLoS Comput. Biol.* **2006**, *2*, e85.

80. Ueda, Y.; Taketomi, H.; Gō, N. Studies on protein folding, unfolding, and fluctuations by computer simulation. II. A three-dimensional lattice model of lysozyme. *Biopolymers* **1978**, *17*, 1531–1548.

81. Laing, C.; Schlick, T. Computational approaches to RNA structure prediction, analysis, and design. *Curr. Opin. Struct. Biol.* **2011**, *21*, 306–318.

82. Urbanc, B.; Borreguero, J.; Cruz, L.; Stanley, H. Ab initio discrete molecular dynamics approach to protein folding and aggregation. *Methods Enzymol.* **2006**, *412*, 314–338.

83. Urbanc, B.; Cruz, L.; Teplow, D. B.; Stanley, H. E. Computer simulations of Alzheimers amyloid-protein folding and assembly. *Curr. Alzheimer Res.* **2006**, *3*, 493–504.

84. Sharma, S.; Ding, F.; Dokholyan, N. V. Probing protein aggregation using discrete molecular dynamics. *Front. Biosci.* **2008**, *13*, 4795–4808.

85. Walsh, D. M.; Hartley, D. M.; Kusumoto, Y.; Fezoui, Y.; Condron, M. M.; Lomakin, A.; Benedek, G. B.; Selkoe, D. J.; Teplow, D. B. Amyloid β-protein fibrillogenesis. *J. Biol. Chem.* **1999**, *274*, 25945–25952.

86. Lambert, M. P.; Barlow, A. K.; Chromy, B. A.; Edwards, C.; Freed, R.; Liosatos, M.; Morgan, T. E.; Rozovsky, I.; Trommer, B.; Viola, K. L. et al. Diffusible, nonfibrillar ligands derived from Abeta1-42 are potent central nervous system neurotoxins. *Proc. Natl. Acad. Sci. U. S. A.* **1998**, *95*, 6448–6453.

87. Walsh, D. M.; Klyubin, I.; Fadeeva, J. V.; Cullen, W. K.; Anwyl, R.; Wolfe, M. S.; Rowan, M. J.; Selkoe, D. J. Naturally secreted oligomers of amyloid beta protein potently inhibit hippocampal long-term potentiation in vivo. *Nature* **2002**, *416*, 535–539.

88. Peng, S.; Ding, F.; Urbanc, B.; Buldyrev, S. V.; Cruz, L.; Stanley, H. E.; Dokholyan, N. V. Discrete molecular dynamics simulations of peptide aggregation. *Phys. Rev. E.* **2004**, *69*, 041908.

89. Urbanc, B.; Cruz, L.; Ding, F.; Sammond, D.; Khare, S.; Buldyrev, S. V.; Stanley, H. E.; Dokholyan, N. V. Molecular dynamics simulation of amyloid β dimer formation. *Biophys. J.* **2004**, *87*, 2310–2321.

90. Bitan, G.; Kirkitadze, M. D.; Lomakin, A.; Vollers, S. S.; Benedek, G. B.; Teplow, D. B. Amyloid beta-protein (Abeta) assembly: Abeta 40 and Abeta 42 oligomerize through distinct pathways. *Proc. Natl. Acad. Sci. U. S. A.* **2003**, *100*, 330–335.

91. Bitan, G.; Vollers, S. S.; Teplow, D. B. Elucidation of primary structure elements controlling early amyloid beta-protein oligomerization. *J. Biol. Chem.* **2003**, *278*, 34882–34889.

92. Urbanc, B.; Betnel, M.; Cruz, L.; Bitan, G.; Teplow, D. Elucidation of amyloid β-protein oligomerization mechanisms: Discrete molecular dynamics study. *J. Am. Chem. Soc.* **2010**, *132*, 4266–4280.

93. Borreguero, J.; Urbanc, B; Lazo, N. D.; Buldyrev, S. D.; Teplow, D. B.; Stanley, H. E. Folding events in the 21–30 region of amyloid β-protein (Aβ) studied in silico. *Proc. Natl. Acad. Sci. U. S. A.* **2005**, *102*, 6015–6020.
94. Urbanc, B.; Betnel, M.; Cruz, L.; Li, H. Structural basis for Aβ1–42 toxicity inhibition by Aβ C-Terminal fragments: Discrete molecular dynamics study. *J. Mol. Biol.* **2011**, *410*, 316–328.
95. Emperador, A.; Meyer, T.; Orozco, M. United-atom discrete molecular dynamics of proteins using physics-based potentials. *J. Chem. Theory Comput.* **2008**, *4*, 2001–2010.
96. Proctor, E. A.; Yin, S.; Tropsha, A.; Dokholyan, N. V. Discrete molecular dynamics distinguishes nativelike binding poses from decoys in difficult targets. *Biophys. J.* **2012**, *102*, 144–151.
97. Emperador, A.; Solernou, A.; Sfriso, P.; Pons, C.; Gelpí, J. L.; Fernández-Recio, J.; Orozco, M. Efficient relaxation of protein–protein interfaces by discrete molecular dynamics simulations. *J. Chem. Theory Comput.* **2013**, *9*, 1222–1229.
98. Cheng, T. M. K.; Blundell, T. L.; Fernandez Recio, J. pyDock: Electrostatics and desolvation for effective scoring of rigid-body protein–protein docking. *Proteins* **2007**, *68*, 503–515.

Section VI

Ensemble Methods

Chapter 13

Use of Ensemble Methods to Describe Biomolecular Dynamics by Small Angle X-Ray Scattering

Giancarlo Tria, Dmitri I. Svergun, and Pau Bernadó

CONTENTS

INTRODUCTION

In this chapter, we present modern computational approaches to use small-angle x-ray scattering (SAXS) in the analysis of flexible proteins. SAXS is a powerful method for analyzing the structure and structural changes of biological macromolecules in solution (Svergun et al. 2013). The method is most often used for purified solutions of globular proteins and complexes where all particles can be considered identical and the experimental scattering can be related to the structure of a single particle (although the scattering is typically isotropic due to the average over particle orientations). In this case, SAXS provides information about the overall shape and can construct three-dimensional (3-D) low-resolution models, either *ab initio* or using rigid body analysis in terms of known high resolution structures of domains or subunits.

For the solutions of nonidentical particles, the measured scattering reflects the average not only over the orientations but also over the different structures of the particles present in solution. In this case, shapes of individual components cannot be reconstructed given only the experimental scattering from the mixture. However, if the scattering intensities from the components are known *a priori*, their volume fractions can be determined. A typical application is the characterization of oligomeric mixtures, where only a few types of particles coexist in solution (e.g., monomer-dimer equilibrium) but much more complicated systems can also be analyzed. In particular, SAXS can be readily employed for the analysis of flexible systems including multidomain proteins with flexible linkers or intrinsically disordered proteins (IDPs).

The major problem of the SAXS data analysis for flexible systems is that the solution may contain an astronomic number of components (individual macromolecules having different conformations). Until recently, it was only possible to qualitatively distinguish between globular and unfolded macromolecules from the SAXS data. The situation has changed with the advent of methods utilizing ensemble analysis, which allows the extraction of quantitative information about the flexibility from the scattering data. In the ensemble approach one from the very beginning admits that the scattering data cannot be represented by a single model. Instead, one generates a pool of possible structures covering the entire conformational space explored by the macromolecule and attempts to fit the experimental data by a subensemble of the general pool. The overall properties of the subensemble compatible with the experiment can be subsequently analyzed and provide quantitative information about flexibility and dynamics of the macromolecule. The results of this analysis are often complemented by other methods providing structural information on such systems (methods such as nuclear magnetic resonance [NMR], circular dichroism [CD], calorimetry, and structure prediction).

In this chapter, after a brief introduction to the theoretical and experimental basics of SAXS, we will provide an overview of the computational ensemble methods employed to study conformational flexibility. Applications of these methods and their synergistic use with complementary techniques will be illustrated by recent results on various flexible systems.

SAXS BASIS

SAXS utilizes elastic scattering of photons by electrons in macromolecules. In a SAXS experiment the dissolved macromolecules are irradiated by a monochromatic x-ray beam at a wavelength λ and the elastically scattered x-rays have the same wavelength. (Blanchet and Svergun 2013). The scattering intensity from each macromolecule $I(s)$ is a function of the scattering vector $s = k_1 - k_0$, where k_0 and k_1 are the wavevectors of the incident and the scattered beam, both having the same magnitude $2\pi/\lambda$ (Figure 13.1a). The scattering from an ensemble of noninteracting identical particles in solution is a continuous isotropic function proportional to the scattering from a single particle averaged over all orientations (Ω) (Feigin and Svergun 1987):

$$I(s) = \langle A(s) \cdot A^*(s) \rangle_\Omega \tag{13.1}$$

where the momentum transfer $s = k_1 - k_0 = 4\pi\sin(\theta)/\lambda$ is a function of the scattering angle (2θ) (note that in some publications, the momentum transfer is

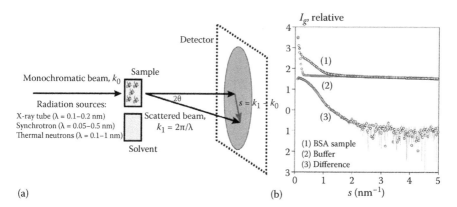

Figure 13.1 (a) Schematic representation of a typical SAS experiment. (b) X-ray scattering patterns from a solution of BSA measured at X33 (DORIS, Hamburg) in 50-mM HEPES, pH 7.5, solvent scattering, and the difference curve. (Reprinted from *J Struct Biol*, 172, Mertens, H. D. and D. I. Svergun. 2010, 128–141, Copyright 2010, with permission from Elsevier.)

denoted as q). A SAXS experiment on macromolecules in solution consists of collecting the intensity curve $I(s)$ for the particle as well as for the solvent in which the particle is dissolved. The difference between the two radially averaged one-dimensional (1-D) curves represents the SAXS curve from the particle. These intensities decrease rapidly with increasing angle and are usually presented in semilogarithmic scale (Figure 13.1b).

Valuable low-resolution structural information can be obtained from the experimental SAXS data without model assumptions. The interatomic (or pairwise) distance distribution, $p(r)$, is obtained as the inverse Fourier transformation of the scattering pattern $I(s)$:

$$p(r) = \Im[I(s)]^{-1} = \frac{r^2}{2\pi^2} \int_0^\infty \frac{s^2 I(s) \sin(sr)}{sr} dr \qquad (13.2)$$

The $p(r)$ function describes in the real space the frequencies of interatomic distance r within a particle weighted by the electron density (Feigin and Svergun 1987). The *maximum particle distance*, D_{max}, estimated iteratively using programs such as GNOM (Semenyuk and Svergun 1991) or ITP (Glatter 1977) when calculating $p(r)$, represents the maxim distance between two mass units of the protein. In addition to D_{max}, other structural parameters can be derived from raw data including *radius of gyration* (R_g), *molecular mass* (MM), and *hydrated particle volume* (V_p). The R_g corresponds to the root-mean-squared distance of each scattering element from the center of mass weighted by their scattering length and provide information about the electron density distribution within a particle. R_g is normally derived from Guinier's approximation $I(s) = I(0)\exp(-(sR_g)^2/3)$ (Guinier 1939) that is valid at small angles ($s \cdot R_g < 1.3$). In addition to the R_g, Guinier's analysis also provides the *forward scattering intensity*, $I(0)$, that is sensitive to the particle composition and concentration (Jacques and Trewhella 2010). Consequently, the MM can be estimated directly from the experimental data, for example, by comparison with the scattering from a standard protein ($MM_{particle} = I(0)_{particle} {}^* MM_{standard}/I(0)_{standard}$). However, although very straightforward, this method is strongly dependent on the solute concentration so that the latter must be accurately measured (Mylonas and Svergun 2007). Alternatively, the *hydrated particle volume* (V_p) allows one to estimate the MM independently of the solute concentration. The V_p is calculated from the scattering profile using Porod's equation (Porod 1982):

$$V_p \cong 2\pi^2 I(0)/Q, \quad Q = \int_{s_{min}}^{s_{max}} s^2 I(s) ds \qquad (13.3)$$

where Q is the so-called Porod invariant, and for proteins, the MM can be estimated as $MM[kDa] \approx V_p[nm^3]/1.7$ with an accuracy of about 20% (Petoukhov et al. 2012).

Monodispersity of the sample is a necessary prerequisite for the analysis of the scattering data and to extract parameters in terms of the single particle structure. Polydisperse systems containing different types of particles can be still characterized as long as information of the species in the mixture is available. If the theoretical scattering intensity from each component in a mixture is known, the measured intensity $I(s)$ is a linear combination (Koch, Vachette, and Svergun 2003)

$$I(s) = \sum_s v_k I_k(s) \tag{13.4}$$

where v_k and $I_k(s)$ are the volume fraction and the scattering intensity, respectively, for the kth component in solution. Using the program OLIGOMER (Konarev et al. 2006) it is possible to find the volume fractions occupied by each component in the mixture. For this, the experimental data from the mixture are fitted by Equation 13.4 and the best fitting values of v_k for known $I_k(s)$ are readily determined by a linear least squares procedure, minimizing the discrepancy between the experimental and computed data.

Obviously, this approach is not applicable for the macromolecular systems with pronounced flexibility like IDPs or proteins with intrinsically disordered regions (IDRs), where the number of accessible conformations (and thus, possible components in Equation 13.4) is astronomical. Traditionally, Kratky plots [$I(s)s^2$ as a function of s] (Glatter and Kratky 1982) have been used to qualitatively identify disordered states and distinguish them from globular particles (Figure 13.2). The Kratky representation has the capacity to enhance particular features of scattering profiles allowing an easier identification of different degrees of compactness (Doniach 2001). The scattering intensity from a globular protein has an asymptotic behavior $\sim 1/s^4$ at higher scattering angles, providing a bell-shaped Kratky plot with a well-defined maximum. In contrast, for a disordered chain the Kratky plot presents a plateau over a specific range of s, which is followed by a monotonic increase at higher angles. The latter behavior is often observed experimentally for unfolded proteins. Kratky plot is a useful tool to identify dynamics, but it provides qualitative information only. Regardless of the specific degree of motion, the quantitative interpretation of protein dynamics using SAXS must be conducted with ensemble methods.

As it was established recently, the SAXS data from dynamic systems where numerous multiple conformations are present in solution do contain quantitative information about the conformations explored, and the principles

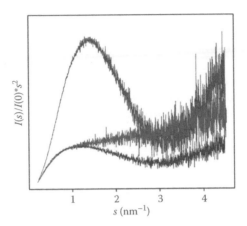

Figure 13.2   Dimensionless Kratky plot for three constructs of Src-kinase: the globular SH3 domain (blue), the fully disordered unique domain (red), and a construct joining both domains (purple). The prototypical features of globular and disordered domains are combined in the partially folded construct. (Reproduced by permission of The Royal Society of Chemistry.)

were proposed to utilize this information (Bernado et al. 2005, 2007, 2009; von Ossowski et al. 2005). The concept of using an ensemble (also called *supertertiary structure* [Tompa 2012]) to describe biomolecular dynamics is widely accepted today in the structural biology community and several computer programs have been developed to address this question. In the rest of the chapter we will focus on the ensemble approach by reviewing the most popular computer programs currently available to quantitatively characterize flexible particles using SAXS in solution. Some practical examples using ensemble approaches will also be presented.

ENSEMBLE APPROACHES FOR SAXS DATA ANALYSIS

Proteins display a large spectrum of motional modes encompassing fast fluctuations around the average globular structure, slow large-scale molecular reorganizations, and the inherent disorder observed in IDPs and IDRs. In all these scenarios, protein dynamics is crucial for biological function including recognition, regulation, and catalysis. In recent years, structural biologists have addressed the challenge of describing dynamic systems in terms of ensembles of reliable conformations guided by experimental data that represents average values for the complete ensemble of conformations (Bernado and Blackledge 2010). SAXS has not been exempt from this tendency and several approaches have been developed to characterize biomolecular mobility. These methods are based on a common strategy that consists of three consecutive steps:

1. Computational generation of a large ensemble describing the conformational landscape available to the protein
2. Computation of the scattering properties for each of these individual conformations
3. Selection of a subensemble of conformations that collectively describe the experimental profile using distinct optimization methods

In the following section, before presenting specific examples, a succinct description of the different approaches available will be presented.

Ensemble Optimization Method

The Ensemble Optimization Method (EOM) methodology has been developed as a general approach to address the structural characterization of IDPs by SAXS (Bernado et al. 2007) and it was the first ensemble method proposed for SAXS. Using this strategy the experimental SAXS profile is assumed to derive from an (undetermined *a priori*) number of coexisting conformational states. A subensemble of conformations is selected by a genetic algorithm (GA) from the scattering patterns computed from a large pool representing the maximum flexibility allowed by the protein topology. In that sense, EOM can be considered as a data-driven optimized ensemble strategy (Figure 13.3). The price to pay in order to optimize the large number of degrees of freedom by EOM using low-content information data is that assumptions are required to generate the pool of conformations. The pool of random conformations is also used to establish a threshold for the later quantitative interpretation of the results in terms of distributions of selected structures. In this sense the EOM approach is not entirely model-independent as it always relies on the structural rules imposed for the generation of conformations.

In the EOM algorithm, a potential solution is represented by an ensemble containing N different conformers of the same molecule. The appropriate ensemble is selected from a pool containing $M>>N$ conformers that should cover the conformational space available to the molecule. Subsequently, GA is then employed to select subsets of configurations that collectively fit the experimental data. The scattering data from the subensemble is computed by averaging the individual scattering patterns assuming that all conformers are equally populated.

$$I^{EOM}(s) = \frac{1}{N} \sum_{n=1}^{N} I_n(s)$$

(13.5)

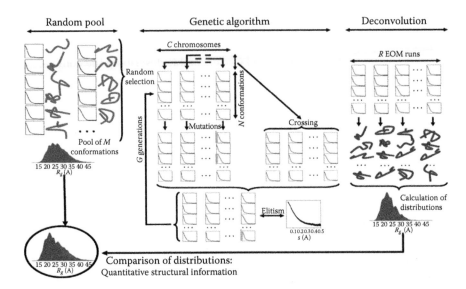

Figure 13.3   Schematic representation of the EOM strategy for the analysis of SAXS data. *M* randomly distributed conformations/curves (left part of the figure) are generated to make the initial pool. *C* chromosomes (ensembles) are selected from the pool and used to feed the genetic operators (mutations, crossing, and elitism) along the genetic algorithm (GA) process that runs for *G* generations. GA is then repeated *R* independent times, and each run provides an ensemble of *N* selected conformations/curves that fit the experimental profile (right part of the figure). The R_g distribution of the selected ($N \times R$) conformations is compared with that derived from the initial pool that is considered as a complete conformational freedom scenario. From this comparison a quantitative structural estimation of the protein conformations coexisting in solution can be derived. (Reproduced by permission of The Royal Society of Chemistry.)

where $I_n(s)$ is the scattering from the *n*th conformer. In order to speed up the optimization, the scattering curves from all the structures in the pool are precomputed, normally using the program CRYSOL (Svergun, Barberato, and Koch 1995), and the subsequent GA operators are applied to find an optimal subensemble of conformations/curves. Details of the optimization algorithm used by EOM are explained in detail in the original study (Bernado et al. 2007).

EOM was conceived to structurally characterize highly disordered states that display an enormous conformational sampling. In these scenarios, the number of conformations used by the EOM to describe the behavior of the protein in solution is much smaller than those sampled by the protein in solution. As a consequence, EOM results are normally reported as distributions of the low-resolution structural parameters, R_g, D_{max}, anisotropy, and interdomain distances. These distributions are compared with those derived from the pool that represent a model with complete conformational freedom in order to delineate the overall properties of biomolecules in solution. Despite the low-resolution nature of the results obtained by EOM, the distributions provided by EOM represent a major advance over traditional approaches that condense all structural characteristics of disordered systems in averaged structural parameters. EOM can also be used in multidomain proteins, where several globular domains are connected through flexible linkers that confer high levels of flexibility to the protein. In this scenario, the EOM approach can provide essential information about the interdomain distance distributions that indirectly report on the structural coupling between folded units, and have direct impact on their biological function.

Minimal Ensemble Search

The bases of the minimal ensemble search (MES) approach are very similar to those described for the EOM, where a large conformational exploration of the molecule is followed by a subensemble selection performed with a GA (Pelikan, Hura, and Hammel 2009). In MES the main emphasis is put in not overfitting the experimental data. One strategy to achieve this aim is to search for the minimal number of conformations/curves that collectively describe the data. The search is then performed by sequentially increasing the number of curves necessary to describe the data from 1 to 5. In contrast to EOM, the relative populations of each of the conformations tested are not considered equivalent and they are also optimized along the process. Comparison of the figure of merit χ_i^2 upon increasing the number of conformations is used to define the minimal number of conformations able to describe the data. Although no rigorous statistical tests are applied, this is a reasonable compromise to avoid overfitting and overinterpreting the data. Inspection of the optimized ensemble with respect to the complete pregenerated pool of conformations allows for the discrimination of the degree of plasticity of the system under study. Subensembles presenting conformations with very different sizes and shapes suggest a high degree of flexibility. Conversely, in cases where all conformations selected are structurally similar, a reduced degree of flexibility is inferred. Examples of these two scenarios are presented in the original study (Pelikan, Hura, and Hammel 2009). Therefore, the spread of the low-resolution structural descriptors R_g and D_{max} of the optimized subensemble are used as a diagnostic of flexibility in a similar way to the width of the distributions derived from the EOM approach.

MES is especially useful for biomolecular systems exhibiting a reduced mobility where the dynamic landscape sampled can be well described with few conformations. In these circumstances MES can quantitatively provide relative populations of the species present in solution.

Basis-Set Supported SAXS

The basis-set supported SAXS (BSS-SAXS) analysis was developed to study the conformational transitions of Hck tyrosine kinase (Yang et al. 2010). Equivalently to the previously described approaches, BSS-SAXS begins with an exhaustive exploration of the conformational space, followed by the precomputation of the individual scattering properties of each of them, and finally, a selection of conformations/curves that collectively reproduce the raw data. The first difference with respect to the other approaches is that the authors preselect the conformations and their associated theoretical SAXS curves prior to the optimization process. In the original study of Hck tyrosine kinase, this process is done in two consecutive steps. First, a large number of conformations generated by the coarse-grained simulation program was clustered in 25 families that were considered as distinct conformational states of the kinase. This number was further reduced to nine by clustering these families presenting very similar theoretical SAXS profiles. Notice that this second step can put conformationally different structures in a common state. Although it can be seen as a limitation, the final objective of any ensemble process optimization is based on the capacity to discriminate between the scattering properties of the precomputed conformations. In BSS-SAXS, the authors performed this discrimination before the optimization process, leading to discrete description of the conformational fluctuations of Hck tyrosine kinase. An important difference with respect to previously described approaches is the algorithm used to optimize the relative populations based on SAXS data, a Bayesian-based Monte Carlo (BSM). The authors assumed independent experimental errors and that a uniform prior that set the probabilities of each state equivalent previous to the BSM. The great advantage of the Bayesian approach is that for each SAXS curve analyzed it yields the fractional population for each of the states along with their uncertainties.

Like MES, BSS-SAXS method is especially useful for biological systems sampling a discrete number of conformational states. The advantage of BSS-SAXS is the rigorous statistical analysis applied. However, BSS-SAXS can only be applied when the pool of conformations represents a realistic description of the conformational sampling of the biomolecule, which is not always straightforward.

Ensemble Refinement of SAXS

The ensemble refinement of SAXS (EROS) approach (Rozycki, Kim, and Hummer 2011) has similarities with previously described methods. As in all previous approaches, a vast exploration of the conformational space is performed; in the EROS case a coarse-grained approach is used. Similarly to

BSS-SAXS, all these conformations are structurally clustered to identify independent states. The number of conformations of each of the clusters defines their relative initial population. This population is refined using a maximum-entropy algorithm in order to prevent overfitting. In practice, a pseudofree-energy function is constructed to minimize the discrepancy with respect to the raw SAXS data. The first term of this pseudofree energy is the traditional χ_i^2 that directly compares the experimental and the ensemble averaged scattering profiles. The second term, which is more novel, is designated as an effective entropy that penalizes variations on the relative populations of each of the clusters with respect to their initial values, which were purely computational. Therefore, the authors assume that an initially built ensemble of conformations is already close to reality and the EROS algorithm slightly refines their relative weight to describe the data attributing big changes to overfitting. In a subsequent study the maximum entropy term was substituted by a simple term that increases the pseudofree energy when the number of conformations required increases (Francis et al. 2011). Rozycki et al. applied the EROS approach to the SAXS curves measured for the ESCRT-III CHMP3 protein that contains a long terminal tail that hosts two small helical regions. Remarkably, at low salt concentration, conformations found by EROS are very similar to a previous crystallographic structure where one of the small helices of the tail is bound to the globular domain. In these experimental conditions, the SAXS curve can be accurately reproduced with six states. Conversely, in the presence of large salt concentration, the small helical fragment dissociates and the protein becomes much more flexible. In these circumstances, a much larger number of states (60) is required to reproduce the SAXS curve.

ENSEMBLE

The program ENSEMBLE derives ensembles of disordered proteins that collectively describe SAXS curves in addition of a variety of NMR parameters including residual dipolar couplings (RDCs), J-couplings, chemical shifts (CSs), paramagnetic relaxation enhancement (PRE) and its ratio, nuclear Overhauser effects, hydrodynamic radius, solvent accessibility restraints, and ^{15}N R_2 relaxation rates (Krzeminski et al. 2013; Marsh and Forman-Kay 2009; Marsh et al. 2007). Some of these parameters are back-calculated from the ensemble and compared with the experimental measurement, whereas others are transformed into structurally relevant parameters such as distances or solvent accessibility surfaces. ENSEMBLE uses a switching Monte Carlo algorithm to select a subensemble from a large pool of initial conformations ($\approx$100,000) that sample extensively the conformational space. This approach addresses the intrinsic problem of underrestraining and consequent overfitting by finding the smallest ensemble that is consistent with all experimental restraints imposed. Like EOM, ENSEMBLE is especially suited for highly flexible proteins such as IDPs. The capacity to integrate a variety of NMR parameters enables the simultaneous description of local conformational preferences and overall size and shape descriptors. The program ENSEMBLE has recently been applied to structurally characterize the intrinsically disordered protein Sic1 and its hexaphosphorylated version, pSic1, by combining several NMR parameters, including CS, PREs, RDCs, and ^{15}N R_2, with SAXS data (Mittag et al. 2010). The resulting ensembles confirmed that both Sic1 and pSic1 contain significant amounts of secondary and tertiary structure despite their disordered nature. Mainly due to the incorporation of SAXS data into the ensemble refinement, it was demonstrated that pSic1 displays a more compact structure than Sic1 that requires conformations sampling R_g values from 10 to 60 Å. Authors attribute this difference to the high positive net charge in Sic1, which is alleviated in pSic1 upon phosphorylation.

APPLICATION OF ENSEMBLE MODELS TO STUDY BIOMOLECULAR DYNAMICS BY SAXS

In the following, several examples of the use of ensemble approaches based on SAXS data to study motions in biomolecules will be described. Here, our aim is not to screen all the examples that can be found in the literature, but to succinctly describe those that we find more relevant to demonstrate the capabilities of the ensemble approach. We have divided this section into subsections related to the specific dynamic scenarios of the systems.

IDPs

IDPs are characterized by the lack of permanent secondary or tertiary structure and consequently sample an astronomical number of conformations. IDPs are emerging as a very important family of proteins that are involved in the vast majority of signaling and cell cycle control processes. Despite its relevance, their structural characterization that should provide insights into the molecular bases of their biological function remains challenging. Important progress in the biophysical and structural characterization of IDPs has been achieved in the last decade mainly using NMR and SAXS.

The development of realistic ensemble models of unstructured states of proteins has been an important subject of research for many years (Zhou 2004). Although these models were initially focused on proteins under denaturing conditions (Shortle 1996), more recently specific structural models of IDPs have been developed in order to structurally characterize this family of proteins

(Bernado et al. 2005). These models are necessarily simple as they have to be fast enough to produce large ensembles with the capacity to sample the conformational landscape as exhaustively as possible. SAXS curves have emerged as a stringent way to validate these structural models. In general, large ensembles of the protein and their associated individual scattering properties are calculated using appropriate programs. Subsequently, the individual profiles are averaged and compared with the experimental curve. If a good agreement between both curves is obtained, one can conclude that the structural model has the overall size and shape properties of the protein in solution. Molecular dynamics (MD) simulations accomplish these conditions for small peptides (Zagrovic et al. 2005), but for larger proteins the computational cost of the conformational sampling hampers this approach.

The program flexible-meccano (FM) has been the most tested structural model for IDPs. FM assembles peptidic units, considered as rigid entities, in a consecutive way. The force field used for this algorithm includes a coil description of the residue-specific Ramachandran space sampled by the amino acids and a coarse-grained description of the side chains that avoids collapse within the chain. This program has been tested for a large number of IDPs and it has successfully described several NMR observables and SAXS data measured for these proteins (Jensen et al. 2009). FM was used to describe the SAXS profile of protein X, a three-helix bundle that has a 60 amino acid long tail attached at the N-terminus. An excellent agreement was obtained between the experimental and the FM-derived SAXS curves. Importantly, the same ensemble is able to reasonably describe the residue-specific conformational sampling probed with RDCs. These observations suggest the capacity of FM to derive general models for IDPs that simultaneously describe local and overall properties. This agreement is not always necessarily obtained as IDPs can have regions that present transiently structured regions that severely modify their properties at residue and global levels. In these situations, the strategy followed consists of the refinement of the secondary structural elements (type, location, and population) using NMR data, mainly RDCs and/or chemical shifts. Subsequently these models can be validated by comparing the theoretical SAXS derived from these models with those measured experimentally. There are a few examples of this strategy such as p53 protein or the K18 construct of tau protein (Bernado and Svergun 2012).

The encouraging results obtained for the structural modeling of several IDPs using FM prompted to the derivation of a specific parametrization of Flory's equation for IDPs (Bernado and Blackledge 2009)

$$R_g = (2.54 \pm 0.01)^* N^{(0.522 \pm 0.01)} \tag{13.6}$$

The exponential value obtained from the parametrization, $\nu = 0.522 \pm 0.01$, is notably smaller than that derived from the dataset of denatured proteins, $\nu =$

0.598 ± 0.028, indicating that IDPs are more compact than chemically dena-tured proteins (Kohn et al. 2004). As some IDPs are expected to have certain populations of secondary or tertiary structure, this parametrization can be used as an interpretative tool, and departures from expected values are indi-cations of compactness of extendedness of the protein (Bernado and Svergun 2012).

Ensemble selection is the best choice to exploit SAXS data on IDPs when high-resolution information from NMR is not available. Among the ensemble methods described in the section "Ensemble approaches for SAXS data analy-sis" EOM is the most popular one as it was designed and tested to study highly flexible systems such as IDPs and IDRs. EOM results are reported as distribu-tions of low-resolution structural parameters that represent the variety of con-formations accessible to the molecule in solution. Three studies that exploit EOM to pinpoint structural changes will be succinctly described in the section "Multidomain Proteins."

The first example is the biophysical characterization of the N-terminal region of vesicular stomatitis virus phosphoprotein (VSV-P_{60}) (Leyrat et al. 2011). This 68 amino acid long protein, containing the recognition element of the nucleo-protein, presents large R_g and D_{max} values indicating that VSV-P_{60} is an IDP in agreement with the poor dispersion of the ^{1}H NMR signals, CD spectroscopy, and size-exclusion chromatography (SEC). Analysis of the chemical shifts and relaxation rates by NMR identify two regions of the protein with transient heli-cal conformations. The EOM analysis of the SAXS curve displays a bimodal dis-tribution of R_g values containing two subpopulations of compact and extended conformers. The bimodal distribution was maintained regardless of the param-eters used for the GA procedure suggesting the robustness of this observation. The structural nature of the two subpopulations was investigated by the use of different cosolutes with distinct structural effects. Initial experiments were per-formed in a 50-mM Arg/Glu buffer that exerts compaction in proteins increas-ing their solubility and stability (Blobel et al. 2011). A decrease in the amount of Arg/Glu to 10 mM induces a reduction of the amount of compact conformations, and the addition of trimethylamine N-oxide (TMAO), a well-known stabilizing cosolute, substantially increases the subpopulation of compact conformations. Conversely, addition of 6M of GdmCl, a desestabilizing agent, induces a uni-modal distribution that is shifted toward extended conformations. The authors hypothesize that the presence of fluctuating helical elements and the distribu-tion of charged residues could preconfigure certain VSV-P_{60} conformational states to facilitate the recognition event with the nucleoprotein.

The effect of temperature jumps in the structure of protein tau (441 resi-dues) has recently been studied by SAXS on wild-type tau and a mutant that mimics a phosphorylated state (Shkumatov et al. 2011). SAXS curves have been measured at 10°C and 50°C and no changes in the apparent R_g were observed,

66 ± 3 and 65 ± 3 Å, respectively. However, SAXS curves measured after a fast temperature jump, from 10°C to 50°C or from 50°C to 10°C, presented a notable compaction manifested in the apparent R_g, 55 ± 3 and 56 ± 3 Å, respectively. This temperature-jump induced compaction could be studied in more detail when the same data were analyzed using ensemble approaches. The EOM analyses clearly show that after temperature jumps tau protein presents a much narrower distribution of conformations than those obtained at fixed temperatures and with maxima displaced toward smaller R_g values (Figure 13.4). Interestingly, the structural compaction is preserved for several hours after the temperature jump until it reaches the conformational equilibrium. The

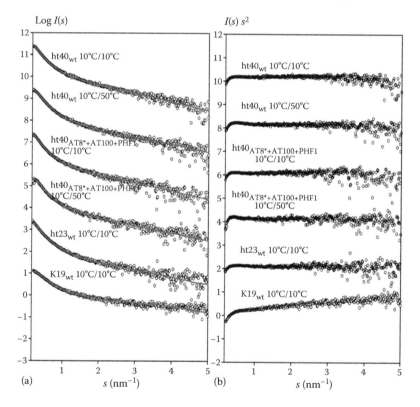

Figure 13.4 SAXS measurements on tau protein constructs. (a) Experimental SAXS curves (open circles) with their corresponding EOM fits (solid lines) for full-length constructs (hTau40wt, hTau40AT8*+AT100+PHF1) are shown at equilibrium (10°C/10°C) and after temperature jumps (10°C/50°C). Experimental data for short constructs (hTau23, K19) are shown at equilibrium temperature condition (10°C/10°C). (b) Kratky plots corresponding to data in (A).

(Continued)

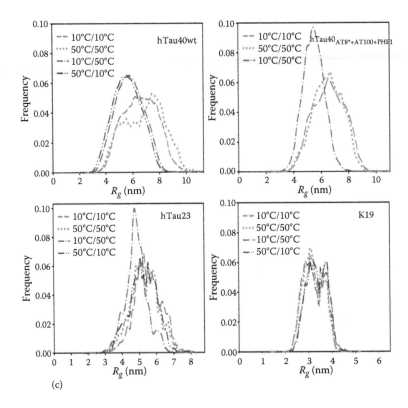

Figure 13.4 (Continued) SAXS measurements on tau protein constructs. (c) Changes in the ensemble dimensions studied by SAXS. Ensemble optimization analysis of the SAXS profile measured for full-length (hTau40wt, hTau40AT8*+AT100+PHF1) and short (hTau23, K19) tau constructs in equilibrium (10°C/10°C, 50°C/50°C) and after temperature-jumps (10°C/50°C, 50°C/10°C). (Shkumatov, A. V., S. Chinnathambi, E. Mandelkow, and D. I. Svergun: Structural memory of natively unfolded tau protein detected by small-angle X-ray scattering. *Proteins.* 2011. 79 (7). 2122–2131. Copyright Wiley-VCH Verlag GmbH & Co. KGaA. Reproduced with permission.)

authors attributed this intriguing effect to a structural memory of the protein that could be related to the more compact nature of hyperphosphorylated tau involved in neurodegeneration in Alzheimer's disease. Importantly, in this and the previous example the use of ensemble approaches has provided a unique picture of the structural perturbations exerted by the addition of cosolvents and temperature variations to IDPs.

The EOM provides overall structural information whereas the exact regions responsible for conformational particularities remain elusive. This lack of resolution can be improved if SAXS data of the full-length protein and from several deletion mutants are simultaneously fitted using the EOM approach. In this way, contributions from different chain fragments to the SAXS data can be separated to identify these regions responsible for the increase or decrease in compactness of the protein. The SAXS resolution is therefore improved by the addition of multiple experimental curves. One has to be careful in using this approach as it is only valid if the structural elements of the full-length protein remain intact in the deletion mutants, which is not easy to detect *a priori*. Despite the potential of this simultaneous fitting approach, it has not been applied in many cases. The most relevant application has been to protein tau (Mylonas et al. 2008). Two tau isoforms were studied, ht40 and ht23, and SAXS data for the full-length and three- and six-deletion mutants for each isoform were used, respectively. The EOM unambiguously identifies the so-called repeat region as the source of residual secondary structure in tau, in perfect agreement with previous NMR studies indicating the presence of turns and extended fragments in this region (Mukrasch et al. 2005, 2007). The ensemble derived from the multiple curve fitting, monitored with the averaged C_α-C_α interresidue distance matrix, identifies a distinct conformational behavior depending on the number of repeats present in the isoforms (Figure 13.5). For the isoform ht23, with three repeats, the maximum separation is found within the repeat domain itself. The full-length ht40 isoform with four domains reveals an enhanced separation between the repeat domain and the preceding region. These results suggest that the different number of turns (one per repeat) may lead to different global arrangements of the chain in that region, enhancing or shortening the average interdomain distances expected from a random coil.

Multidomain Proteins

Modular or multidomain proteins, comprising two or more folded domains tethered by linkers, are common in nature (Levitt 2009). Due to their special architecture, multidomain proteins can often adopt several conformations in solution, and two different scenarios with respect to their mobility can be envisioned. In some cases the rigid (globular) domains move among distinct arrangements in a concerted way, facilitated by the flexibility of the linkers. A typical example of this situation is the collapse from an extended conformation of calcium binding protein calmodulin (Trewhella et al. 1990). In other situations, the inherent flexibility of the linker induces the protein to sample an astronomical number of conformations. Therefore, linkers behave as IDRs that provide a large capacity for spatial exploration to globular (functional) domains. Several biological advantages of this arrangement can be envisioned with respect to the situation where the domains are presented unconnected (Dunker et al. 2002).

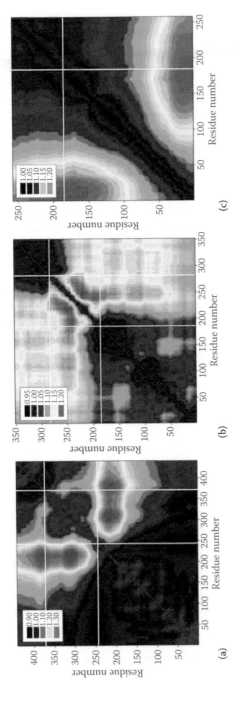

Figure 13.5 Cα-Cα distance matrices of ht40 (a), ht23 (b), and K23 (c) using EOM multiple-curve fitting. Each point of the plot shows the ratio of the average Cα-Cα distance of the selected structures to all the structures from the pool. The legend shows the ratio represented by each color. White lines indicate the residues of the repeat domain for ht40 and ht23 and the N- and C-terminal connections for K23. (Reprinted with permission from Mylonas et al. 2007, s245–s249. Copyright 2007, American Chemical Society.)

Identification of the dynamic regime is the first step that must be achieved as the modeling strategies in each scenario are different. Kratky plots ($I(s)s^2$ as a function of s) have been used to qualitatively identify disordered states and distinguish them from globular particles. In the specific case of multidomain proteins, they present a dual behavior and consequently SAXS profiles and Kratky plots present contributions from both regions. The relative weight of each contribution will depend on the amount of protein in globular and disordered states. Also the absence of a plateau in the Porod-Debye representation has been attributed to the presence of protein flexibility (Rambo and Tainer 2011). In an extensive study using synthetic SAXS, data derived from models of rigid and highly flexible proteins identified several features that prompt toward the assignment of extensive protein dynamics (Bernado 2010). Protein dynamics causes a general smoothing of the SAXS profiles that is enhanced in their Kratky representations. Additionally, correlation peaks in the $p(r)$ function also disappear in the presence of dynamics. In the case where the data is analyzed as a single conformation by *ab initio* or rigid-body approaches, the resulting models display a systematic reduction of the resolution (*ab initio* modeling) and the presence of isolated domains within the rigid body model. The analyses of the results derived from ensemble methods such as EOM and MES are also an excellent diagnostic of high flexibility. In EOM, R_g distributions displaying a sharp peak indicate the presence of major rigid conformation. In MES, when the selected conformations fall within a restricted region of the R_g vs RMSD map also indicate the presence of a rigid particle. In the following, relevant examples of the application of ensemble methods for multidomain proteins in both scenarios—restricted motions and high flexibility—will be presented.

Concerted Motions in Multidomain Proteins

Certain proteins present conformational jumps between discrete conformations. In some cases these concerted motions are mediated by hinge points along the sequence, normally connecting globular regions of the protein, that allow a certain degree of flexibility that is translated to the rest of the protein, causing notable changes in the size and shape of the particle. This breathing phenomenon is normally inherent to the fold of the protein that is a continuous exchange between two or more extreme conformations but that is probably sampling a large number of intermediates. The study of these motions is challenging as along the process of crystallization only one of the available conformations in solution is preserved. Some proteins display structural rearrangements involving large portions of the protein. Very often these rearrangements are mediated by external stimuli such as the interaction with other biomolecules or posttranslational modifications. Ensemble methods in SAXS have notably contributed to the molecular bases of both kinds of concerted motions. Two examples will be succinctly described.

In a recent study, the N-terminal of the Nup192, a major component of the nuclear pore complex (NPC) inner ring was crystallized displaying an α-helical fold with three domains named D1, D2, and D3 (Sampathkumar et al. 2013). Despite the high resolution of the structure, the agreement to the SAXS curve was unsatisfactory, indicating that the structure of the Nup192 in solution was different or it displayed some degree of plasticity. The analysis of the B-factors of the x-ray structure indicated a larger degree of mobility in D1 than in the other domains. A simplistic elastic network model suggested a long range motion of D1 and D3 relative to the central D2 and identified two residues of this domain, in the border with domains D1 and D3, as the hinge points responsible for this motion. In order to identify the species involved in this conformational equilibrium, the authors performed a high-temperature MD simulation, and small subensembles of the resulting 110,000 conformations were optimized using the MES approach. The authors concluded that a 3:2 mixture of two conformations, one open and one closed, was sufficient to explain the SAXS profile. Importantly, these two conformations were experimentally observed using electron microscopy analysis. The authors hypothesized that the inherent flexibility of Nup192 could be relevant for its function as a part of a large assembly such as the NPC.

A traditional example of concerted motions directly related with enzymatic activity is the the Src family of nonreceptor protein tyrosine kinases. The Src family is considered to experience notable structural rearrangements when going from inactive to active triggered by the formation or breakage of intramolecular interactions. Their general architecture is composed by two binding domains, SH3 and SH2, respectively, and a highly conserved catalytic domain, all interconnected by disordered regions. Current models of conformational regulation are based on high-resolution crystallographic structures of both forms. The inactive form is a rather compact form of the involved domains whereas partly disassembled forms are observed by crystallography for the active state. This model has been challenged recently for the Src and the Hck kinases based on SAXS data and ensemble analysis.

The EOM approach was applied to SAXS data measured for two forms of Src-kinase in constitutively inactive and active states (Bernado et al. 2008). The inactive form of the protein presented a compact structure very close to that found in crystallography according to the narrow R_g distribution in EOM analysis. The same approach on the constitutively active form, hosting a point mutation impeding a key intramolecular contact, presented a virtually equivalent R_g distribution to the inactive protein with the addition of a small shoulder at slightly larger R_g. This observation suggested equilibrium between at least two conformations. Furthermore, using the crystallographic structures, it was observed that the major form of the protein ($\approx 85\%$) is in a closed conformation that is in principle inactive and only ≈15% of the conformations adopted open (active) conformations. However, this last approach yielded systematic deviations in the description of the experimental curve, suggesting the presence of additional conformations in solution. This last hypothesis raised for Src-kinase study was demonstrated in Hck kinase, another member of Src-kinase family, using the Basis Set Supported SAXS approach (Yang et al. 2010). Using a clustering procedure based on the structural SAXS-profile similarities, the authors identified nine different states that were used for the conformational study of Hck-kinase in the absence/presence of two types of external peptide ligands that mimic two intramolecular anchoring points (Figure 13.6). For the free protein the SAXS data can be mainly described as a compact state (≈82%), although the presence of minor contributions from partly and fully disassembled states is required. In the presence of a peptide that breaks the interaction between the SH2 and the kinase domain, a significant shift in the state distribution from compact to disassembled states is observed. The fraction of the assembled state decreases dramatically from ≈82% to ≈22%, while that of the disassembled state increases from ≈8% to ≈29% when compared to the wild-type Hck. In the presence of a peptide that breaks the connection between the SH3 domain and the proline-rich linker connecting the SH2 and the kinase domains, the compact state becomes only ≈7% of the population whereas the disassembled states reaches ≈50%. Two additional correlated partly disassembled states appear in these circumstances representing ≈39% of the conformations. Interestingly, when both peptides are simultaneously added, the assembled state disappears and the fully disassembled states are the only representatives of the conformational state sampled by Hck. This study nicely demonstrates how ensemble methods based on reduced number of states offers a structure-based approach to characterize concerted motions in multidomain proteins. This is especially relevant when the relative populations of these states can be rationally modulated using external signals.

Highly Flexible Multidomain Proteins

Highly flexible multidomain proteins represent a scenario close to IDPs in as much as astronomical numbers of conformations are accessible in solution. Here, in addition to the structural parameter distributions, the interdomain distance distribution can be derived. This is an important piece of structural information that is difficult to obtain with other methods. In many cases, the length and its conformational features of the linker are related with the coupled activity of individual domains. In addition, the interdomain distance distribution provides indirect evidence of residual structuration in the linker. In these cases, the distance between the domains is highly influenced by the structural features present in the linker. This effect has been observed in the SAXS study of the ribosomal protein L12, a dimeric two-domain protein

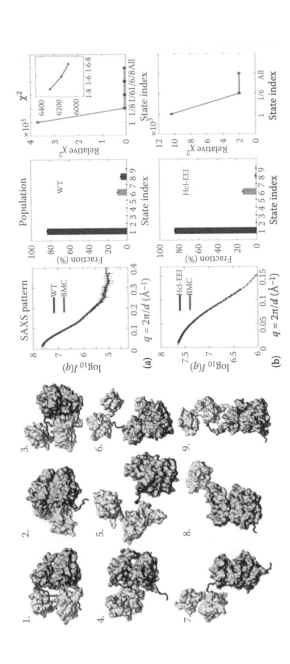

Figure 13.6 (Left) Multiple assembly conformational states adopted by the multidomain Hck kinase from a CG model. Representative structures of these nine scattering states, ranging in architecture from fully to partially assembled to disassembled states. The catalytic domain is in blue, SH2 in green, and SH3 in yellow. States 1 and 8 are similar to the assembled Hck (pdb entry 1QCF) and the partially active c-Src (pdb entry 1Y57) structures, respectively. (Right) Analysis of wild-type Hck and high-affinity C-tail mutant (Hck-YEEI) in solution. The scattering pattern (left), the assembly state population (center), and the relative χ^2 scores for different combinations (right) are shown. (a) SAXS data for the wild-type Hck (black) and BMC fitted scattering profile (red) are shown. Relative χ^2 scores show that the inclusion of states 6 and 8 improves the fit to the SAXS data. (b) SAXS data for the Hck-YEEI mutant (black) and BMC fitted scattering (red) are shown. Inclusion of state 6 greatly improves agreement with the SAXS data (right). (From Yang, S. C., L. Blachowicz, L. Makowski, and B. Roux. 2010. Multidomain assembled states of Hck tyrosine kinase in solution. *Proc Natl Acad Sci USA* 107 (36):15757–15762. Copyright 2010, National Academy of Sciences, U.S.A.)

connected by a 20 amino acid long linker (Bernado et al. 2010). The EOM analysis of the SAXS curve provides information of the 3-D space sampled by the C-terminal domains with respect to the dimeric N-terminal one and indicates that the overall shape of L12 is larger and more anisotropic than expected for a random linker. Interestingly, this analysis indicates an asymmetric behavior of both linkers that is hypothesized to be relevant for the biological function when docked on the ribosome. A rotational normal mode analysis of the EOM-derived ensemble indicates a small degree of motional coupling between both domains that suggests a transient structuration of the linker. This is not surprising taking into account that the linker is highly enriched in alanine residues and it is consequently prone to form transient α-helical structures, and this secondary structural element in the linker has been observed in several crystallographic structures of the protein.

A nice study on how ensemble methods can provide detailed pictures of highly flexible proteins is the SAXS study of high-mobility group protein B1 (HMGB1) (Stott et al. 2010). HMGB1 consists of two tandem HMG-box domains joined by a linker with an acidic C-terminal tail. NMR studies showed that the acidic tail regulates HMGB1-deoxyribonucleic acid (DNA) recognition by interacting intramolecularly to the DNA-binding surfaces of both HMG-boxes. From these studies, mainly sensitive to binding events at residue level, it was proposed that HMGB1 is in a dynamic equilibrium between a collapsed tail-bound and open tail-free state. This scenario has been studied by the combined use of SAXS and NMR for the full-length protein and two deletion mutants lacking different number residues of the acidic tail. The EOM analysis of a SAXS curve of HMGB1 at 0-mM salt displayed a relatively narrow R_g distribution shifted toward compact structures when compared to the pool of random conformations. The presence of the open conformation could not be detected by this analysis, suggesting the predominance of collapsed conformers in solution. Upon the increase of the ionic strength, the R_g distribution became slightly wider due to the breakage of some of the anchoring points of the tail, which are mainly electrostatic, but not all of them as fully open conformations could not be observed. Conversely, deletion mutants lacking intramolecular interactions displayed R_g distributions that were very similar to the disordered state represented in the pool. The collapsed nature of HMGB1 driven by the acidic tail was validated by NMR relaxation experiments. Despite the residue level information offered by NMR, the overall arrangement of domains could only be achieved by the use of the EOM strategy.

Ensemble Methods in Nucleic Acids

Over the years SAXS in solution has also been used for structural characterization of nucleic acids. Nucleic acids tend to assume multiple conformations when observed in a native-like environment providing them with the conformational plasticity that facilitates their biological activity in recognition and catalysis. Consequently, ensemble approaches are the most suited strategy for their structural and dynamic characterization. In a recent study, Kazantsev et al. (2011) have elegantly shown how three phylogenetically distinct bacterial ribonuclease P ribonucleic acids (RNAs) from *Escherichia coli* (*Eco*), *Agrobacterium tumefaciens* (*Atu*), and *Bacillus stearothermophilus* (*Bst*) are better represented by small ensembles of well-folded conformations than by a single well-defined conformation. In fact, standard *ab initio* reconstructions hardly define unique and interpretable envelopes with a normalized spatial discrepancy (NSD) tending to 1 (*Bst* = 0.823, *Eco* = 0.842, and *Atu* = 1.02) suggesting the presence of intrinsic flexibility in the three molecules. Normal modes were used to explore the conformational landscape of these RNAs, and more than 1000 structures were calculated by perturbing these biomolecules using the five lowest elastic modes and all bimodal perturbations. These structures were submitted to the EOM approach to define the conformational variability accessible in solution. The EOM results prove that the three RNAs behave differently in solution (Figure 13.7). On one hand, a single EOM-refined structure is sufficient to have a very nice description of the SAXS curve of *Bst* RNA (χ^2). When increasing the number of conformations a minor improvement in the fit was obtained. On the other hand, for both *Eco* and *Atu* RNAs a multiconformation ensemble representation is required to describe the data, with χ^2 going from 13.4 and 35.5 to 2.26 and 1.87, respectively, when passing from single models to small ensembles. The ensemble approach confirms the initial hypothesis in which *Eco* and *Atu* RNAs were supposed to be more flexible to explain actual biological data. This study highlights the power of the joint use of ensemble SAXS representations with advanced modeling techniques in the study of nucleic acids.

An excellent example of the application of ensemble methods to address the study of large and flexible RNA molecules is the recent characterization of the internal ribosome entry site (IRES) fragment of the hepatitis C virus (HCV) (Perard et al. 2013). IRES fragments recruit ribosomes to initiate the cap-independent translation of viral RNA. Some secondary structure elements have been described in this 332-nucleotide RNA, although its complete structure remains elusive due to its intrinsic dynamics. The authors build an atomic model of the HCV IRES joining the previously crystallized elements with canonical fragments in agreement with the secondary structural map. This model was energetically minimized but was unable to describe the SAXS curve measured (χ_i = 10). Alternative single-conformation models were tested by submitting this initial model to harmonic normal mode displacements followed by MD simulations. This hybrid procedure ensured an exhaustive exploration of the conformational space potentially available by this RNA with 8000 structures. One of these structures notably improved the agreement to the

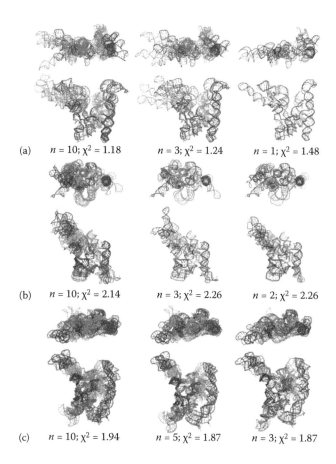

(a) $n = 10; \chi^2 = 1.18$ $n = 3; \chi^2 = 1.24$ $n = 1; \chi^2 = 1.48$

(b) $n = 10; \chi^2 = 2.14$ $n = 3; \chi^2 = 2.26$ $n = 2; \chi^2 = 2.26$

(c) $n = 10; \chi^2 = 1.94$ $n = 5; \chi^2 = 1.87$ $n = 3; \chi^2 = 1.87$

Figure 13.7   Conformational ensembles derived from the EOM fit of SAXS data for *Bst* (a), *Eco* (b), and *Atu* (c) RNAs. Typical selected solutions depending on the sizes of the ensembles used for *Bst* and *Eco* RNA from the pools of 1051 NMA-perturbed conformations and for *Atu* RNA from a pool of 13,735 filtered MD-perturbed conformations. (Extracted from Kazantsev, A. V., R. P. Rambo, S. Karimpour, J. Santalucia, Jr., J. A. Tainer, and N. R. Pace. 2011. Solution structure of RNase P RNA. *RNA* 17 (6):1159–1171. Cold Spring Harbor Laboratory Press. With permission.)

SAXS curve ($\chi_i = 2.8$), although still systematic deviations were observed along the momentum transfer range measured. To improve the description, the EOM approach was used to find subensembles able to describe the curve. A minimal set of five conformations was able to provide a good description of the SAXS curve ($\chi_i = 1.92$). These structures displayed a moderate conformational variability mainly along one of the motional modes of the molecule. Interestingly, the resulting structures were incompatible with the cryo-EM densities of the

HCV IRES bound to the ribosomal 40S subunit and the elF3. However, very simple articulations can transform solutions present in the SAXS-derived ensemble to the structure found in the bound state.

Biomolecular Complexes

In recent years, genome-sequencing initiatives have provided an almost complete list of the macromolecules present in several organisms. This crucial knowledge has proved unsatisfactory to describe the complexity of biological activity. Virtually all molecular processes (i.e., transcription and replication) are performed by macromolecular assemblies, and the regulation of metabolic pathways is governed by an intricate network of transient protein–protein and protein–nucleic acid interactions. Large-scale interaction discovery experiments have provided an extensive chart of interacting partners in various organisms (Russell and Aloy 2008). However, understanding biological processes at the molecular level requires knowledge of 3-D arrangements of biomolecular complexes.

Two biomolecular complex families can be distinguished on the basis of their stability *in vivo*. Complexes with large-contact interfaces are characterized by strong binding affinities and their partners are rarely found in isolation. These arrangements force relatively fixed conformations that provide optimal orientations to ensure high efficiency. Large macromolecular machines (i.e., RNA polymerases or ribosomes) are among this family. A radically different model is required for signaling and regulation, in which interactions must be dynamic and versatile to rapidly respond to continuous changes induced by external stimuli (Russel et al. 2009; Stein et al. 2009). Often governed by small interfaces with a limited number of contacts, these low-affinity interactions are normally transient and their partners are very often found in isolation. Robust approaches have been developed to study both families of complexes by SAXS (Blobel et al. 2009; Petoukhov and Svergun 2005; Williamson et al. 2008). In many circumstances dynamic phenomena can be crucial for the biological function of these complexes and therefore it must be taken into account in modeling procedures driven by SAXS data. Some examples on how ensemble methods have been used to study dynamic biomolecular complexes will be described in the following.

Dynamics in Biomolecular Complexes

A recent study on the complex formation between the proliferating cell nuclear antigen (PCNA) and ubiquitin exemplifies the use of ensemble methods in biomolecular complexes (Tsutakawa et al. 2011). The monoubiquitylation of PCNA in response to DNA damage leads to the recruitment of specialized translesion polymerases to the damage locus. Macromolecular x-ray crystallography of PCNA-Ub showed that the interaction surface was coincident with that used for the translesion polymerase. This observation suggests a certain degree of mobility of the PCNA-Ub complex that was addressed by SAXS and ensemble methods. To avoid species polydispersity, ubiquitin was fixed by either split-function or by chemical cross-linking with trimeric PCNA mutant K164C providing triubiquitinated complexes with almost equivalent SAXS properties. The crystallographic structure of monoubiquitinated PCNA does not describe properly SAXS data. A large ensemble of conformations of the triubiquitinated PCNA, built using high-temperature MD simulations, was used for the MES analysis of the SAXS curves. An important improvement in the description of the SAXS data was obtained when three conformations of ubiquitin were considered in the subensemble, suggesting an important level of flexibility of ubiquitin molecules when covalently attached to PCNA. A more detailed picture of the binding sites was endeavored by the authors by combining multiscale computational procedures including Brownian dynamics, docking with Rosetta, loop modeling, and MD simulations. This complex procedure identified three sites in PCNA that could accommodate covalently linked ubiquitin. These three positions were used to generate 130 models of the three-ubiquitinated complexes that were used as a pool for a MES calculation. An important improvement in the level of agreement to the SAXS curves was obtained providing a detailed picture of the dynamics of the complex where approximately 25%–30% of the time the ubiquitin moieties are positioned on the back face of the ring in the position indicated by the x-ray crystal structure, about 25%–30% of the time they are in a flexible position in which the ubiquitin is interacting with the PCNA, and about 40%–50% of the time they are interacting with the side of the PCNA ring at the subunit-subunit interface. This is another example that illustrates the synergy between ensemble approaches based on SAXS data with high-level computational tools (Figure 13.8).

In a recent series of studies, the structural characterization of the membrane budding endosomal sorting complexes required for transport I and II (ESCRT-I and -II) supercomplexes from yeast have been addressed (Boura et al. 2011, 2012). ESCRT-I directly binds to ubiquitin and functions with ESCRT-II to bud membranes away from the cytosol by an unknown mechanism. They characterized the individual component, ESCRT-I and ESCRT-II, and their complex in solution using SAXS, electronic paramagnetic resonance (EPR) and single-molecule Förster resonance energy transfer (smFRET). Both proteins are heterotetramers that contain IDRs connecting globular domains of known structure. For each individual complex, a large ensemble of structures sampling the complete conformational space was created. Then, following the EROS approach, a subensemble of them was optimized using SAXS or the combination of SAXS, EPR, and smFRET. For ESCRT-I an acceptable fit was obtained when using two conformations of the complex displaying the two long flexible arms either in a closed or open conformation. In both conformations the arms harboring globular

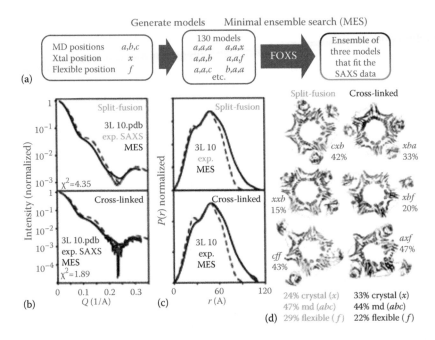

Figure 13.8 MES ensemble of both discrete and flexible positions of ubiquitin relative to PCNA for the split-fusion (green) and cross-linked (blue) PCNA–Ub. (a) Schematic showing MES methodology. One hundred thirty models were generated where the three ubiquitins per PCNA homotrimer were placed at the crystallographic (x) position, the MD-identified positions (a,b,c), or the BILBOMD-generated ensemble (f). Ensembles of three models were then compared to the experimental SAXS data with FOXS. (b) The scattering curve of the best MES ensemble fits the experimental scattering data better than the crystal structure (3L10). (c) $P(r)$ plots showing the good fit of the MES ensemble to the experimental data. (d) Ensemble of three models that with the best fit the experimental scattering curve are shown in ribbon models. Relative populations of each position in x, $a/b/c$, or f show that ubiquitin adopts both the crystallographic and computationally determined discrete positions indicating that it is flexible in solution. (From Tsutakawa, S. E., A. W. Van Wynsberghe, B. D. Freudenthal, C. P. Weinacht, L. Gakhar, M. T. Washington, Z. H. Zhuang, J. A. Tainer, and I. Ivanov. 2011. Solution X-ray scattering combined with computational modeling reveals multiple conformations of covalently bound ubiquitin on PCNA. *Proc Natl Acad Sci U S A* 108 (43):17672–17677. Copyright 2011 National Academy of Sciences, U.S.A.)

domains are placed in the same side of the highly elongated shape of the protein. Simultaneous fit of SAXS and EPR data required the presence of six conformations that resembled the open and close conformations previously found. The ensemble created for ESCRT-II, a Y-shaped protein that only contains a flexible linker, using a replica exchange Monte Carlo (REMC) approach essentially reproduced the SAXS data without refinement. The simultaneous fit of SAXS and smFRET with EROS slightly improved both data sets with respect to the initial models by reducing the population of highly extended conformation. Structural models of the supercomplex were also built using the REMC method. The simultaneous fit of SAXS, EPR, and smFRET yielded a subensemble of 15 different conformations that eliminated highly extended models to account for the averaged R_g measured. The selected conformations of the individual components in the complex are quantitatively similar to these found in isolation. The vast majority of the conformations adopted a crescent shape that is typical of proteins that target membranes to induce curvature.

Nucleoprotein complexes, formed by a protein and a nucleic acid, can display high levels of plasticity, especially when ssRNA is involved. This situation is exemplified in a recent study of a 34-nt-long regulatory RNA (DsrA) with the hexameric RNA chaperone host factor Q (Hfq) (Ribeiro Ede et al. 2012). These two molecules form a high-affinity (nM range) equimolar complex that has been addressed by SAXS, NMR, and modelling tools. Analysis of the SAXS curve of the complex already suggests that the RNA displays an elongated shape upon binding to hexameric Hfq. This observation was further confirmed applying EOM. A large pool of the structural model of DsrA:Hfq were built using the crystallographic structure of the protein with a RNA represented by dummy residues with RanCh. The selected subensemble presents a relatively narrow distribution of R_gs, suggesting confined flexibility. This reduced amount of potential flexibility enabled the construction of rigid-body models of the nucleoprotein complex that were in agreement with the hydrodynamic measurements performed with dynamic light scattering (DLS).

Transient Biomolecular Complexes

Despite the biological relevance of transient ($K_d > 100 \mu M$) biomolecular complexes, retrieving valuable structural information for these systems is a challenge. The main reason is the coexistence of different species, the complex, and its subunits, in variable relative concentrations. Therefore, appropriate sample conditions are required to ensure a detectable amount of the complex, and an experimental technique sensitive enough to the population of the targeted complex must be used. In this context, SAXS is an excellent technique to address these challenging systems as it requires relatively high concentrated samples. Experimentalists often try to reduce the complexity of the samples by different approaches: engineering proteins to stabilize or disrupt contact, use cross-linking agents (Hu et al. 2012), or by using size exclusion chromatography coupled to SAXS, although in certain circumstances equilibriums are unavoidable. In these cases, several species coexist in solution and the analysis of the data in terms of structures necessarily requires ensemble approaches and the capacity to model all the species present.

There are several examples where transient homo- or heterooligomerization has been addressed by SAXS making use of the available high-resolution structures of the individual components. This is the case in the study of the transient complex of methyltransferase (Mtr) and corrinoid iron-sulfur protein (CFeSP), an assembly responsible for methyl transfer (Ando et al. 2012). SAXS data were measured for the individual components and for 15 titration experiments, where increasing amounts of CeFeSP (up to 150 μM) were added to a 50-μM sample of Mtr. SAXS profiles as well as the apparent R_g presented substantial changes upon CeFeSP addition. The authors made use of the available x-ray structures of the individual components and the 2:1 CeFeSP:Mtr complex. In addition a 1:1 complex derived from the x-ray structure was used. For each of the four species the theoretical SAXS curve was computed and used to determine the relative population using the program OLIGOMER. Results unambiguously indicated that in these conditions the 1:1 complex was formed and the 2:1 complex was not necessary to explain the data. That was a surprising observation in apparent contradiction with the x-ray structure found. To unravel these contradictory results, a series of curves was collected in the crystallization conditions revealing that the 2:1 complex was present. Interestingly, both complexes 1:1 and 2:1 were required to explain the complete data set. Importantly, the single value decomposition (SVD) of both data sets indicates that three and four species were required to explain the data. SVD analysis is a crucial preliminary analysis step in order to identify the number of species that have to be used in the ensemble approach.

A similar approach was used to explain the oligomerization properties of STAT-5 protein (Bernado et al. 2009). The three SAXS curves measured in the range of 1.15 to 4.6 mg/ml indicated displayed concentration dependent effects, and the SVD analysis indicated the presence of two species. Two approaches were used to obtain molecular insights into STAT-5 oligomerization. First, the three SAXS curves were simultaneously fitted as a combination of the monomeric structure and all dimeric structures found by crystallography for other STAT proteins in phosphorylated and unphosphorylated states. For each dimeric arrangement, the only fitted parameter was the dissociation constant, K_d, that defines the relative population of monomers and dimers at any given concentration. This analysis confirmed that the antiparallel dimeric structure found in x-ray crystallographic studies for STAT-5 is prevalent in solution, governed by a K_d of ≈80 μM. Interestingly, the authors tried the same approach but using a large pool of 5000 potential dimers computed using a protein docking program. For each of the docking poses a global χ_i^2 was obtained. From this analysis, simulating the situation where the dimer was unknown, five different clusters of dimers were obtained. Interestingly the most populated one, and the only displaying a certain degree of symmetry, was very similar to the crystallographic structure. Despite the uncertainty of the exact dimeric arrangement, this analysis allows a very precise determination of the dissociation constant, $K_d = 86 \pm 11$ μM, and the R_g of the dimer in solution, $R_g = 46.4 \pm 1.8$ Å.

Systems with Complex Dynamics

In this chapter, we have circumscribed biomolecular dynamics within specific scenarios. However, many situations can be envisioned where these scenarios do not describe motions probed by biomolecules and multiple scenarios coexist. Proper description of these phenomena will require advanced computational tools to sample motions of different amplitude and timescale. A recent study about the nucleoprotein complex between translin and a ssRNA molecule exemplifies this situation (Perez-Cano et al. 2013). Translin is a highly conserved RNA and DNA binding protein that plays an essential role in eukaryotic cells. Human translin functions as an octamer, but in the octameric crystallographic structure, the residues responsible for nucleic acid binding are inside the empty cage of the dimer of tetramer assembly and they are not accessible. A SAXS curve measured for translin was not compatible either with the x-ray structure or with the cryo-EM reconstruction, which is notably more open. Using an exhaustive normal mode exploration and ensemble methods based on SAXS data, the authors concluded that translin is in a continuous exchange between a major open form (≈80%) and a closed one similar to the x-ray structure (≈20%). The presence of this open form explains the mechanism of the entrance of single-stranded nucleic acids in the protein cavity. The analysis of the SAXS data measured for the complex of translin with a 24-nucleotide ssRNA indicates that the nucleic acid is placed in the interior of the cavity. Molecular modeling of the complex was highly challenging as two dynamic scenarios were present in the complex: on one hand, conformational fluctuations experienced by translin between open and closed conformations; on the other, the presence of an inherently flexible protein placed in a fluctuating cavity with eight symmetrical binding sites available. Through an elaborate computational strategy the authors modeled thousands of potential structures for the complex by docking rigid dinucleotides to the four conformations of translin found for the free state using MES. Ensemble approaches indicate that the relative population between open and closed conformations in translin were not modified by the ssRNA. The improvement in the fit of the SAXS data using EOM ($\chi_i = 1.62$) with respect to the MES ($\chi_i = 1.73$) suggests that ssRNA is sampling several conformations within a cavity composed of eight symmetrical binding sites. However, this scenario of high flexibility of the ssRNA in the complex cannot be demonstrated due to the low-resolution nature of SAXS data (Figure 13.9).

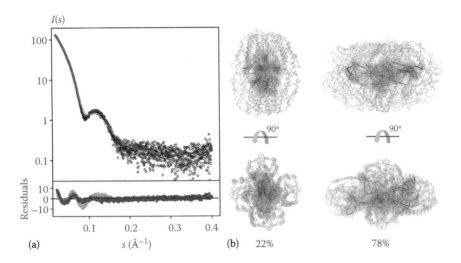

Figure 13.9 Structural and dynamic models of translin–ssRNA complexes according to SAXS data. (a) Experimental fit of the best single-conformation of the complex (in green) and the dynamic structural model of translin–RNA interaction (in red) to the experimental SAXS curve (black circles). Point-by-point deviations for each model are displayed at the bottom with the same color code. (b) The best equilibrium of conformers was found to be formed by an ensemble of compact and open translin forms bound to structurally different RNA molecules, with a fitting to SAXS data of $\chi i = 1.62$, notably better than the one found for the best single conformation ($\chi i = 2.44$). Translin is shown in gray ribbon, while the docked RNA conformations are shown in ribbon and surface, colored according to their population within the global ensemble (red is higher population; blue is lower). Two different views are displayed. (Perez-Cano, L., E. Eliahoo, K. Lasker, H. J. Wolfson, F. Glaser, H. Manor, P. Bernado, and J. Fernandez-Recio, Conformational transitions in human translin enable nucleic acid binding. *Nucleic Acids Res,* 41 (21): 9956–9966, 2013, by permission of Oxford University Press.)

FINAL REMARKS

Quantitative characterization of flexible macromolecular objects using the SAXS data is a relatively new area of application of this technique. The novel approaches operating in terms of ensemble distributions allow one to conduct meaningful analysis of these challenging systems. Ensemble analysis can cover a broad spectrum of flexible systems, from multidomain or multisubunit objects tethered by linkers to intrinsically disordered states. The new methods can also tackle macromolecular interactions revealing possible changes of flexibility upon binding and provide indications about transiently formed conformations (Marsh, Teichmann, and Forman-Kay 2012). However, it should always be kept in mind that SAXS data physically correspond to an average of the scattering patterns from many independent macromolecules. Construction of 3-D models from this data is a difficult and ambiguous task even for rigid monodisperse systems. Ensemble analyses, where an additional conformational averaging is involved, are even more susceptible to overinterpretation and overfitting of the data. To increase the information content, SAXS studies of flexible systems are often combined with complementary techniques, mainly NMR but also with FRET, CD, AFM, and other methods. Computational methods with the capacity to distort 3-D arrangements allowing the exploration of the conformational space are crucial tools for the proper derivation of dynamic information from SAXS data. We envision an increasing number of studies integrating SAXS and other biophysical data into advanced computational methods to tackle relevant biological questions.

ACKNOWLEDGMENTS

Agence Nationale de la Recherche (SPIN-HD-ANR-CHEX-2011) and ATIP-Avenir program are acknowledged for the financial support (PB).

REFERENCES

Ando, N., Y. Kung, M. Can, G. Bender, S. W. Ragsdale, and C. L. Drennan. 2012. Transient B12-dependent methyltransferase complexes revealed by small-angle X-ray scattering. *J Am Chem Soc* 134 (43):17945–17954.

Bernado, P. 2010. Effect of interdomain dynamics on the structure determination of modular proteins by small-angle scattering. *Eur Biophys J* 39 (5):769–780.

Bernado, P., and M. Blackledge. 2009. A self-consistent description of the conformational behavior of chemically denatured proteins from NMR and small angle scattering. *Biophys J* 97 (10):2839–2845.

Bernado, P., and M. Blackledge. 2010. Proteins in dynamic equilibrium. *Nature* 468: 1046–1048.

Bernado, P., L. Blanchard, P. Timmins, D. Marion, R. W. Ruigrok, and M. Blackledge. 2005. A structural model for unfolded proteins from residual dipolar couplings and small-angle x-ray scattering. *Proc Natl Acad Sci U S A* 102 (47):17002–17007.

Bernado, P., K. Modig, P. Grela, D. I. Svergun, M. Tchorzewski, M. Pons, and M. Akke. 2010. Structure and dynamics of ribosomal protein L12: An ensemble model based on SAXS and NMR relaxation. *Biophys J* 98 (10):2374–2382.

Bernado, P., E. Mylonas, M. V. Petoukhov, M. Blackledge, and D. I. Svergun. 2007. Structural characterization of flexible proteins using small-angle X-ray scattering. *J Am Chem Soc* 129 (17):5656–5664.

Bernado, P., Y. Perez, J. Blobel, J. Fernandez-Recio, D. I. Svergun, and M. Pons. 2009. Structural characterization of unphosphorylated STAT5a oligomerization equilibrium in solution by small-angle X-ray scattering. *Protein Sci* 18 (4):716–726.

Bernado, P., Y. Perez, D. I. Svergun, and M. Pons. 2008. Structural characterization of the active and inactive states of Src kinase in solution by small-angle X-ray scattering. *J Mol Biol* 376 (2):492–505.

Bernado, P., and D. I. Svergun. 2012. Structural analysis of intrinsically disordered proteins by small-angle X-ray scattering. *Mol Biosyst* 8 (1):151–167.

Blanchet, C. E., and D. I. Svergun. 2013. Small-angle X-ray scattering on biological macromolecules and nanocomposites in solution. *Annu Rev Phys Chem* 64:37–54.

Blobel, J., P. Bernado, D. I. Svergun, R. Tauler, and M. Pons. 2009. Low-resolution structures of transient protein–protein complexes using small-angle X-ray scattering. *J Am Chem Soc* 131 (12):4378–4386.

Blobel, J., U. Brath, P. Bernado, C. Diehl, L. Ballester, A. Sornosa, M. Akke, and M. Pons. 2011. Protein loop compaction and the origin of the effect of arginine and glutamic acid mixtures on solubility, stability and transient oligomerization of proteins. *Eur Biophys J* 40 (12):1327–1338.

Boura, E., B. Rozycki, H. S. Chung, D. Z. Herrick, B. Canagarajah, D. S. Cafiso, W. A. Eaton, G. Hummer, and J. H. Hurley. 2012. Solution structure of the ESCRT-I and -II supercomplex: Implications for membrane budding and scission. *Structure* 20 (5):874–886.

Boura, E., B. Rózycki, D. Z. Herrick, H. S. Chung, J. Vecer, W. A. Eaton, D. S. Cafiso, G. Hummer, and J. H. Hurley. 2011. Solution structure of the ESCRT-I complex by small-angle X-ray scattering, EPR, and FRET spectroscopy. *Proc Natl Acad Sci U S A* 108 (23):9437–9442.

Doniach, S. 2001. Changes in biomolecular conformation seen by small angle X-ray scattering. *Chem Rev* 101 (6):1763–1778.

Dunker, A. K., C. J. Brown, J. D. Lawson, L. M. Iakoucheva, and Z. Obradović. 2002. Intrinsic disorder and protein function. *Biochemistry* 41 (21):6573–6582.

Feigin, L. A., and D. I. Svergun. 1987. *Structure Analysis by Small-Angle X-Ray and Neutron Scattering*. New York: Plenum Press.

Francis, D. M., B. Rozycki, D. Koveal, G. Hummer, R. Page, and W. Peti. 2011. Structural basis of p38alpha regulation by hematopoietic tyrosine phosphatase. *Nat Chem Biol* 7 (12):916–924.

Glatter, O. 1977. A new method for the evaluation of small-angle scattering data. *J Appl Cryst* 10:415–421.

Glatter, O., and O. Kratky. 1982. *Small Angle X-Ray Scattering*. London: Academic Press.

Guinier, A. 1939. La diffraction des rayons X aux tres petits angles; application a l'etude de phenomenes ultramicroscopiques. *Ann Phys (Paris)* 12:161–237.

Hu, S-H., A. E. Whitten, G. J. King, A. Jones, A. F. Rowland, D. James, and J. Martin. 2012. The weak complex between RhoGAP protein ARHGAP22 and signal regulatory protein 14-3-3 has 1:2 stoichiometry and a single peptide binding mode. *PLoS One* 7 (8):e41731.

Jacques, D. A., and J. Trewhella. 2010. Small-angle scattering for structural biology— Expanding the frontier while avoiding the pitfalls. *Protein Sci* 19 (4):642–657.

Jensen, M. R., P. R. L. Markwick, S. Meier, S. Griesinger, M. Zweckstetter, S. Grzesiek, P. Bernadó, and M. Blackledge. 2009. Quantitative determination of the conformational properties of partially folded and intrinsically disordered proteins using NMR dipolar couplings. *Structure* 17 (9):1169–1185.

Kazantsev, A. V., R. P. Rambo, S. Karimpour, J. Santalucia, Jr., J. A. Tainer, and N. R. Pace. 2011. Solution structure of RNase P RNA. *RNA* 17 (6):1159–1171.

Koch, M. H., P. Vachette, and D. I. Svergun. 2003. Small-angle scattering: A view on the properties, structures and structural changes of biological macromolecules in solution. *Q Rev Biophys* 36 (2):147–227.

Kohn, J. E., I. S. Millett, J. Jacob, B. Zagrovic, T. M. Dillon, N. Cingel, R. S. Dothager, S. Seifert, P. Thiyagarajan, T. R. Sosnick et al. 2004. Random-coil behavior and the dimensions of chemically unfolded proteins. *Proc Natl Acad Sci U S A* 101 (34):12491–12496.

Konarev, P. V., M. V. Petoukhov, V. V. Volkov, and D. I. Svergun. 2006. ATSAS 2.1, a program package for small-angle scattering data analysis. *J Appl Crystallogr* 39:277–286.

Krzeminski, M., J. A. Marsh, C. Neale, W. Y. Choy, and J. D. Forman-Kay. 2013. Characterization of disordered proteins with ENSEMBLE. *Bioinformatics* 29 (3):398–399.

Levitt, M. 2009. Nature of the protein universe. *Proc Natl Acad Sci U S A* 106 (27):11079–11084.

Leyrat, C., M. R. Jensen, E. A. Ribeiro, Jr., F. C. Gerard, R. W. Ruigrok, M. Blackledge, and M. Jamin. 2011. The N(0)-binding region of the vesicular stomatitis virus phosphoprotein is globally disordered but contains transient alpha-helices. *Protein Sci* 20 (3):542–556.

Marsh, J. A., and J. D. Forman-Kay. 2009. Structure and disorder in an unfolded state under nondenaturing conditions from ensemble models consistent with a large number of experimental restraints. *J Mol Biol* 391 (2):359–374.

Marsh, J. A., C. Neale, F. E. Jack, W. Y. Choy, A. Y. Lee, K. A. Crowhurst, and J. D. Forman-Kay. 2007. Improved structural characterizations of the drkN SH3 domain unfolded state suggest a compact ensemble with native-like and non-native structure. *J Mol Biol* 367 (5):1494–1510.

Marsh, J. A., S. A. Teichmann, and J. D. Forman-Kay. 2012. Probing the diverse landscape of protein flexibility and binding. *Curr Opin Struct Biol* 22 (5):643–650.

Mittag, T., J. Marsh, A. Grishaev, S. Orlicky, H. Lin, F. Sicheri, M. Tyers, and J. D. Forman-Kay. 2010. Structure/function implications in a dynamic complex of the intrinsically disordered Sic1 with the Cdc4 subunit of an SCF ubiquitin ligase. *Structure* 18 (4):494–506.

Mukrasch, M. D., J. Biernat, M. von Bergen, C. Griesinger, E. Mandelkow, and M. Zweckstetter. 2005. Sites of tau important for aggregation populate β-structure and bind to microtubules and polyanions. *J Biol Chem* 280 (26):24978–24986.

Mukrasch, M. D., P. Markwick, J. Biernat, M. von Bergen, P. Bernadó, C. Griesinger, E. Mandelkow, M. Zweckstetter, and M. Blackledge. 2007. Highly populated turn conformations in natively unfolded tau protein identified from residual dipolar couplings and molecular simulation. *J Am Chem Soc* 129 (16):5235–5243.

Mylonas, E., A. Hascher, P. Bernado, M. Blackledge, E. Mandelkow, and D. I. Svergun. 2008. Domain conformation of tau protein studied by solution small-angle X-ray scattering. *Biochemistry* 47 (39):10345–10353.

Mylonas, E., and D. I. Svergun. 2007. Accuracy of molecular mass determination of proteins in solution by small-angle X-ray scattering. *J Appl Cryst* 40:s245–s249.

Pelikan, M., G. L. Hura, and M. Hammel. 2009. Structure and flexibility within proteins as identified through small angle X-ray scattering. *Gen Physiol Biophys* 28 (2):174–189.

Perard, J., C. Leyrat, F. Baudin, E. Drouet, and M. Jamin. 2013. Structure of the full-length HCV IRES in solution. *Nat Commun* 4:1612.

Perez-Cano, L., E. Eliahoo, K. Lasker, H. J. Wolfson, F. Glaser, H. Manor, P. Bernado, and J. Fernandez-Recio. 2013. Conformational transitions in human translin enable nucleic acid binding. *Nucleic Acids Res* 41 (21):9956–9966.

Petoukhov, M. V., D. Franke, A. V. Shkumatov, G. Tria, A. G. Kikhney, M. Gajda, C. Gorba, H. D. T. Mertens, P. V. Konarev, and D. I. Svergun. 2012. New developments in the ATSAS program package for small-angle scattering data analysis. *J Appl Crystallogr* 45 (2):342–350.

Petoukhov, M. V., and D. I. Svergun. 2005. Global rigid body modelling of macromolecular complexes against small-angle scattering data. *Biophys J* 89 (2):1237–1250.

Porod, G. 1982. General theory. In *Small-Angle X-Ray Scattering*, Edited by O. Glatter and O. Kratky, 17–51. London: Academic Press.

Rambo, R. P., and J. A. Tainer. 2011. Characterizing flexible and intrinsically unstructured biological macromolecules by SAS using the Porod-Debye law. *Biopolymers* 95 (8):559–571.

Ribeiro Ede, Jr., A., M. Beich-Frandsen, P. V. Konarev, W. Shang, B. Vecerek, G. Kontaxis, H. Hammerle, H. Peterlik, D. I. Svergun, U. Blasi, and K. Djinovic-Carugo. 2012. Structural flexibility of RNA as molecular basis for Hfq chaperone function. *Nucleic Acids Res* 40 (16):8072–8084.

Rozycki, B., Y. C. Kim, and G. Hummer. 2011. SAXS ensemble refinement of ESCRT-III CHMP3 conformational transitions. *Structure* 19 (1):109–116.

Russel, D., K. Lasker, J. Phillips, D. Schneidman-Duhovny, J. A. Velazquez-Muriel, and A. Sali. 2009. The structural dynamics of macromolecular processes. *Curr Opin Cell Biol* 21 (1):97–108.

Russell, R. B., and P. Aloy. 2008. Targeting and tinkering with interaction networks. *Nat Chem Biol* 4 (11):666–673.

Sampathkumar, P., S. J. Kim, P. Upla, W. J. Rice, J. Phillips, B. L. Timney, U. Pieper, J. B. Bonanno, J. Fernandez-Martinez, Z. Hakhverdyan et al. 2013. Structure, dynamics, evolution, and function of a major scaffold component in the nuclear pore complex. *Structure* 21 (4):560–571.

Semenyuk, A. V., and D. I. Svergun. 1991. GNOM—A program package for small-angle scattering data processing. *J Appl Crystallogr* 24:537–540.

Shkumatov, A. V., S. Chinnathambi, E. Mandelkow, and D. I. Svergun. 2011. Structural memory of natively unfolded tau protein detected by small-angle X-ray scattering. *Proteins* 79 (7):2122–2131.

Shortle, D. 1996. The denatured state (the other half of the folding equation) and its role in protein stability. *FASEB J* 10 (1):27–34.

Stein, A., R. A. Pache, P. Bernado, M. Pons, and P. Aloy. 2009. Dynamic interactions of proteins in complex networks: A more structured view. *FEBS J* 276 (19):5390–5405.

Stott, K., M. Watson, F. S. Howe, J. G. Grossmann, and J. O. Thomas. 2010. Tail-mediated collapse of HMGB1 is dynamic and occurs via differential binding of the acidic tail to the A and B domains. *J Mol Biol* 403 (5):706–722.

Svergun, D. I., C. Barberato, and M. H. J. Koch. 1995. CRYSOL—A program to evaluate X-ray solution scattering of biological macromolecules from atomic coordinates. *J Appl Crystallogr* 28:768–773.

Svergun, D. I., M. H. J. Koch, P. A. Timmins, and R. P. May. 2013. *Small Angle X-Ray and Neutron Scattering from Solutions of Biological Macromolecules*. New York: Oxford University Press.

Tompa, P. 2012. On the supertertiary structure of proteins. *Nat Chem Biol* 8 (7):597–600.

Trewhella, J., D. K. Blumenthal, S. E. Rokop, and P. A. Seeger. 1990. Small-angle scattering studies show distinct conformations of calmodulin in its complexes with two peptides based on the regulatory domain of the catalytic subunit of phosphorylase kinase. *Biochemistry* 29 (40):9316–9324.

Tsutakawa, S. E., A. W. Van Wynsberghe, B. D. Freudenthal, C. P. Weinacht, L. Gakhar, M. T. Washington, Z. H. Zhuang, J. A. Tainer, and I. Ivanov. 2011. Solution X-ray scattering combined with computational modeling reveals multiple conformations of covalently bound ubiquitin on PCNA. *Proc Natl Acad Sci U S A* 108 (43):17672–17677.

von Ossowski, I., J. T. Eaton, M. Czjzek, S. J. Perkins, T. P. Frandsen, M. Schulein, P. Panine, B. Henrissat, and V. Receveur-Brechot. 2005. Protein disorder: Conformational distribution of the flexible linker in a chimeric double cellulase. *Biophys J* 88 (4):2823–2832.

Williamson, T. E., B. A. Craig, E. Kondrashkina, C. Bailey-Kellogg, and A. M. Friedman. 2008. Analysis of self-associating proteins by singular value decomposition of solution scattering data. *Biophys J* 94 (12):4906–4923.

Yang, S. C., L. Blachowicz, L. Makowski, and B. Roux. 2010. Multidomain assembled states of Hck tyrosine kinase in solution. *Proc Natl Acad Sci U S A* 107 (36):15757–15762.

Zagrovic, B., J. Lipfert, E. J. Sorin, I. S. Millett, W. F. van Gunsteren, S. Doniach, and V. S. Pande. 2005. Unusual compactness of a polyproline type II structure. *Proc Natl Acad Sci U S A* 102 (33):11698–11703.

Zhou, X. 2004. Polymer models of protein stability, folding and interactions. *Biochemistry* 43 (8):2141–2154.

Chapter 14

Bridging Experiments and Simulations

Structure Calculations with a Dynamical Touch

Florian Heinkel, Alexander Cumberworth, and Jörg Gsponer

CONTENTS

INTRODUCTION

The central task of structural biology is to determine the three-dimensional structure of bioactive macromolecular compounds and draw structure-to-function connections by "reading" the structural details. Three experimental methods play a dominant role in achieving this task: x-ray crystallography, cryoelectron microscopy, and nuclear magnetic resonance.

In the early twentieth century the course of modern structural biology was set when Max von Laue realized that the magnitude of the wavelength of x-ray beams was in the range of the spacings between atoms in crystals and molecules. This realization lead to the development of x-ray diffraction as a technique for structure determination, beginning with his description of the diffraction behavior of a copper(II) sulfate crystal together with his technicians Walter Friedrich and Paul Knipping in 1912 (Friedrich, Knipping, and von Laue 1912). Bringing the famous Bragg's law into play, William Lawrence Bragg and his father William Henry Bragg made the connection between the diffraction pattern and the structure of atoms in crystal lattices of simple salt crystals shortly after (Bragg 1913a, b). Although the basic instruments needed to structurally characterize the more complicated macromolecular biopolymers that define life were available, the world of biological science had to wait until the 1950s for those tools to be applied by Rosalid Franklin, Raymond Gosling, James Watson, and Francis Crick to determine the helical structure of deoxyribonucleic acid (DNA) (Franklin and Gosling 1953; Watson and Crick 1953). After solving the phase problem that becomes severe for large, highly structurally heterogeneous, and nonsymmetric objects, the first three-dimensional structures of proteins were reported for myoglobin and hemoglobin by John Kendrew in 1958 (Kendrew et al. 1958, 1960) and Max Perutz in 1960 (Perutz et al. 1960). They mark the hour of birth for modern structural biology that is now able to describe macromolecules at atomic resolution. To date, x-ray crystallography has produced countless three-dimensional structures of proteins, dominates the entries in the Protein Data Bank (PDB) (>85%) and has given valuable insight into many biological processes.

Although used for the structural characterization of biological samples for nearly a century, electron microscopy (EM) has not been of much interest for the analysis of bioactive macromolecules at the nanometer scale because of radiation damage and resolution limitations. However, fairly recent technical advances have given rise to a renaissance of this method (Kourkoutis, Plitzko, and Baumeister 2012). Not long after being able to produce vitreous ice (Bruggeller and Mayer 1980; Dubochet and McDowall 1981) and prepare specimens in it, thereby keeping them hydrated while protecting them from the destructive impact of accelerated electrons, the first cryo-electron micrographs of biological samples were observed (Adrian et al. 1984). With the methods for three-dimensional reconstruction from two-dimensional images, the available image processing power and the EM device design constantly evolving, cryo-EM is now one of the three major forces in structural biology. Cryo-EM is slowly expanding its applicability from sampling impressively large systems and gaining large-scale mechanistic insight (Davies et al. 2012; Raddi et al. 2012) toward high-resolution models of smaller protein complexes (Mao et al. 2013; Wiedenheft et al. 2011).

The youngest method for structural characterization of biological macromolecules at atomic resolution is nuclear magnetic resonance (NMR) spectroscopy. The underlying principle of nuclei absorbing and emitting electromagnetic radiation when they are exposed to an exterior magnetic field was first discovered by Isidor Rabi in 1938 (Rabi et al. 1938) and described in bulk matter in 1945 by Edward Purcell (Purcell, Torrey, and Pound 1946) and Felix Bloch (Bloch, Hansen, and Packard 1946). After developing the ability to measure subtle differences in the resonance frequency, which depend on the chemical environment of a particular nucleus (Arnold, Dharmatti, and Packar 1951), NMR was applied to analytical chemistry and the characterization of small molecules. The discovery of the nuclear Overhauser effect (NOE) (Anderson and Freeman 1962) ultimately led to NMR spectroscopy being used as a method to determine the structure of proteins, the first being done by Wüthrich and coworkers (Williamson, Havel, and Wüthrich 1985) in 1985, whereby NOE-derived distance restraints were applied to a protein model of fixed bond lengths and angles. Today, NOE-derived distance restraints as well as other restraints derived from NMR measurands such as chemical shifts, J-couplings, or residual dipolar couplings are routinely used in structure calculations (Schwieters et al. 2003). While NMR is generally limited in terms of the molecular weight of the specimen due to overlapping spectra and relaxation effects, its power lies in looking at solvated proteins rather than proteins packed in crystals or frozen in ice, and in providing a large arsenal of tools to investigate not only the structure but also the dynamics of biomolecules. However, many of the NMR measurands that report on biomolecular dynamics are not straightforward to interpret in structural terms, and therefore are not included routinely in structure calculations.

As a result of the noninclusion of dynamics data in structure calculations, the vast majority of entries in the PDB are determined under the assumption that a single, average structure can satisfy all or most of the experimental restraints. Furthermore, it is assumed this average structure is a good representation of the experimentally probed state and that this state is the native state of the protein. Such approaches may contribute to a misleading picture of biomolecules being rather static and well represented by a single conformation. It is true that if one looks at the energy landscape, certain biomolecules have a narrow, highly populated native state well and that a single conformation can provide a good representation (Vendruscolo, Paci, Karplus et al. 2003). However, when the structures of biomolecules are probed by the experimental methods introduced above, measured quantities are averages of billions of molecules in different energetic and conformational states and because acquiring data is often not instantaneous, one is sampling each molecule in all the conformations it adopts during measuring time. The entity of structures is called the ensemble and its width can be very large for highly dynamic proteins (Lindorff-Larsen et al. 2004). Consistently, flexibility of proteins can be very important for their biological function (Boehr, Dyson, and Wright 2006; Henzler-Wildman and Kern 2007) and even high-energy states that are not part of the native well play a nonnegligible role for some proteins (Vendruscolo, Zurdo et al. 2003). Hence, for dynamic proteins with a broad native well, an average structure can be of very little structural significance and hard to interpret, or there may even be no single structure that satisfies the experimental data (Bonvin and Brünger 1995; Clore and Schwieters 2006; Richter et al. 2007; Torda, Scheek, and van Gunsteren 1989).

Computer simulations can be used to investigate the dynamics of biomolecules at the atomic level of detail. Molecular dynamics (MD) simulations are the most popular type of simulations used. Systems are modeled classically using molecular mechanics and empirical data to derive force fields that describe how all atoms in a given system interact with each other. Starting with some initial coordinates and momenta, Newton's laws can be integrated over time to provide a trajectory of the system; in other words, molecular movies at an atomic resolution can be obtained. The first MD simulation of a biomolecule, bovine pancreatic trypsin inhibitor, was performed in 1977 by McCammon, Gelin, and Karplus (McCammon, Glein, and Karplus 1977). The simulation was carried out in a vacuum with a rather crude force field and only a few picoseconds of data were able to be collected due to the limitations of the available computer power. However, it represented a huge step forward in the quest to understand protein dynamics; over the following years, new and improved force fields have been developed, solvent models have been created and refined, and most importantly, computational power has increased exponentially as predicted by Moore's law. MD simulations are now used routinely to refine experimentally determined structures, study protein folding and ligand docking, and investigate all types of system-specific questions regarding links between structure, dynamics, and function (Karplus 2003). However, the results of MD simulations classically suffer from the fact that the force fields used are at best good approximations of the "real" force field that acts on biomolecules in nature, as well as convergence problems due to limitations in the number of systems and the timescale that can be simulated, although significant improvements have recently been achieved in this regard (Shaw et al. 2010).

To overcome the limitations of both classical structure calculations and MD simulations, the idea to combine both methods in order to generate structure ensembles that are consistent with the measured experimental data was born. In this approach, the average over a number of different structures—the calculated ensemble—is restrained to fit experimental data thereby allowing the protein to adopt several conformations. This is consistent with the real situation of structural averaging during measurement (Bonvin, Boelens, and Kaptein 1994). In principle, data from all experimental methods described above can be used. However, NOE distance parameters in combination with additional data from NMR experiments probing dynamics on different timescales ranging from picoseconds to seconds are most widely applied (Vendruscolo and Dobson 2005). The width of the ensemble and thereby the nature of dynamics described by this approach depends on the experimental data that is used and which time window it probes. Therefore, multiple experimental methods probing different regimes of motion have to be used to get a complete picture of the dynamic properties of a protein on various timescales and describe the energy landscape of the protein accurately. In this chapter, we provide the the-oretical background for different types of ensemble calculations and illustrate their use in a few examples.

RESTRAINED SIMULATIONS

Formulation of Restraints in MD Simulations

Generally, the force fields used in MD simulations consist of bonded and non-bonded energy terms V_b and V_{nb}, respectively.

$$V(\mathbf{r}) = V_b + V_{nb} \tag{14.1}$$

The total potential $V(\mathbf{r})$ is recalculated during the course of the simulation after a given time step Δt. The force applied to each atom F_i at all time points during the simulation is calculated as the first derivative of this potential energy with respect to the coordinates each atom, $\mathbf{r}_i$.

$$F_i = -\frac{\partial V}{\partial \mathbf{r}_i} \tag{14.2}$$

In restrained MD simulations, a pseudoenegy or restraint potential term $V_{\text{restraint}}(\mathbf{r},t)$ is added to the force field that penalizes deviations between simulated and measured sets of parameters A_i^{sim} and A_i^{exp}. Hence, at the core of the restraint potential is a reaction coordinate ρ that can be defined in the simplest case as

$$\rho = \sum_i \left(A_i^{\text{sim}} - A_i^{\text{exp}} \right)^x \tag{14.3}$$

There are multiple ways to ensure that ρ reaches a value of 0 as time progresses. One important method was first introduced by Paci and Karplus (1999) based on time-dependent perturbation proposed by P. Ballone and S. Rubini (unpublished result). In their formalism, the perturbation (restraint)

$V_{\text{restraint}}$ calculated over M replicas depends on the time through the reaction coordinate:

$$V_{\text{restraint}}(\mathbf{r},t)=\begin{cases} \dfrac{M\alpha}{2}(\rho-\rho_a)^2 & \text{for } \rho(t)>\rho_a \\ 0 & \text{for } \rho(t)\leq\rho_a \end{cases} \tag{14.4}$$

$$\rho_a(t)=\min_{0\leq\tau\leq t}\rho(\tau) \tag{14.5}$$

This potential penalizes a conformational evolution that corresponds to higher values of $\rho(t)$ than its minimum by applying a harmonic force on the atoms that violate the restraint to drive them back toward an acceptable conformation. The weight of the restraint that sets the magnitude of the applied forces and how stringently a system is driven to fit the experimental parameters is controlled by the force constant α, which has to be determined individually for each system studied. It must be selected so that it is high enough to allow the restraints to be satisfied. In practice, the alpha value is steadily increased in an initial simulation in order to find its upper bound. The evolution of the half-harmonic potential during this simulation is illustrated in Figure 14.1. Parameters are back-calculated and compared to the experimental ones across the ensemble in the form of the reaction coordinate ρ at each step of the simulation. A simulated conformation with a new minimal value of

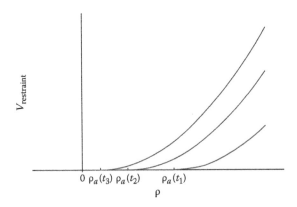

Figure 14.1 Behavior of the half harmonic potential $V_{\text{restraint}}(\mathbf{r},t)$ versus ρ over time. The potential is shown for three different time points $t_1 > t_2 > t_3$. As new minima of ρ are sampled, ρ_a gets smaller and the potential with respect to the origin becomes more narrow.

$\rho(t)$ is always favorable as it falls in the left half of the potential where no penalty force is applied. In this case, the calculated $\rho(t)$ becomes the new standard $\rho_a(t)$ so that the comparison value at a given time $\rho_a(t)$ is always minimal. If the minimal value of $\rho(t)$ at a given force constant α has converged but is not yet zero, α is increased. As time progresses from t_1 to t_3, $\rho_a(t)$ gets smaller because new minima are sampled and the right half of the potential gets narrower with respect to the origin.

This restraint potential term is included in the general expression for the potential energy of the MD force field:

$$V(\mathbf{r}, t) = V_b + V_{nb} + V_{restraint} \tag{14.6}$$

The first derivative of the restraint potential gives the penalty forces applied to each atom i, $F_{i,restraint}$.

$$F_{i,restraint} = -\frac{\partial V_{restraint}}{\partial \mathbf{r}_i} \tag{14.7}$$

All restrained simulations discussed here are commonly combined with some form of simulated annealing (Dedmon et al. 2005; Gsponer et al. 2008; Khorvash, Lamour, and Gsponer 2011; Lindorff-Larsen et al. 2005; Richter et al. 2007). Simulated annealing structure calculations involve raising and lowering the temperature of the system in order to increase sampling efficiency and take a broader range of the conformational space into account. NOE restraints are added here as simple flat-bottom harmonic distance restraints to prevent the system from falling apart at the higher temperatures (or to assist in solving the structure if initial coordinates are not known beforehand).

Time- and Ensemble-Averaging

Any macroscopic observable $\langle A \rangle$ that results from measurements of the bulk is equal to an ensemble average of A_i over the N microscopic states of the system:

$$\langle A \rangle = \frac{1}{N} \sum_{i=1}^{N} A_i \tag{14.8}$$

In structure calculations, this observable could be distances, angles, dihedral angles, or parameters that report on the amplitude of motions in the structure. Depending on the type of observable and the probability distribution of A_i, a single structure may not be able to correctly reproduce $\langle A \rangle$. To account for that, the simulated parameter A_i^{sim} in Equation 14.3 is taken as an average A_i^{sim} over a number of calculated structures. This can be either accomplished by averaging over the structures of one simulated system over time (time averaging) (Torda, Scheek, and van Gunsteren 1989, 1990), or by averaging over the structures of several identical systems simulated in parallel (ensemble averaging) (Kemmink and Creighton 1993). The true time average at time t of a simulated parameter for a single system evolving over time is given as

$$\overline{A_i^{sim}(t)} = \frac{1}{t}\int_0^t A_i^{sim}(t')dt' \tag{14.9}$$

Here, t denotes a fixed time for which the average is calculated, whereas t' is the integrated variable. The problem with this formulation of the average is that as time increases, the value becomes less and less sensitive to fluctuations in the MD simulation. To overcome that issue, a decaying memory function based on the time constant τ can be applied that smoothens out increasing insensitivity (Torda, Scheek, and van Gunsteren 1989).

$$\overline{A_i^{sim}(t)} = \frac{1}{\tau(1-e^{\frac{-t}{\tau}})}\int_0^t e^{\frac{-t'}{\tau}} A_i^{sim}(t-t')dt' \tag{14.10}$$

As an MD simulation increments the timescale by Δt instead of being continuous, t' has to be replaced by an explicit time point t_k for application. The integral becomes a simple sum over all N_t time-points.

$$\overline{A_i^{sim}(t)} = \frac{1}{\tau(1-e^{\frac{-t}{\tau}})}\sum_{k=0}^{N_t} e^{\frac{-t_k}{\tau}} A_i^{sim}(t-t_k) \tag{14.11}$$

For ensemble averaged calculations of $\overline{A_i^{sim}}$, Equation 14.8 can be directly applied as a sum of the number of computed parameters A_i over all replicas M in the simulated ensemble.

$$\overline{A_i^{sim}} = \frac{1}{M}\sum_{i=1}^{M} A_i \tag{14.12}$$

One crucial factor for success of the simulation and whether the determined ensemble resembles the actual dynamic properties of the investigated system is the choice of the ensemble size M (or the length of t in a time-averaging approach). If one chooses the ensemble to be too small, it will not be able to sample a big enough variety of conformations and dynamic properties are overlooked. In this case, the system is overrestrained (or underfitted). The other extreme case is the choice of too big an ensemble and the problem of underrestraining (or overfitting). Here, too many replicas in the ensemble (or too long a t) increases the number of degrees of freedom in the back-calculation to a point where a better fit of simulated and experimental parameters does not necessarily resemble a better conformational fit but is just caused by an increased number of fitted parameters. It has been proposed that the number of replicas M to be used in the calculated ensemble to obtain an accurate representation of the experimental ensemble highly depends on the observable and has to be estimated separately in each case (Richter et al. 2007).

Timescale of the Dynamic Properties to be Represented by the Ensemble

The timescale of dynamics that the calculated structures reproduce depends on the timescale that the experimentally measured observable probes. In theory, the observable can be any experimentally determined quantity that probes dynamics and can be back-calculated from the atomic coordinates in a simulation. Fortunately, the timescale on which protein dynamics occur can be mostly covered by a whole toolbox of NMR experiments that are able to monitor conformational change of various origins within a distinct time window with high sensitivity (Kleckner and Foster 2011). Its ability to link protein dynamics and function has been proven for important biological features such as enzyme catalysis, protein folding, sparsely populated excited conformational states, and thermodynamics. The methods range from real-time (RT) NMR that includes hydrogen exchange (HX) and looks at conformational change occurring in the slow time-regime of milliseconds and above, probing folding and large-scale structural rearrangements such as whole domain movement, to nuclear spin relaxation (NSR) and residual dipolar coupling (RDC), both probing dynamics in the fast region of the dynamics spectrum covering side chain rotation, loop movement, and molecular tumbling. To investigate intermediate structural fluctuations, methods range from paramagnetic relaxation enhancement (PRE) and rotating frame relaxation dispersion (RF RD) in the microseconds frame to Carr-Purcell Meiboom-Gill relaxation dispersion (CPMG RD), lineshape (LS) analysis, and exchange spectroscopy (EXSY) that cover the time window from milliseconds to seconds. Figure 14.2 provides an overview of the available methods with the according time frames and probed conformational changes. The main question we discuss in the following is how data from some of these NMR experiments are specifically implemented as restraints in MD simulations and how these are performed in principle to give insight into the dynamical behavior of proteins.

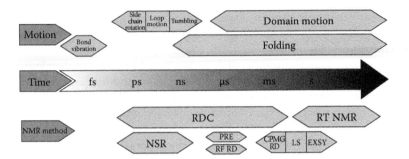

Figure 14.2 A timescale of protein dynamics probed by NMR experiments. Motional regimes are shown in the top part above the timescale and the NMR methods in the bottom part below the timescale. The width of the bars for motional regimes and NMR methods indicates the timescale on which they are occurring or that they are probing, respectively. The NMR methods shown are nuclear spin relaxation (NSR), residual dipolar coupling (RDC), paramagnetic relaxation enhancement (PRE), rotating frame relaxation dispersion (RF RD), Carr-Purcell Meiboom-Gill relaxation dispersion (CPMG RD), lineshape analysis (LS), exchange spectroscopy (EXSY), and real-time (RT) NMR.

NMR DATA USED IN RESTRAINED MD SIMULATIONS

ps-ns Dynamics from ^{15}N Relaxation Experiments (S^2-Order Parameters)

S^2-order parameters allow some insight into the dynamics of a protein on the picosecond to nanosecond timescale. Processes contributing to dynamics on this timescale include bond vibrations, side chain rotamer interconversions, and backbone torsion angle rotations. The S^2-order parameter quantifies the amplitude of the internal motions of bond vectors investigated; it can also be thought of as indicating the level of heterogeneity of the bond vector across the ensemble. The parameter can take on a value from 0 to 1, with 0 indicating that the bond vector is isotropically distributed across the ensemble and 1 indicating that the bond vector is exactly the same in all copies. Hence, this parameter can be used in restrained MD simulations to generate ensembles whose distributions of bond vectors match those found in experiments and correspond to motions on the picosecond to nanosecond timescale. Experimentally, these parameters can be obtained from the results of NSR measurements. Because the local chemical environment can affect the rate at which an excited nucleus returns to its ground state, measuring NSR provides another perspective on what is going on at the atomic level. More specifically, oscillating magnetic fields created by the movements of polarized and charged groups can actually stimulate an excited nucleus to return to its ground state. One of the ways that these oscillating fields can be created is through chemical shift anisotropy. When bond vectors change their orientation relative to a magnetic field, they experience a different local magnetic field. The oscillations caused by ps-ns timescale motions occur with a frequency in the range needed to lead to relaxation enhancement of excited nuclei. Dipolar coupling is another important contributor to such oscillating fields. It refers to an interaction that occurs between two nuclear spins that can change the local magnetic field as the distance between the two nuclei changes or the orientation of the vector changes relative to the main magnetic field.

Usually, two relaxation rate constants and the heteronuclear ^{15}N NOE are measured for each N-H bond vector in NSR experiments. R_1 and R_2 denote the longitudinal and transverse relaxation rates, respectively, and R_{NOE} the heteronuclear ^{15}N NOE. These relaxation constants are determined by the spectral density functions $J(\omega)$ (Chen, Brooks, and Wright 2004). The spectral density function $J(\omega)$ is simply the Fourier transform of the bond vector's angular autocorrelation function, $C(t)$

$$J(\omega) = 2 \int_0^\infty C(t) \cos(\omega t) \mathrm{d}t \qquad (14.13)$$

where ω is a Larmor frequency (or sum/difference of Larmor frequencies). In the model-free analysis it is assumed that the internal motions are independent and much faster than the motions of the entire molecule. This allows the angular autocorrelation function to be broken into two pieces:

$$C(t) = C_0(t)C_I(t) \qquad (14.14)$$

where $C_0(t)$ and $C_I(t)$ are the overall tumbling and internal motion correlation functions, respectively. The internal correlation function can be modeled with the order parameters using a number of different equations. In the simplest case, it is modeled as follows:

$$C_I(t) = S^2 + (1 - S^2)e^{\frac{-t}{\tau_e}} \qquad (14.15)$$

where S^2 is the squared generalized order parameter and τ_e is the relaxation time constant. Starting from some initial values of the parameters (in the case of the internal correlation function, S^2 and τ_e), theoretical values for the three experimentally determined relaxation quantities can be determined. By minimizing the following function

$$\chi^2 = \sum_{j=1}^M \frac{(R_j - \hat{R}_j)^2}{\sigma_j^2} \qquad (14.16)$$

where M is the number of relaxation parameters measured for each bond vector (usually three), R_j is the jth experimentally measured relaxation constant, $\hat{R}_j$ is the corresponding theoretically calculated relaxation constant, and σ_j is the standard deviation associated with the experimentally measured relaxation constant, accurate experimental values for the generalized order parameter and the time constant can be obtained.

In order to implement the generalized order parameter as a restraint in simulations, it must be able to be back-calculated from atomic coordinates as found in trajectories. Assuming the bond distances are constant (and that we are examining NH bond vectors), as is the case in most simulations for hydrogen-containing bonds, the following equation provides the solution (Best and Vendruscolo 2004):

$$S_{ij}^{2,\text{sim}} = \frac{3}{2(r_{ij}^{\text{min}})^4}\left(\langle x_{ij}^2\rangle^2 + \langle y_{ij}^2\rangle^2 + \langle z_{ij}^2\rangle^2 + 2\langle x_{ij}y_{ij}\rangle^2\right.$$

$$\left. + 2\langle x_{ij}z_{ij}\rangle^2 + 2\langle y_{ij}z_{ij}\rangle^2\right) - \frac{1}{2}$$

(14.17)

where x_{ij}, y_{ij}, and z_{ij} are components of the bond vector $\vec{r}_{ij}$ and r_{ij}^{min} is the equilibrium length of the bond. The back-calculated order parameter can then be combined with the experimental value to create the reaction coordinate:

$$\rho(t) = \frac{1}{N_{S^2}} \sum_{(i,j)\in D} \left(S_{ij}^{2,\text{sim}} - S_{ij}^{2,\text{exp}}\right)^2$$

(14.18)

where N_{S^2} is the number of bond vectors being used and D is the set of atom pairs associated with the vectors. The reaction coordinate can then be introduced in the restraining potential V_{S^2} similar to Equations 14.4 and 14.5 to determine when the restraints are to be applied. At each step of the simulation, an alignment to minimize the root-mean-square deviation (RMSD) of each structure in the ensemble to a reference is carried out. The alignment allows the overall rotational and translational motions to be eliminated.

An investigation into the substrate binding mechanism of calmodulin, a protein important in calcium-mediated signal transduction, provides an example of S^2-order parameters being applied to a system of interest to achieve a more accurate picture of the structural ensemble (Gsponer et al. 2008). Calmodulin is made up of two homologous domains, the C-terminal domain (CTD) and the N-terminal domain (NTD), each containing two EF-hand helix-loop-helix motifs, as well as a linker region between them. Without calcium bound, calmodulin rests in a closed state not able to bind its substrates. Binding of calcium ions to the C- and N-terminal motifs causes the structured domains to shift to an open state capable of binding substrates.

Structure calculations were performed on both an uncomplexed state (Ca^{2+}-CaM) and a substrate-complexed state (Cam-MLCK). The number of replicas used for the ensemble was chosen to be 16 according to a study by Richter et al. (Richter et al. 2007). It was shown here that S^2 restraints were more susceptible to overrestraining, and so derived restraints should be averaged over a larger number of replicas than NOEs. In order to prevent dampening of the large, interdomain motions by the S^2 order restraints, two separate alignments—one for each structured domain—had to be carried out.

The calculated ensembles can be viewed in Figure 14.3. Detailed analyses of the calculated ensembles revealed that correlated motions within the Ca^{2+}-CaM state direct the structural fluctuations toward complex-like substates. Specifically, it was shown that while the calculated ensembles of the CTD have average values of certain structural properties that are consistent with those found in the X-ray structure of the unbound state, they also contain conformations with structural properties that are more similar to the complexed state, possibly allowing for more efficient binding to the substrate.

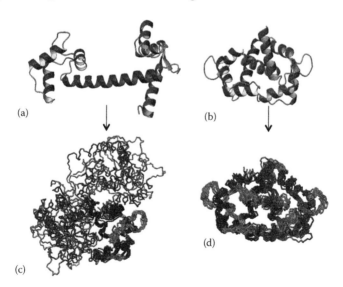

(a)   (b)   (c)   (d)

Figure 14.3   Cartoon representation of the crystal structure of Ca^{2+}-CaM (a) and CaM-MLCK (b). Ribbon diagram of the ensemble representing the Ca^{2+}-CaM state, aligned to the CTD domain (c) and the CaM-MLCK state (d). (From Gsponer, J. et al., *Structure [London, England: 1993]* 16 (5): 736–746, 2008.)

μs-ms Dynamics from PRE Experiments

PRE NMR spectroscopy is a technique that makes use of the magnetic dipolar interaction of nuclei such as protons with the free electron of a paramagnetic probe. It is able to sample sparsely populated states of low lifetimes and internal motion in the intermediate μs-ms time frame, which includes fast and larger structural rearrangements such as domain movement and folding processes. A paramagnetic center in the sample causes a significant change in the relaxation behavior of the protein nuclei within a certain distance. It has an effect on the longitudinal relaxation rate R_1 as well as the transverse relaxation rate R_2. However, the R_2 relaxation effects are considered to be more reliable in the PRE interpretation (Clore and Iwahara 2009). In case the system investigated does not already contain a paramagnetic probe such as a metal ion bound as a cofactor, it has to be artificially introduced at a solvent-exposed site (Kosen 1989). This can be achieved by site-specifically introducing a cysteine that can then be cross-linked with either a modified chelator like EDTA in complex with the desired paramagnetic metal ion (Ikegami et al. 2004) or a nitroxide-stable spin radical (Battiste and Wagner 2000). PRE data can be used as MD restraints in different ways including an S^2-like manner that takes fast local fluctuations into account. In the following, we focus on distance restraints derived from PRE data.

The central measured quantity of a PRE experiment is Γ_2, the difference between the relaxation rate of a ^{1}H nucleus with and without the paramagnetic centre present, $R_{2,\mathrm{para}}$ and $R_{2,\mathrm{dia}}$, respectively. It essentially quantifies the paramagnetic contribution to the transverse relaxation and can be readily obtained from the intensities of the NMR signals in a two-time-point experiment for both the paramagnetically labeled and the unlabeled sample (Clore and Iwahara 2009).

$$\Gamma_2 = R_{2,\mathrm{para}} - R_{2,\mathrm{dia}} = \frac{1}{T_b - T_a} \ln \frac{I_{\mathrm{dia}}(T_b) I_{\mathrm{para}}(T_a)}{I_{\mathrm{dia}}(T_a) I_{\mathrm{para}}(T_b)} \qquad (14.19)$$

T_a and T_b denote the two time points and I_{dia} and I_{para} the intensities of the ^{1}H signals with and without the paramagnetic center present, respectively. As in NOE experiments, the magnitude of the PRE effect scales with the inverse of the sixth power of the distance of the nucleus to the paramagnetic probe, r^{-6}. This relation is not trivial, as dipole-dipole interactions and Curie-spin relaxation can occur. In the simplest description used for systems where the Curie-spin contribution is negligible, the Solomon-Bloembergen equation connects the transverse relaxation effect of a free electron on a proton Γ_2 to the length r of the vector if this vector is fixed (Kosen 1989):

$$\Gamma_2 = \frac{K}{r^6}\left(4\tau_C + \frac{3\tau_C}{1+\omega_H^2\tau_C^2}\right) \tag{14.20}$$

K is a constant depending on the gyromagnetic ratio of the interacting particles, r is the ^{1}H to spin-label distance, ω_H is the Larmor frequency of a proton, and τ_C is the correlation time for the dipole-dipole interaction between the electron and the ^{1}H nucleus. Because the gyromagnetic ratio of an unpaired electron is so much larger (1000×) than that of a proton, PRE experiments are able to look at interactions over a distance of up to 35 Å as opposed to 6 Å for ^{1}H-^{1}H NOE where K is much smaller. This equation is only an approximation, as the vector between the free electron of the paramagnetic probe and the proton is really exposed to fast structural fluctuations. These internal structural fluctuations can be included in a model-free extension of the Solomon-Bloembergen equation similar to the S^2-order parameters described earlier (Iwahara, Schwieters, and Clore 2004). However, it has been shown these dynamics occur on a fast enough timescale so that they do not significantly affect the apparent correlation time τ_C, while r is seen as a statistical average over the possible conformations (Krugh 1979). To account for this averaging, an error of up to ± 5 Å is added to all experimental distances that are to be used in the simulations afterwards (Battiste and Wagner 2000; Gillespie and Shortle 1997). The correlation time τ_C can be approximated to be the global correlation time of the entire molecule (Battiste and Wagner 2000) or just an estimated value (Gillespie and Shortle 1997) as its influence on the calculated distance within its error bounds is minor. The obtained distance r is the experimental parameter d_{ij}^{exp} used in PRE-restrained MD simulations, while i and j denote the indices of the proton and the paramagnetic atom, respectively. The distance back-calculated from the simulated conformations d_{ij}^{sim} is easily obtained as an average over the distances r_{ij} in all replicas M of the ensemble (Lindorff-Larsen et al. 2004).

$$d_{ij}^{sim} = \left(\frac{1}{M}\sum_{k=1}^{M}\frac{1}{r_{ij,k}^6}\right)^{-1/6} \tag{14.21}$$

d_{ij}^{sim} and d_{ij}^{exp} can then be used as the central parameters of the restraint in the reaction coordinate ρ, as shown in Equation 14.22.

$$\rho(t) = \frac{1}{N_{PRE}}\sum_{(i,j)\in D}\left(d_{ij}^{exp}-d_{ij}^{sim}\right)^2 \tag{14.22}$$

N_{PRE} denotes the number of PRE restraints applied and D the set of existing proton-spin-label pairs. The reaction coordinate can then be introduced in the restraining potential V_{PRE}, as shown in Equations 14.3 and 14.4 to determine when the restraints are applied.

This method has been applied to determine an ensemble of structures of the intrinsically disordered protein α-synuclein (αSyn) that plays a role in Parkinson's disease (Dedmon et al. 2005). In a slightly different implementation, PRE experimental data has also been used to characterize a denatured structural ensemble of bovine acyl-coenzyme A binding protein (Lindorff-Larsen et al. 2004).

One interesting feature of PRE as a tool for examining dynamics is its ability to probe sparsely populated states that otherwise would not be visible in an NMR spectrum (Iwahara and Clore 2006). For illustration, we will consider a two-state system with a major conformational state A that is highly populated and a minor conformational state B that is only sparsely populated and invisible in a normal NMR spectrum. Hence, PRE is only directly measurable on the signal for the proton in state A. The two states are exchanging at a combined exchange rate $k_{ex} = k_{AB} + k_{BA}$.

$$A \underset{k_{BA}}{\overset{k_{AB}}{\rightleftharpoons}} B$$

In state A, a probed nucleus is much further away from the paramagnetic center than in state B. The PRE intensity of the proton in state A is $\Gamma_{2,A}$, while the intensity in state B is $\Gamma_{2,B}$. As Γ_2 scales with r^{-6} and as the proton in B is closer to the paramagnetic probe, its PRE intensity $\Gamma_{2,B}$ is much greater than its intensity in state A, $\Gamma_{2,A}$.

$$\Gamma_{2,B} \gg \Gamma_{2,A} \tag{14.23}$$

In the slow exchange limit, when k_{ex} is small compared to the PRE rates, this doesn't affect the apparent PRE rate measured on the proton in major state A, Γ_2^{app} and it is equal to $\Gamma_{2,A}$. However, in the fast exchange limit, when the exchange rate is much greater with respect to the PRE rate

$$k_{ex} \gg (\gamma_{2,B} - \gamma_{2,A}) \tag{14.24}$$

the apparent PRE rate measured on the proton in major state A, Γ_2^{app}, becomes a weighted average of the PRE rates of the proton in the two states. In other words, information about the minor state B is included as a footprint in the signal for state A. p_A and p_B denote the populations of state A and B, respectively:

$$\Gamma_2^{app} = p_A \Gamma_{2,A} + p_B \Gamma_{2,B} \tag{14.25}$$

If one knows the approximate distance of the proton in state A from where the paramagnetic probe was introduced and thereby $\Gamma_{2,A}$, PRE can then be used to characterize an otherwise invisible, sparsely populated state by calculating $\Gamma_{2,B}$.

PRE and its structural interpretation by MD simulations have been used to study sparsely populated states in protein–protein and protein–DNA complexes. One feature of these complexes is very specific binding interfaces. However, the three-dimensional space that free molecules have to sample in order to find their specific binding orientation and to form the interaction is huge. Specific complexes would be rare if not for a mechanism to lower the degrees of freedom in sampling of orientations. This mechanism is based on the formation of low-affinity non-specific complexes once two molecules meet. Once these low-affinity complexes are formed, sampling of relative orientations can occur, allowing the specific position to be found. Initially formed complexes are sparsely populated and have short lifetimes. However, they can be sampled using PRE. Here, DNA fragments containing a specific protein binding sequence and random flanking sequences were paramagnetically labeled at different positions and their PRE effect on the binding protein was investigated. It was found that the protein was sampling a variety of nonspecific interactions in addition to the specific complex (Iwahara and Clore 2006). This supports the hypothesis of the protein sliding along the DNA in a low-affinity interaction until it finds its specific binding sequence, thereby reducing the search for the specific complex interaction to a one-dimensional problem. A similar approach has been used to investigate the situation for protein–protein interactions where nonspecific encounter complexes lower the dimensionality of conformational sampling to two. In a study by Fawzi et al., two distinct populations of encounter complexes have been identified for the N-terminal domain of enzyme I (EIN) and the histidine phosphocarrier protein (HPr) as a model system. One of them was found to be sterically occluded by the specific complex and one of them able to coexist along with the specific complex (Fawzi et al. 2010).

ms-ks Dynamics from Hydrogen Exchange Experiments

In hydrogen exchange experiments (HX), protein dynamics on a timescale of milliseconds and greater can be monitored. These include large-scale conformational changes such as domain motion or protein folding and unfolding. The method makes use of the different spin properties of ^1H (protium, H) and ^2H (deuterium, D) nuclei. The basis of the experiment is a solvent exchange from water to deuterium oxide that causes protons bound to the protein to exchange with the deuterium ions in the solvent. This process can be either acid-catalyzed by H^+ and D^+ at a low pH or base-catalyzed by OH^- and OD^- at a high pH. As deuterium nuclei are invisible in a hydrogen NMR spectrum, the exchange between the two can be monitored over time by taking a series of spectra after the swap of the solvent in which the hydrogen signals are decaying due to the replacement by deuterium. The more exposed the hydrogens are to the solvent, the faster they exchange. The more protected they are due to burial or hydrogen bonding, the slower they exchange (Englander and Kallenbach 1983; Hibbert and Emsley 1990). The behavior of the signal intensity of proton signals over time gives the apparent exchange rate k^{exp}. It can be interpreted with a microscopic two-state model (Hvidt 1964) in which the hydrogens are described as being in an equilibrium between a closed state in which no exchange is possible and an open state in which they are capable of exchanging with the solvent. The interconversion between the open and the closed state is described by the rate constants k^{open} and k^{close} that are dependent on the three-dimensional structure of the protein whereas the intrinsic solvent-exchange rate k^{int} can be determined on the basis of pH, temperature, and the nearby amino acid sequence (Bai et al. 1993).

$$XH_{closed} \xrightleftharpoons[k^{close}]{k^{open}} XH_{open} \xrightarrow[k^{int}]{D^+} XD + H^+$$

In principle, there are two limiting cases for the value of k^{int} that affect the interpretation of the model (Hvidt and Nielsen 1966). In the EX1 limit, the intrinsic exchange rate is much greater than the rate of the proton becoming protected again, ($k^{int} \gg k^{close}$). Hence, every event of an amide proton becoming unprotected also leads to exchange.

$$k^{exp} \approx k^{open} \tag{14.26}$$

EX1 experiments are typically carried out at high pH or in the presence of denaturants. The EX2 limit, which denotes a slow intrinsic exchange compared to the rate of the proton becoming protected again ($k^{int} \ll k^{close}$), is experimentally more commonly probed because it is often more easily accessible. In this case, the experimentally observed exchange rate corresponds to the equilibrium between the open and the closed state and the exchange of the proton.

$$k^{exp} \approx k^{int} \frac{k^{open}}{k^{close}} \tag{14.27}$$

In the EX2 limit, the observed exchange rate k_i^{exp} for each amide proton is directly related to the difference in free energy ΔG_i:

$$\Delta G_i = RTln\frac{k_i^{closed}}{k_i^{open}} = RTln\frac{k_i^{int}}{k_i^{exp}} \tag{14.28}$$

The HX experimental protection factor for the ith residue P_i^{exp} can be defined as

$$P_i^{exp} = \frac{k_i^{int}}{k_i^{exp}} \tag{14.29}$$

and worked into the difference in free energy for each amide proton ΔG_i:

$$\Delta G_i = RTlnP_i^{exp} \tag{14.30}$$

This difference in free energy can either correspond to a global exchange—when the proton is only exchanging in the completely unfolded state, or local exchange—when the exchange occurs upon local structural fluctuations. To distinguish between the two, structural perturbation dependence (denaturant, temperature, mutagenesis) has to be experimentally determined (Bai and Englander 1996; Neira et al. 1997). Given the local case, hydrogen bonding and burial of the residues are the two predominant origins of amide protection (Woodward and Hilton 1980), which is why they are considered in the parameterization for the back-calculation of protection factors for simulated structures (Vendruscolo, Paci, Dobson et al. 2003). This back-calculation is purely empirical as opposed to accurate descriptions that are available for the S^2-order parameters or the PRE:

$$lnP_i^{sim}(C) = \beta_c N_i^c(C) + \beta_h N_i^h(C) \tag{14.31}$$

Here $N_i^c(C)$—the number of contacts—accounts for burial as the number of heavy atoms within a certain range and $N_i^h(C)$ for hydrogen bonding as being the number of hydrogen bonds that a certain proton forms in a conformation C. The β values are empirical weighting factors that take into account the difference in the influence of the two effects. To get a value that is comparable to the experimental situation of conformational averaging, an average over the protection factors of several conformations has to be taken:

$$\overline{lnP_i^{sim}} = \beta_c \overline{N_i^c} + \beta_h \overline{N_i^h} \tag{14.32}$$

In practice, this corresponds to the arithmetic mean over M simulated replicas as in Equation 14.11:

$$\overline{lnP_i^{sim}} = \frac{1}{M} \sum_{k=1}^{M} lnP_i^{sim}(C_k) = \frac{1}{M} \sum_{k=1}^{M} (\beta_c N_i^c(C_k) + \beta_h N_i^h(C_k)) \tag{14.33}$$

To eventually be able to determine an ensemble that satisfies the HX restraints, the simulated protection factors lnP_i^{sim} and the experimentally determined protection factors lnP_i^{exp} are included in the reaction coordinate ρ as described in Equation 14.3. N_{HX} denotes the number of protection factors used in the calculation:

$$\rho(t) = \sum_{i=1}^{N_{HX}} \left(\overline{lnP_i^{sim} - lnP_i^{exp}} \right)^2 \tag{14.34}$$

The reaction coordinate is then included in the pseudo-energy term as described in Equations 14.4 and 14.5 (Best and Vendruscolo 2006).

Simulations that use HX data as restraints enable the sampling of regions of the conformational space that correspond to rare fluctuations taking place on the sub-second or longer timescales. Therefore, this approach generates ensembles that contain those rare structures for regions where hydrogen exchange takes place. It has been successfully used to determine structures representing the exchange-competent intermediate state (EIS) of the bacterial immunity protein Im7 (Gsponer et al. 2006). This EIS was shown to share many structural and biophysical properties with the kinetic folding intermediate of Im7. As intermediates on the folding path of proteins are more heterogeneous than their native counterparts, their characterization must involve the determination of structural ensembles that accurately represent the conformational fluctuations of proteins in these states; simulations with HX data as restraints allow this to be done. Similarly HX-restrained simulations can be used to investigate dynamics that could underlie transitions between different functional or pathological states of proteins. For instance, HX-restrained simulations were used to investigate long timescale fluctuations in human prion protein (PrP^C) (Khorvash, Lamour, and Gsponer 2011). A structural conversion of the initially soluble globular protein (PrP^C) into a misfolded form (PrP^{Sc}) that can aggregate and accumulate in the brain underlies infectious and lethal neurodegenerative prion diseases. PrP^{Sc} is rich in β-sheets (43%) while PrP^C has a high α-helical content (42%) and only a few residues in β-sheets (3%). Hence, structural plasticity of PrP^C plays an important role in prion pathogenesis. By calculating the structural ensembles representing the exchange-competent state of human PrP^C (Figure 14.4), it was found that partial unfolding of one of the α-helices in PrP^C leads to the formation of new β-strands in some structures of the exchange-competent state. Overall, these restrained simulations provided new structural insights into the long timescale fluctuations of PrP^C that are more difficult to look at using classical simulations.

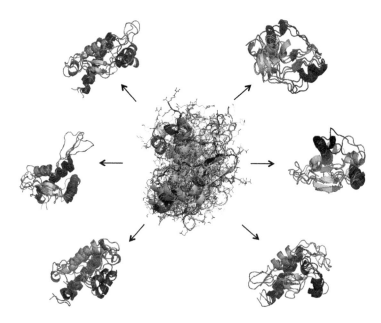

Figure 14.4 Structural ensemble representing the exchange-competent state of PrP^C (center). Representative structures of the six largest clusters of the exchange-competent state are shown outside. (From Khorvash, M. et al., *Biochemistry* 50 (47): 10192–10194, 2011.)

CONCLUDING REMARKS

In this chapter we have shown that our understanding of energy landscapes and sparsely populated states of proteins can be hugely expanded by applying experimental data as restraints in computational descriptions of macromolecules. These states are worth looking at because they can be of high biological significance. Although limited to NMR data in our description, every experimental technique that yields data giving insight into protein dynamics that can be somehow back-calculated from atomic coordinates can theoretically be implemented as a restraint in MD simulations. This gives us the tools to investigate protein dynamics further as existing methods are improved and new experiment-to-theory bridges are built.

REFERENCES

Adrian, M., J. Dubochet, J. Lepault et al. 1984. Cryo-electron microscopy of viruses. *Nature* 308 (5954): 32–36.

Anderson, W. A., and R. Freeman. 1962. Influence of a second radiofrequency field on high-resolution nuclear magnetic resonance spectra. *The Journal of Chemical Physics* 37 (1): 85–103.

Arnold, J. T., S. S. Dharmatti, and M. E. Packard. 1951. Chemical effects on nuclear induction signals from organic compounds. *The Journal of Chemical Physics* 19 (4): 507.

Bai, Y., and S. W. Englander. 1996. Future directions in folding: The multi-state nature of protein structure. *Proteins* 24: 145–151.

Bai, Y., J. S. Milne, L. Mayne et al. 1993. Primary structure effects on peptide group hydrogen exchange. *Proteins: Structure, Function, and Bioinformatics* 17 (1): 75–86.

Battiste, J. L., and G. Wagner. 2000. Utilization of site-directed spin labeling and high-resolution heteronuclear nuclear magnetic resonance for global fold determination of large proteins with limited nuclear Overhauser effect data. *Biochemistry* 39 (18): 5355–5365.

Best, R. B., and M. Vendruscolo. 2004. Determination of protein structures consistent with NMR order parameters. *Journal of the American Chemical Society* 126 (26): 8090–8091.

Best, R. B., and M. Vendruscolo. 2006. Structural interpretation of hydrogen exchange protection factors in proteins: Characterization of the native state fluctuations of CI2. *Structure (London, England: 1993)* 14 (1): 97–106.

Bloch, F., W. Hansen, and M. Packard. 1946. The nuclear induction experiment. *Physical Review* 70 (7–8): 474–485.

Boehr, D. D., H. J. Dyson, and P. E. Wright. 2006. An NMR perspective on enzyme dynamics. *Chemical Reviews* 106 (8): 3055–3079.

Bonvin, A. M., R. Boelens, and R. Kaptein. 1994. Time- and ensemble-averaged direct NOE restraints. *Journal of Biomolecular NMR* 4 (1): 143–149.

Bonvin, A. M., and A. T. Brünger. 1995. Conformational variability of solution nuclear magnetic resonance structures. *Journal of Molecular Biology* 250 (1): 80–93.

Bragg, W. L. 1913a. The diffraction of short electromagnetic waves by a crystal. *Proceedings of the Cambridge Philosophical Society* 17: 43.

Bragg, W. L. 1913b. The structure of some crystals as indicated by their diffraction of x-rays. *Proceedings of the Royal Society of London* A89 (610): 248–277.

Bruggeller, P., and E. Mayer. 1980. Complete vitrification in pure liquid water and dilute aqueous solutions. *Nature* 288 (5791): 569–571.

Chen, J., C. L. Brooks, and P. E. Wright. 2004. Model-free analysis of protein dynamics: Assessment of accuracy and model selection protocols based on molecular dynamics simulation. *Journal of Biomolecular NMR* 29: 243–257.

Clore, G. M., and J. Iwahara. 2009. Theory, practice, and applications of paramagnetic relaxation enhancement for the characterization of transient low-population states of biological macromolecules and their complexes. *Chemical Reviews* 109 (9): 4108–4139.

Clore, G. M., and C. D. Schwieters. 2006. Concordance of residual dipolar couplings, backbone order parameters and crystallographic B-factors for a small alpha/beta protein: A unified picture of high probability, fast atomic motions in proteins. *Journal of Molecular Biology* 355 (5): 879–886.

Davies, K. M., C. Anselmi, I. Wittig et al. 2012. Structure of the yeast F1Fo-ATP synthase dimer and its role in shaping the mitochondrial cristae. *Proceedings of the National Academy of Sciences of the United States of America* 109 (34): 13602–13607.

Dedmon, M. M., K. Lindorff-Larsen, J. Christodoulou et al. 2005. Mapping long-range interactions in alpha-synuclein using spin-label NMR and ensemble molecular dynamics simulations. *Journal of the American Chemical Society* 127 (2): 476–477.

Dubochet, J., and A. W. McDowall. 1981. Vitrification of pure water for electron microscopy. *Journal of Microscopy* 124 (3): 3–4.

Englander, S. W., and N. R. Kallenbach. 1983. Hydrogen-exchange and structural dynamics of proteins and nucleic-acids. *Quarterly Reviews of Biophysics* 16: 521–655.

Fawzi, N. L., M. Doucleff, J.-Y. Suh et al. 2010. Mechanistic details of a protein-protein association pathway revealed by paramagnetic relaxation enhancement titration measurements. *Proceedings of the National Academy of Sciences of the United States of America* 107 (4): 1379–1384.

Franklin, R. E., and R. G. Gosling. 1953. Molecular configuration in sodium thymonucleate. *Nature* 171 (4356): 740–741.

Friedrich, W., Knipping P., and von Laue, M. 1912. Interferenz-Erscheinungen bei Röntgenstrahlen. *Sitzungsberichte der Mathematisch-Physikalischen Classe der Königlich-Bayerischen Akademie der Wissenschaften zu München* 1912: 303.

Gillespie, J. R., and D. Shortle. 1997. Characterization of long-range structure in the denatured state of staphylococcal nuclease. II. Distance restraints from paramagnetic relaxation and calculation of an ensemble of structures. *Journal of Molecular Biology* 268 (1): 170–184.

Gsponer, J., J. Christodoulou, A. Cavalli et al. 2008. A coupled equilibrium shift mechanism in calmodulin-mediated signal transduction. *Structure (London, England: 1993)* 16 (5): 736–746.

Gsponer, J., H. Hopearuoho, S. B.-M. Whittaker et al. 2006. Determination of an ensemble of structures representing the intermediate state of the bacterial immunity protein Im7. *Proceedings of the National Academy of Sciences of the United States of America* 103 (1): 99–104.

Henzler-Wildman, K., and D. Kern. 2007. Dynamic personalities of proteins. *Nature* 450 (7172): 964–972.

Hibbert, F., and J. Emsley. 1990. Hydrogen bonding and chemical reactivity. *Advances in Physical Organic Chemistry* 26: 255–379.

Hvidt, A. 1964. A discussion of the pH dependence of the hydrogen-deuterium exchange of proteins. *Comptes-rendus des travaux du Laboratoire Carlsberg* 34: 299.

Hvidt, A., and S. O. Nielsen. 1966. Hydrogen exchange in proteins. *Advances in Protein Chemistry* 21: 287–386.

Ikegami, T., L. Verdier, P. Sakhaii et al. 2004. Novel techniques for weak alignment of proteins in solution using chemical tags coordinating lanthanide ions. *Journal of Biomolecular NMR* 29 (3): 339–349.

Iwahara, J., and G. M. Clore. 2006. Detecting transient intermediates in macromolecular binding by paramagnetic NMR. *Nature* 440 (7088): 1227–1230.

Iwahara, J., C. D. Schwieters, and G. M. Clore. 2004. Ensemble approach for NMR structure refinement against (1)H paramagnetic relaxation enhancement data arising from a flexible paramagnetic group attached to a macromolecule. *Journal of the American Chemical Society* 126 (18): 5879–5896.

Karplus, M. 2003. Molecular dynamics of biological macromolecules: A brief history and perspective. *Biopolymers* 68: 350–358.

Kemmink, J., and T. E. Creighton. 1993. Local conformations of peptides representing the entire sequence of bovine pancreatic trypsin inhibitor and their roles in folding. *Journal of Molecular Biology* 234 (3): 861–878.

Kendrew, J. C., G. Bodo, H. M. Dintzis et al. 1958. A three-dimensional model of the myoglobin molecule obtained by x-ray analysis. *Nature* 181 (4610): 662–666.

Kendrew, J. C., R. E. Dickerson, B. E. Strandberg et al. 1960. Structure of myoglobin: A three-dimensional Fourier synthesis at 2 Å resolution. *Nature* 185 (4711): 422–427.

Khorvash, M., G. Lamour, and J. Gsponer. 2011. Long-timescale fluctuations of human prion protein determined by restrained MD simulations. *Biochemistry* 50 (47): 10192–10194.

Kleckner, I. R., and M. P. Foster. 2011. An introduction to NMR-based approaches for measuring protein dynamics. *Biochimica et Biophysica Acta* 1814 (8): 942–968.

Kosen, P. A. 1989. Spin labeling of proteins. *Methods in Enzymology* 177: 86–121.

Kourkoutis, L. F., J. M. Plitzko, and W. Baumeister. 2012. Electron microscopy of biological materials at the nanometer scale. *Annual Review of Materials Research* 42 (1): 33–58.

Krugh, T. R. 1979. Spin-label-induced nuclear magnetic resonance relation studies of enzymes. In *Spin Labeling. II. Theory and Applications*, ed. L. Berliner, 339–372. Academic Press, New York.

Lindorff-Larsen, K., R. B. Best, M. A. Depristo et al. 2005. Simultaneous determination of protein structure and dynamics. *Nature* 433 (7022): 128–132.

Lindorff-Larsen, K., S. Kristjansdottir, K. Teilum et al. 2004. Determination of an ensemble of structures representing the denatured state of the bovine acyl-coenzyme a binding protein. *Journal of the American Chemical Society* 126 (10): 3291–3299.

Mao, Y., L. Wang, C. Gu et al. 2013. Molecular architecture of the uncleaved HIV-1 envelope glycoprotein trimer. *Proceedings of the National Academy of Sciences of the United States of America* 110 (30): 12438–12443.

McCammon, J. A., B. R. Gelin, and M. Karplus. 1977. Dynamics of folded proteins. *Nature* 267: 585–590.

Neira, J. L., L. S. Itzhaki, D. E. Otzen et al. 1997. Hydrogen exchange in chymotrypsin inhibitor 2 probed by mutagenesis. *Journal of Molecular Biology* 270: 99–110.

Paci, E., and M. Karplus. 1999. Forced unfolding of fibronectin type 3 modules: An analysis by biased molecular dynamics simulations. *Journal of Molecular Biology* 288 (3): 441–459.

Perutz, M. F., M. G. Rossmann, A. N. N. F. Cullis et al. 1960. Structure of haemoglobin: A three-dimensional Fourier synthesis at 5.5-Å resolution, obtained by x-ray analysis. *Nature* 185 (4711): 416–422.

Purcell, E., H. Torrey, and R. Pound. 1946. Resonance absorption by nuclear magnetic moments in a solid. *Physical Review* 69 (1–2): 37–38.

Rabi, I., J. Zacharias, S. Millman et al. 1938. A new method of measuring nuclear magnetic moment. *Physical Review* 53 (4): 318.

Raddi, G., D. R. Morado, J. Yan et al. 2012. Three-dimensional structures of pathogenic and saprophytic Leptospira species revealed by cryo-electron tomography. *Journal of Bacteriology* 194 (6): 1299–1306.

Richter, B., J. Gsponer, P. Várnai et al. 2007. The MUMO (minimal under-restraining minimal over-restraining) method for the determination of native state ensembles of proteins. *Journal of Biomolecular NMR* 37 (2): 117–135.

Schwieters, C. D., J. J. Kuszewski, N. Tjandra et al. 2003. The Xplor-NIH NMR molecular structure determination package. *Journal of Magnetic Resonance* 160 (1): 65–73.

Shaw, D. E., P. Maragakis, K. Lindorff-Larsen et al. 2010. Atomic-level characterization of the structural dynamics of proteins. *Science* 330 (6002): 341–346.

Torda, A. E., R. M. Scheek, and W. F. van Gunsteren. 1989. Time-dependent distance restraints in molecular dynamics simulations. *Chemical Physics Letters* 157 (4): 289–294.

Torda, A. E., R. M. Scheek, and W. F. van Gunsteren. 1990. Time-averaged nuclear Overhauser effect distance restraints applied to tendamistat. *Journal of Molecular Biology* 214 (1): 223–235.

Vendruscolo, M., and C. M. Dobson. 2005. Towards complete descriptions of the free-energy landscapes of proteins. *Philosophical Transactions. Series A, Mathematical, Physical, and Engineering Sciences* 363 (1827): 433–450.

Vendruscolo, M., E. Paci, C. M. Dobson et al. 2003. Rare fluctuations of native proteins sampled by equilibrium hydrogen exchange. *Journal of the American Chemical Society* 125: 15686–15687.

Vendruscolo, M., E. Paci, M. Karplus et al. 2003. Structures and relative free energies of partially folded states of proteins. *Proceedings of the National Academy of Sciences of the United States of America* 100 (25): 14817–14821.

Vendruscolo, M., J. Zurdo, C. E. MacPhee et al. 2003. Protein folding and misfolding: A paradigm of self-assembly and regulation in complex biological systems. *Philosophical Transactions. Series A, Mathematical, Physical, and Engineering Sciences* 361 (1807): 1205–1222.

Watson, J. D., and F. H. C. Crick. 1953. Molecular structure of nucleic acids: A structure for deoxyribose nucleic acid. *Nature* 171 (4356): 737–738.

Wiedenheft, B., G. C. Lander, K. Zhou et al. 2011. Structures of the RNA-guided surveillance complex from a bacterial immune system. *Nature* 477 (7365): 486–489.

Williamson, M. P., T. F. Havel, and K. Wüthrich. 1985. Solution conformation of proteinase inhibitor IIA from bull seminal plasma by 1H nuclear magnetic resonance and distance geometry. *Journal of Molecular Biology* 182 (2): 295–315.

Woodward, C. K., and B. D. Hilton. 1980. Hydrogen isotope exchange kinetics of single protons in bovine pancreatic trypsin inhibitor. *Biophysical Journal* 32: 561–575.

Index

Page numbers followed by f and t indicate figures and tables, respectively.

Printed and bound by CPI Group (UK) Ltd, Croydon, CR0 4YY

21/10/2024

01777044-0019